Finite Element Method
Basic Technique and Implementation

Pin Tong
and
John N. Rossettos

DOVER PUBLICATIONS, INC.
Mineola, New York

Copyright

Copyright © 1977 by The Massachusetts Institute of Technology
All rights reserved.

Bibliographical Note

This Dover edition, first published in 2008, is an unabridged republication of the work originally published by The MIT Press, Cambridge, Massachusetts, in 1977. This edition is published by special arrangement with The MIT Press, 55 Hayward Street, Cambridge, MA 02142.

Library of Congress Cataloging-in-Publication Data

Tong, Pin.
    Finite-element method : basic technique and implementation / Pin Tong and John N. Rossettos.
        p. cm.
    Reprint. Originally published: Cambridge, Mass. : MIT Press, 1977.
    Includes bibliographical references and index.
    ISBN-13: 978-0-486-46676-7
    ISBN-10: 0-486-46676-0
    1. Finite element method. I. Rossettos, John N. (John Nicholas) II. Title.

TA347.F5T66 2008
620.001'51825—dc22

                                                         2007047874

Manufactured in the United States of America
Dover Publications, Inc., 31 East 2nd Street, Mineola, N.Y. 11501

This book is dedicated to our parents

Contents

Preface — vii

1 The Finite-Element Method — 1

2 The Finite-Element Method for Poisson's Equation — 40

3 Assembly and Solution for Large Systems — 72

4 Implementation of Assembly and Solution Schemes for Large Systems on High-Speed Computers — 104

5 Applications to Solid Mechanics — 127

6 Interpolation Functions, Numerical Integration, and Higher-Order Elements — 161

7 Bending of Beams and Plates — 208

8 Hybrid Methods — 228

9 Selected Topics and Recent Developments — 249

Appendix A Notation and Matrix Algebra — 283
Appendix B Rectangular Elements — 291
Appendix C Triangular Elements With Straight Edges — 297
Appendix D Variational Methods — 301
References — 315
Index — 323

Preface

Although the finite-element method was originally developed to solve problems in the particular areas of structural mechanics and elasticity, the mathematical basis of the technique renders it applicable to problems to be found throughout applied mathematics, continuum mechanics, engineering, and physics. The method is eminently suitable for numerical solution on modern computers. The present text is aimed at clearly elucidating this broader scope of the finite-element method, as well as the practical aspect of computation. It bridges the gap between the rigorous mathematical foundations of the method and practical applications.

The book is intended for students and for practicing engineers and scientists. It is written so that it can be used by students in applied mathematics as well as those in engineering departments. The text provides material for a first course in the finite-element method that can be followed by beginning graduate students as well as some senior undergraduates. The illustrative examples and homework problems have been designed so that the book may be used for self-study. The typical undergraduate engineering education, including the usual matrix algebra, systems of equations, numerical methods, and the rudiments of ordinary and partial differential equations, should suffice as a prerequisite (an appendix on matrix notation and algebra is included in this book). The undergraduate engineering background in such topics as solid mechanics, fluid mechanics, and heat transfer would be helpful, although the manner of treating physical topics in this work is aimed at making them understandable to mathematics majors. Therefore, mathematics departments can use the text to advantage in special courses in analytic and numerical approximation techniques.

The initial notes for the text were developed for a first-year graduate course offered at the Massachusetts Institute of Technology by the senior author. Based upon that experience, we decided to develop a course emphasizing both the general methodology of the finite-element technique and its actual implementation on the high-speed computer. This was prompted by the necessity to deal with the complex problems arising in modern engineering and science. Since numerical methods are often the "solution of last resort" for complicated problems, it is important for the applied mathematician, the engineer, and the scientist not only to be equipped with the knowledge of the fundamental principles of such methods, but also to be able to utilize the computer to the greatest extent possible—that is to say, with maximum efficiency.

Therefore in this text we attempt to provide the rational development of the method from its mathematical foundations, developing the technique in clearly understandable stages. By understanding the basis of each step, the reader can apply the method to his or her own original problems in applied mathematics, engineering, and science. To encourage such facility, we develop variational approaches of boundary value problems from the start, and the finite-element procedure is developed from such approaches. An appendix on the necessary variational calculus is provided. Several chapters detail the implementation of the technique on the computer by discussing how to solve large systems of algebraic equations, how to arrange the sequence of numerical computations to save execution time, and how to store and retrieve information for reducing computer core storage requirements.

In ch. 1 an introduction, including the history of the method together with its physical and mathematical background, is given. An overview of the underlying philosophy provides an important perspective on the method. Initial discussions of the displacement model, the interpolation-function concept, and convergence are followed by illustrative examples which apply the finite-element method to the solution of ordinary differential equations related to the Sturm-Liouville problem, which arises most frequently in problems in elasticity, fluid mechanics, and heat transfer. In ch. 2, the solution of problems involving partial differential equations is introduced, along with a thorough finite-element analysis of Poisson's equation. In particular, the evaluation of element matrices using triangular and rectangular elements is explained. A basis for the assembly of the element matrices into a master matrix is indicated, and is followed by a simple and complete example (using triangular elements) illustrating the various steps with actual numbers. A step-by-step procedure of assembling the master matrix —a procedure that is especially suitable for large systems, and ideal for computer implementation—is explained in detail in ch. 3, and again illustrated by an example. The application of boundary or constraint conditions to the master matrix is also demonstrated; and the final solution to an illustrative example is obtained and compared with exact results. Various numerical techniques for the solution of large systems of equations are described in ch. 4 with a view toward establishing which computer algorithms are most efficient and accurate; great advantage is taken of the sparseness of the matrices resulting from the use of the finite-element method. In ch. 4 we also discuss techniques of storing information that result in a reduction of core storage requirements and computation sequences that minimize the number of operations on the computer.

Application to problems in elasticity and the bending of beams and plates is given in chs. 5 and 7; here additional exposure is given to the ideas discussed in earlier chapters, together with some new features inherent in bending problems. Chapter 6 is devoted to general interpolation functions, numerical integrations typical of finite-element applications, and higher-order elements. Chapter 8 contains the application of the hybrid finite-element model to second- and fourth-order partial differential equations. In ch. 9 several topics have been selected from the areas of elastic vibrations, heat transfer, and fluid flow. Recent developments in finite-element technique are also described, such as the application of superelements to problems involving singularities and to problems describable in terms of infinite domains. In various chapters of the text some repetition and spelling out of certain operations with key matrices is done purposely in order to make each chapter reasonably self-contained. This detailed exposition will enhance the self-study features of the book.

The emphasis in the text is on the displacement model, but the hybrid-model concept is also introduced, and is shown to evolve naturally from the mathematical foundations of the method. This text is the only book on the subject which contains a complete chapter on hybrid finite-element models and their applications.

The authors want to express their gratitude to many of their teachers, colleagues, friends, and students. We are particularly grateful to Professors Y. C. Fung and T. H. H. Pian. Their deep insight into many facets of engineering and their broad knowledge of mechanics and applied mathematics, together with their lasting contributions to these fields, guided us throughout the development of this book. Special appreciation is expressed to the wives of the authors. They have been a constant source of inspiration and encouragement, and have been willing to make sacrifices so that this book could be written. Jean Tong, with four children and attendant chores, often stayed up at night at the strenuous task of writing many cumbersome equations for the book; and Joy Rossettos steadfastly and patiently typed chapters. A deep acknowledgment is offered to the young children of the authors, who had to sacrifice precious playtime with their fathers. The MIT Press staff has been very cooperative.

Pin Tong
John N. Rossettos

Finite Element Method
Basic Technique and Implementation

1 The Finite-Element Method

1.1 Introduction. History of the Method

The finite-element method first appeared in the 1950s as a technique for handling problems in solid mechanics. It was an outgrowth of the so-called matrix methods of structural analysis (Argyris 1955, 1958; Livesley 1964). The matrix methods, in turn, had been developed during the decade 1945–1955 as a result of the search for systematic methods of analyzing complex structures containing large numbers of components.

The procedure fundamental to all the matrix methods of structural analysis is the expression of relations between the displacements and internal forces at the nodal points of individual structural components in the form of a system of algebraic equations with either nodal displacements, nodal internal forces, or both nodal displacements and internal forces as unknowns. Depending on the unknowns selected, the method is called the *displacement*, the *force*, or the *mixed* method. The system of equations can be written most conveniently in matrix notation, and solutions of these equations can be efficiently obtained by high-speed digital computers.

The name "finite-element method" was first introduced by Clough (1960). For aircraft structural analysis Turner, Clough, Martin, and Topp (1956) had applied the displacement method to plane-stress problems by subdividing the structure into triangular or rectangular "elements." In their formulation the behavior of each element was represented by an element-stiffness matrix relating the forces at the nodal points of an element to the nodal displacements. In contrast to the conventional matrix analysis, in which the relations between forces and displacements for each structural component are derived exactly, the solution for a structure subdivided into elements utilizes only *approximate* displacement functions within each element. The comprehensive treatise on energy theorems and matrix methods by Argyris (1955) had already spelled out in detail the derivation of the element-stiffness matrix for a rectangular panel under plane stress. Also, long before this, Courant (1943) had applied the "element" principle to obtain an approximate solution of St-Venant torsion by assuming a linear distribution of the warping function among the triangular elements of an assemblage. These earlier finite-element approaches were based on the principle of virtual work or the principle of minimum potential energy; they may be viewed as extensions of the Rayleigh-Ritz method, wherein piecewise linear functions are assumed that are continuous only for the approximate solution. The finite-element method itself, however, is much more versatile than the conventional Rayleigh-Ritz method.

During the decade 1960–1970, formulations of finite-element methods in terms of different types of variational principles were developed by Besseling (1963), Melosh (1963), Jones (1964), Gallagher (1964), Pian (1964a), Fraeijs de Veubeke (1964), Herrmann (1965), Prager (1967, 1968), Tong and Pian (1969), and Tong (1970). Indeed, because the assumed functions for the field variables are only piecewise continuous, it has been necessary to develop new and modified variational principles—ones that allow for the discontinuity of the field variables at the interelement boundaries.

Formulations of finite-element methods need not, however, be based only on the variational approach. Oden (1969) was able to write the equations for the finite-element analysis of thermoelasticity problems by starting from the so-called energy-balance method. Szabo and Lee (1969) arrived at the finite-element solution of plane elasticity problems by using the Galerkin method. For certain boundary-value problems involving regular-polygon shaped boundaries for which regular finite-element arrangements can be used, the resulting system of equations is sometimes identical to that of the conventional finite-difference formulation (Pian 1971b). The finite-element method, however, is always more advantageous if applied to irregular domains and to nonhomogeneous media.

In the short period of a decade and a half since the concept of "finite elements" became known in structural mechanics, research and development in the field has mushroomed. Finite-element methods have been applied to such problems in continuum mechanics as the twisting of prismatic bars, steady-state heat conduction, potential flow in ideal fluids, etc. Its users have grown in number from the original few aeronautical structural analysts to large numbers of engineers in civil, mechanical, naval architectural, and nuclear engineering.

1.2 The Physical Concept

The finite-element method starts from an approximate, though systematic, description of a continuum in terms of a finite, though large, number of coordinates or degrees of freedom. The concept has its origins in the theory of structures. A physical structure is usually built up from many structural

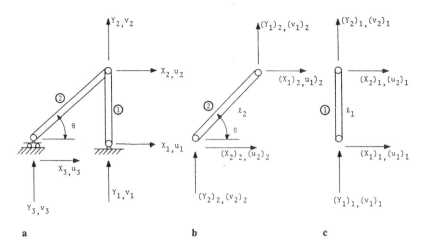

Fig. 1.1 A pin-joint truss: (a) assembled; (b, c) truss elements.

elements. The elements, however, are connected to each other at only a finite number of points. For each element, structural properties such as the force-displacement relations are uniquely defined in terms of the degrees of freedom allowed at the nodes and at some other special points within the element. By various well-known techniques, the behavior of the assembled structure can then be described. For example, in fig. 1.1, the pin-jointed truss assembly consists of two elements, each of which is assumed to carry only axial forces. Let these elements have equal cross-sectional area A and a modulus of elasticity E. The lengths of the two elements are respectively l_1 and l_2. The force-displacement relations for element 1 (fig. 1.1c) are given in matrix form as

$$\begin{Bmatrix} X_1 \\ Y_1 \\ X_2 \\ Y_2 \end{Bmatrix}_1 = \frac{AE}{l_1} \begin{bmatrix} 0 & 0 & 0 & 0 \\ 0 & 1 & 0 & -1 \\ 0 & 0 & 0 & 0 \\ 0 & -1 & 0 & 1 \end{bmatrix} \begin{Bmatrix} u_1 \\ v_1 \\ u_2 \\ v_2 \end{Bmatrix}_1, \tag{1.1}$$

and for element 2 they are

$$\begin{Bmatrix} X_1 \\ Y_1 \\ X_2 \\ Y_2 \end{Bmatrix}_2 = \frac{AE}{l_2} \begin{bmatrix} c^2 & cs & -c^2 & -cs \\ cs & s^2 & -cs & -s^2 \\ -c^2 & -cs & c^2 & cs \\ -cs & -s^2 & cs & s^2 \end{bmatrix} \begin{Bmatrix} u_1 \\ v_1 \\ u_2 \\ v_2 \end{Bmatrix}_2, \tag{1.2}$$

where $c = \cos\theta$, $s = \sin\theta$.

To clarify the physical meaning of these relations, consider for instance the horizontal component of force on element 2 at its node 1, namely $(X_1)_2$, which is given in (1.2). According to Hooke's law this component is equal to

$$(X_1)_2 = F_a \cos\theta = \frac{AE}{l_2}\delta_a \cos\theta,$$

where F_a is the axial force and δ_a is the axial elongation of element 2:

$$\delta_a = (u_1)_2 \cos\theta + (v_1)_2 \sin\theta - (u_2)_2 \cos\theta - (v_2)_2 \sin\theta,$$

so that these relations yield

$$(X_1)_2 = \frac{AE}{l_2}\{[(u_1)_2 - (u_2)_2]\cos^2\theta + [(v_1)_2 - (v_2)_2]\cos\theta \sin\theta\},$$

which is what is given in (1.2) as the product of the first row of the square matrix with the column vector of displacement components. The other force components in (1.1) and (1.2) are obtained in similar fashion.

The two elements can be assembled as follows. We first introduce relations between the so-called global displacement components u_1, v_1, u_2, v_2, u_3, v_3 in fig. 1.1 and the local displacement components for each element, namely, $(u_1)_i$, $(v_1)_i$, $(u_2)_i$, $(v_2)_i$, where i = 1, 2. is the element number. Then

$$u_1 = (u_1)_1, \qquad v_1 = (v_1)_1,$$
$$u_2 = (u_2)_1 = (u_1)_2, \qquad v_2 = (v_2)_1 = (v_1)_2,$$
$$u_3 = (u_2)_2, \qquad v_3 = (v_2)_2.$$

Equilibrium at a node requires that the applied load components be equal to the sum of the load components, on each element at that node. When the relations between global and local displacement components, and eqs. (1.1) and (1.2), are also taken into account, equilibrium then requires that

$$X_1 = (X_1)_1 = 0,$$
$$Y_1 = (Y_1)_1 = \frac{AE}{l_1}v_1 - \frac{AE}{l_1}v_2,$$
$$X_2 = (X_2)_1 + (X_1)_2 = (X_1)_2 = \frac{AE}{l_2}[c^2u_2 + csv_2 - c^2u_3 - csv_3],$$
$$Y_2 = (Y_2)_1 + (Y_1)_2 = \frac{AE}{l_1}(v_2 - v_1) + \frac{AE}{l_2}[csu_2 + s^2v_2 - csu_3 - s^2v_3],$$
$$X_3 = (X_2)_2 = \frac{AE}{l_2}[-c^2u_2 - csv_2 + c^2u_3 + csv_3],$$

$$Y_3 = (Y_2)_2 = \frac{AE}{l_2}[- csu_2 - s^2v_2 + csu_3 + s^2v_3].$$

These equations can be written in matrix form, giving the matrix force-displacement relations for the assembled structure:

$$\begin{Bmatrix} X_1 \\ Y_1 \\ X_2 \\ Y_2 \\ X_3 \\ Y_3 \end{Bmatrix} = AE \begin{bmatrix} 0 & 0 & 0 & 0 & 0 & 0 \\ 0 & 1/l_1 & 0 & -1/l_1 & 0 & 0 \\ 0 & 0 & c^2/l_2 & cs/l_2 & -c^2/l_2 & -cs/l_2 \\ 0 & -1/l_1 & cs/l_2 & 1/l_1 + s^2/l_2 & -cs/l_2 & -s^2/l_2 \\ 0 & 0 & -c^2/l_2 & -cs/l_2 & c^2/l_2 & cs/l_2 \\ 0 & 0 & -cs/l_2 & -s^2/l_2 & cs/l_2 & s^2/l_2 \end{bmatrix} \begin{Bmatrix} u_1 \\ v_1 \\ u_2 \\ v_2 \\ u_3 \\ v_3 \end{Bmatrix}.$$

(1.3)

To generalize this technique to a continuum, we divide the continuum into a finite number of small, nonoverlapping regions which are called *elements* (fig. 1.2). The common boundaries of different elements are called *interelement boundaries*. We may visualize the elements as being connected only at a certain number of discrete points on their common boundaries. These points plus a number of specially chosen points inside the element will be called *nodes*. A finite number of degrees of freedom is associated with each node, and the behavior of the elements will be characterized by these degrees of freedom, which are the generalized coordinates of mechanics. They usually represent some physical quantities, such as displacements, stresses, etc., at

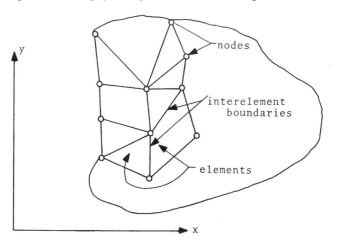

Fig. 1.2 Elements in a continuum.

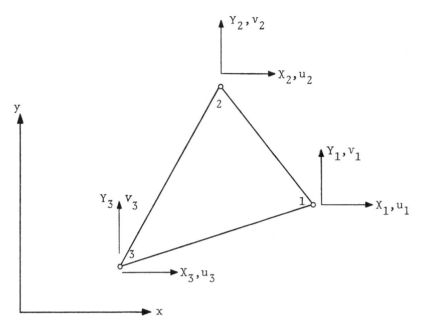

Fig. 1.3 A triangular element.

the nodes. A typical triangular element for plane elasticity, with the vertices chosen as the nodes, is shown in fig. 1.3.

The plane-stress deformation behavior of the triangular region 123 is to be characterized by the relationship between the displacements (u, v) at its vertices and the generalized forces (X, Y) namely, by

$$\begin{Bmatrix} X_1 \\ Y_1 \\ X_2 \\ Y_2 \\ X_3 \\ Y_3 \end{Bmatrix} = \begin{bmatrix} k_{11} & & & & & \\ k_{21} & k_{22} & & & \text{sym} & \\ k_{31} & k_{32} & k_{33} & & & \\ k_{41} & k_{42} & k_{43} & k_{44} & & \\ k_{51} & k_{52} & k_{53} & k_{54} & k_{55} & \\ k_{61} & k_{62} & k_{63} & k_{64} & k_{65} & k_{66} \end{bmatrix} \begin{Bmatrix} u_1 \\ v_1 \\ u_2 \\ v_2 \\ u_3 \\ v_3 \end{Bmatrix}. \quad (1.4)$$

The determination of the k_{ij}'s is one of the basic tasks of the finite-element method. Different elements will normally have the same generalized coordinates at their common nodes. The behavior of the continuum—which is approximated by the assemblage of elements—is to be studied in terms of the entire array of generalized coordinates.

It should be emphasized that the force-displacement relation in (1.1) or (1.2), characterizing a truss element, is exact. However, the use of (1.4) to characterize the triangular region of a plane elastic material is only approximate.

1.3 The Mathematical Concept

In classical continuum mechanics, a field problem is usually described by a set of differential equations with proper boundary conditions, or by the extremum (in many cases, the maximum or the minimum) of a variational principle, if it exists, or by some form of variational statement (incomplete variational principle). For example, the deflection u of a string under a tension N can be described by the second-order differential equation

$$N\frac{d^2u}{dx^2} = p(x)$$

and boundary conditions

$$u = 0 \quad \text{at} \quad x = 0, L;$$

or it can be formulated in terms of the minimum of the functional

$$\Pi = \int_0^L \left[\frac{1}{2}N\left(\frac{du}{dx}\right)^2 + p(x)u\right]dx, \tag{1.5}$$

where u is continuous and equal to zero at $x = 0, L$. As another example, a one-dimensional transient heat conduction problem for the temperature distribution T can be described by the differential equation

$$\frac{\partial T}{\partial t} = \alpha\frac{\partial^2 T}{\partial x^2},$$

where α is the thermal diffusivity, and the boundary conditions are

$$T(0, t > 0) = 1, \qquad \frac{\partial T}{\partial x}(L, t) = 0,$$

and the initial condition is

$$T(x, 0) = 0;$$

or it can be formulated in terms of a variational statement,

$$\delta \Pi = \int_0^L \left[\frac{\partial T}{\partial t} \delta T + \alpha \frac{\partial T}{\partial x} \frac{\partial \delta T}{\partial x} \right] dx = 0, \tag{1.6}$$

or in the form

$$\delta \Pi = \int_0^L \left[\frac{\partial T}{\partial t} \delta T + \frac{\alpha}{2} \delta \left(\frac{\partial T}{\partial x} \right)^2 \right] dx = 0 \tag{1.7}$$

where $\delta T = 0$ at $x = 0$ for $t > 0$, $T(x, 0) = 0$, and $T(0, t > 0) = 1$. Under proper restrictions, the two approaches are mathematically equivalent.

The solution sought for in continuum mechanics usually possesses high-order differentiability, satisfies the differential equation everywhere, and satisfies all the boundary conditions exactly. Even in the classical approximation theories (the Rayleigh-Ritz method or the Galerkin method), only very smooth functions are considered. The situation is different in the finite-element method. The solutions are defined exclusively in terms of a finite number of degrees of freedom and are *piecewise-smooth* functions. The differentiability of these functions is usually lower than the highest order of derivatives in the field equations.

For instance, the two problems mentioned above are governed by second-order differential equations. The finite-element solutions will usually be continuous. Their first partial derivatives will be piecewise continuous and their second partial derivatives will not be defined mathematically at many points. The typical one-dimensional finite-element solution for the string problem shown in fig. 1.4 is a simple piecewise-linear function. The solution u between any two points $x = x_i$ and x_{i+1} is approximated by

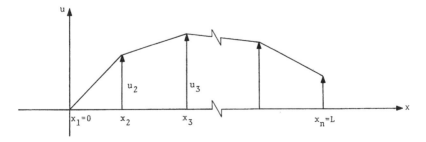

Fig. 1.4 A one-dimensional finite-element solution for a second-order differential equation.

$$u = u_i \left[1 - \frac{x - x_i}{x_{i+1} - x_i} \right] + u_{i+1} \frac{x - x_i}{x_{i+1} - x_i}. \tag{1.8}$$

The first derivative of u has a discontinuity at $x = x_i$. [Here u_i has the physical meaning of being the value of u at $x = x_i$. The u_i are the generalized coordinates of the problem; they are determined by minimizing (1.5)].

For a case in which the governing equation is of fourth order (such as beam or plate bending, or the biharmonic equation), the finite-element solution usually has continuous first partial derivatives only. This case will be examined in detail in ch. 7.

One may visualize the continuum as an assembly of the elements (fig. 1.2 for a two-dimensional domain, or fig. 1.4 for a one-dimensional domain). Each element is itself a continuum. The solution within an individual element is uniquely defined when a proper set of boundary conditions has been prescribed. As an example, for the problem of the string, let the domain $0 \leq x \leq L$ be divided into segments at the points x_i, $i = 1, 2, \cdots, n$ (fig. 1.4). If deflections are prescribed at x_i and x_{i+1}, the deflection for the whole segment $x_i \leq x \leq x_{i+1}$ is uniquely defined. Now, if the deflections prescribed for an element happen to be the same as the solution for this region of the original continuum, the *solution* within the element is, of course, the same as that in this region of the continuum. Thus in searching for the solution of the continuum, we first seek solutions in the individual elements, in terms of their appropriate boundary values. These boundary values at the element level can be thought of as *imaginary* boundary conditions for the purpose of defining the solution within the elements (in the sense that such values depend on the solution of the problem for the entire field). As will be seen later, values for the dependent variables within the element are expressed in terms of values at the nodes of the element boundary by means of an interpolation scheme.

In establishing the solution for the string, imaginary boundary conditions for the element may be chosen as the values of the deflections u_i at the nodal points. The solution within any element $x_i \leq x \leq x_{i+1}$ can then be constructed in terms of the deflections of the two nodes u_i and u_{i+1}. Finally, the deflections of all the nodes are determined by the conditions that the nodal deflections at $x = x_1 = 0$ and $x = x_n = L$ are zero and that each element is in equilibrium with its adjacent elements.

As another example, in an elasticity problem, the imaginary boundary conditions may be chosen to be the values of the displacement field at the boundary of each element, with the requirement that adjoining elements have the same displacements at their common boundary. For any set of the imaginary boundary displacements, there exists a corresponding solution for the displacements, stresses, etc., within each element. The solutions for all the elements, taken together, will clearly provide a continuous displacement field over the entire domain, because the displacement solution for each element will be a continuous function within the element and, by construction, will match the solutions of neighboring elements over all common boundaries. However, the stress (or strain) field from the finite-element solution (which is proportional to the first partial derivatives of the displacements) will, in general, not be continuous over the interelement boundaries. Therefore, the proper value of the boundary displacements must be determined by requiring that all the elements be in equilibrium with their neighbors (i.e., that first partial derivatives match up at boundaries). For an element with a portion of its boundary coinciding with that of the original continuum, the boundary value over that portion must coincide with the prescribed conditions for the continuum. For example, over portions of the boundary where the original prescribed conditions are displacements in the solution, the boundary displacements should be the same as the prescribed displacements. If original prescribed conditions are boundary tractions, the boundary tractions derived from the solution of the element should have the same value as the prescribed tractions.

As another example, suppose we wish to solve the potential problem, which is expressed by the Laplace equation $\nabla^2 \phi = 0$. For any ϕ that is continuous over the boundaries of the elements, we can find the corresponding solution ϕ within each of the elements. The proper values at the boundaries are determined by the condition that the normal gradient of the function for two adjacent elements be the same over the common boundary and that the solution for the elements satisfy the boundary conditions of the original continuum. In establishing the solution for each element using a given set of the imaginary boundary conditions or in satisfying the rules required for the determination of the values in the imaginary boundary conditions, we may try either to satisfy the differential equation directly or to seek the extremum of the appropriate functional.

The construction of a finite-element solution follows a similar procedure. We express the values for the set of imaginary boundary conditions of an element approximately, in terms of a finite number of unknown parameters—usually the generalized coordinates associated with the nodes on the element boundaries. The generalized coordinates of different elements are generally the same when they are associated with a common node. We attempt also to make the boundary values for adjacent elements identical along their entire common boundaries. (The solutions for all elements are then said to be *compatible*.) The solution for each element, which is to be called the local solution, is then obtained analytically, either exactly or approximately. In other words, the element is characterized in terms of a finite number of unknown generalized coordinates whose proper values we must then determine. In practice, the most commonly used approach involves a variational principle. That is, the approximate solution within the element is obtained from the extremum of a certain functional for the element and in terms of the nodal generalized coordinates. These unknowns are then determined from the extremum of another functional (or from the vanishing of the first variation of an incomplete functional) for the entire domain. The functional for the entire domain is usually so set up that the conditions for the extremum of the functional are equivalent to the required conditions for the determination of the value of the imaginary boundary conditions. Evaluating the unknown generalized coordinates from the extremum conditions of the functional is thus equivalent to requiring the approximate satisfaction of those rules in the original continuum problem. It is clear that the approximate solution generally has a very low order of continuity across the interelement boundaries; i.e., its higher derivatives are not defined there.

In the string problem, for example, an approximate solution within the element $x_i \leq x \leq x_{i+1}$ can first be constructed in terms of its nodal deflections by interpolation in the form

$$u = u_i\left(1 - \frac{x - x_i}{x_{i+1} - x_i}\right) + u_{i+1}\frac{x - x_i}{x_{i+1} - x_i} + \alpha\frac{(x_{i+1} - x)(x - x_i)}{(x_{i+1} - x_i)^2}, \qquad (1.9)$$

where u_i and u_{i+1} are the nodal deflections at the nodes x_i and x_{i+1}, respectively. The unknown α can be first determined in terms of u_i and u_{i+1} by minimizing the functional

$$\pi_i = \int_{x_i}^{x_{i+1}} \left[\frac{1}{2}N\left(\frac{du}{dx}\right)^2 + p(x)u\right]dx$$

with respect to α. (For a first-order approximation, α can be set equal to zero.) Then the functional Π of (1.5) can be written as

$$\Pi = \sum_{i=1}^{n-1} \pi_i, \qquad (1.10)$$

with π_i expressed in terms of the nodal values u_i and u_{i+i}. Finally, the unknowns u_i of all the nodes are determined by minimizing Π with respect to u_i subject to the condition that $u_1 = u_{n+1} = 0$.

In the elasticity example, we interpolate the boundary displacements in terms of the nodal generalized coordinates. For example, the boundary displacements along the edges of the triangle in fig. 1.3 are interpolated in terms of the nodal displacements. For the edge between nodes 1 and 2, the displacements u and v may be expressed by

$$u = u_1\left(1 - \frac{s}{l_{12}}\right) + u_2\frac{s}{l_{12}},$$
$$v = v_1\left(1 - \frac{s}{l_{12}}\right) + v_2\frac{s}{l_{12}},$$
(1.11)

where s is measured from node 1 along the edge, and l_{12} is the distance between the two nodes. (It is common practice in the finite-element method to interpolate the boundary displacements of each element in such a way that the displacements will be exactly compatible with those of neighboring elements.) An approximate solution within the element can now be obtained in terms of the nodal unknowns from the extremum of a certain functional, which might be the potential energy or the complementary energy for the element. This allows the approximate potential energy of each individual element to be expressed in terms of its nodal unknowns. Finally, by finding the minimum of the potential energy for the entire domain—which is the sum of the potential energies of all elements in (1.10)—with respect to all these nodal unknowns, we can determine their values. In this process, the equilibrium equations and the original prescribed boundary conditions are satisfied approximately. The first partial derivatives of the displacements of different elements in the direction normal to the element boundaries are generally discontinuous across the interelement boundaries.

1.4 The Underlying Philosophy

At first it may not appear very sophisticated to establish the solution for the continuum by first constructing the solution for each individual element in terms of some imaginary boundary conditions with unknown values and then determining values of the imaginary boundary conditions by imposing particular requirements. However, the basic premise of the finite-element method is that each element is small, so that the variation of the boundary value over its entire boundary will be small. Thus the boundary condition can be approximated by a set of values of the generalized coordinates at the boundary nodes and by some smooth function interpolated between them along the boundaries. For any such set of boundary values a simple approximate solution within the element can also be easily established. The error in the approximation depends upon the size and the number of degrees of freedom of the element. For a sufficiently small element the error will be small. The relationships among the as yet unknown generalized coordinates are expressed by a set of algebraic equations, and the solution of a problem in a continuum is thus reduced to an algebraic process. To improve the approximation one can either divide the domain into smaller elements, or use more degrees of freedom per element, or do both.

The finite-element method is, then, a technique of using local, approximate solutions to build up a solution for an entire domain. One of the most important aspects of the local, approximate representation of a function is that it can also be used to build up a complete description of the function. For instance, any smooth function can be described approximately within a small neighborhood in terms of its value, slope, etc., at one or several points. In the finite-element method we focus on a single element. The local, approximate solution for that element will be formulated and constructed in terms of a finite number of generalized coordinates, loading magnitudes, etc.; That solution will then be used for all other elements in the entire domain. Even for a different problem of a similar type (e.g., elasticity problems of different load, geometry, or boundary conditions), the same or similar elements can be used. In constructing the local, approximate solution, we note that within a small region all smooth functions look like polynomials. We can therefore always express the element boundary conditions by polynomials with unknown coefficients, or equivalently by the nodal generalized coordinates with polynomials as interpolation functions, such as (1.10) along an edge of a triangular element (fig. 1.3). Any external loads can be expressed in terms of polynomials with known coefficients. Within the element, the solution can also be approximated by polynomials or simply by interpolating the boundary value inside the element by polynomials, such as (1.8) for one-dimensional problems, where u_i and u_{i+1} are the boundary values of the element.

The degree of approximation now becomes a question of how good we can make the approximation of a function by polynomials over one element. This, of course, depends on how smooth the original function is, how large the element is, and what degree of complete polynomials are used (the last condition is usually dictated by the number of degrees of freedom of the element). Approximating by polynomials is equivalent to using truncated Taylor-series expansions to represent a function over an element. Thus, for a region where the function is very smooth, or its Taylor-series expansion converges rapidly and has a large radius of convergence, we may use large elements and/or fewer degrees of freedom for an element (which is equivalent to truncating the Taylor series at low order) and still have a good approximation. On the other hand, over a region where the function varies rapidly or its Taylor series has a small radius of convergence, we must use small elements and/or many degrees of freedom per element. In practice, we may not know exactly how the actual solution behaves. However, the *mesh arrangement* (finite-element subdivision) can be estimated and planned in the manner outlined above to obtain economically a good approximate solution.

1.5 Some General Considerations

In the actual construction of a finite-element solution, there are many ways of specifying the boundary values (imaginary boundary conditions) and the approximate solution (interpolation function) within an element. In elasticity problems, for example, we may use the displacement field, the boundary tractions, or a combination of the two, to express the element boundary conditions; the approximate solution within the element can be constructed in terms of displacements, stresses, or both. The different choices will result in different finite-element models. (Pian and Tong 1969, 1972).

As noted previously, we are relatively free to choose the shape of the elements, the nodal points, and the type and the number of degrees of freedom. The choice of element must be based on a compromise between accuracy and convenience. One of the considerations in choosing the shape is that the original domain be fitted nicely by the elements: usually they are in the shape of polygons such as triangles, rectangles, or general quadrilaterals in a two-dimensional domain, and polyhedrons such as tetrahedrons or hexahedrons in a three-dimensional domain. The elements can also be in a shape which can be transformed into a polygon or polyhedron by a simple coordinate transformation (such as the so-called isoparametric transformation). As for choosing the nodes, the first criterion is the ease with which the geometry of the element can be described from the nodal locations. Another criterion is ease of construction and better representation of the local, approximate solution which will be expressed in terms of the generalized coordinates of the nodes. For example, in a one-dimensional problem in which the region is divided into small segments, the end points of segments are natural choices as nodes, as shown in fig. 1.4. However, one may also use a number of points within the segment as the nodal points in order to construct a more accurate local, approximate solution. In a two-dimensional domain that has been divided into triangular elements, the three vertices or the midpoints of the three sides of the triangle are commonly used as nodes. In addition, we may choose a number of nodal points within and/or on the sides of the triangle. The type and number of degrees of freedom at each node are chosen mainly for good representation and easy construction of the approximate solution for the element; this consideration is particularly crucial when we use an interpolation function as the approximate solution.

One of the most important considerations in a good representation is to ensure the compatibility of the boundary values of adjacent elements along their entire common boundaries. When we choose to have the solution of the element in terms of its boundary displacements, as in the elasticity example in sec. 1.3, we would like to make sure that the boundary displacements, which are in terms of the generalized coordinates and interpolation functions, are the same at the interelement boundaries. If we choose to have the element solution in terms of the equilibrium boundary tractions, we would like the tractions to be completely matched over the entire common boundaries of adjacent elements.

1.6 The Displacement Model

So far, general concepts associated with various finite-element methods have been discussed. Most of the existing finite-element models can be categorized by the way boundary values (imaginary boundary conditions) are expressed over the element boundaries and the procedures used to construct approximate solutions within the element for such values (Pian and Tong 1969, 1972). One of the simplest finite-element models is the so-called displacement model (following the terminology of solid mechanics), in which the approximate solution is obtained by interpolating the function from the boundary to the interior of the element. For example, (1.8) can be viewed as an approximate solution to the string problem in the element $x_i \leq x \leq x_{i+1}$; however, u actually interpolates the nodal values u_i and u_{i+1} by a linear function. More generally, if we express boundary values in elasticity problems in terms of displacements, and if the displacements within the element are represented by interpolation functions, then the strains, stresses, etc., can be derived from the approximate displacements. For example, for the displacements given by (1.11) over the boundaries of the triangle in fig. 1.3, the displacement within the triangle can be approximated by simple interpolation in the form

$$u = u_1 f_1(x, y) + u_2 f_2(x, y) + u_3 f_3(x, y),$$
$$v = v_1 f_1(x, y) + v_2 f_2(x, y) + v_3 f_3(x, y). \tag{1.12}$$

The f_i are called *interpolation functions* and are chosen so that the u and v of (1.12) satisfy (1.11) over the side with nodes 1 and 2, and similar equations over the other sides.

For the displacement model, which will be used throughout most of this book, the finite-element method can be developed from a much simpler point of view. We approximate a general function u by

$$u = \sum_i u_i f(\mathbf{x}). \tag{1.13}$$

(Throughout this book vectors, as here, and matrices are denoted by boldface symbols.)

Here the u_i are unknown parameters, and the $f_i(\mathbf{x})$ are interpolation functions. The unknown parameters are determined from the stationary value of variational functionals such as (1.5) or from variational statements such as (1.6) or (1.7), much as in the Rayleigh-Ritz method. The two methods differ by the way in which the unknown parameters and the interpolation functions are chosen. In the Rayleigh-Ritz method, $f_i(\mathbf{x})$ is usually a very smooth function, and u_i is just its amplitude. In the finite-element method, u_i is usually either the value of u or a derivative of u at some discrete point in the domain, and $f_i(\mathbf{x})$ is chosen so that it vanishes everywhere except for a sub-

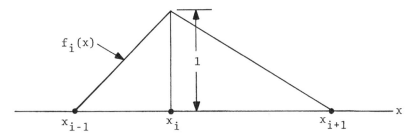

Fig. 1.5 A piecewise-linear representation of the function u of (1.8).

region of the continuum. This subregion is usually a collection of a few elements. The function $f_i(x)$ is usually only piecewise smooth. For a problem governed by differential equations of order n, it is only required that $f_i(x)$ be $(n-1)$-times differentiable (Tong and Pian 1967; Fix and Strang 1969). For example, a piecewise-linear representation of the function u of (1.8) is shown in fig 1.5. Here

$$f_i(x) = \begin{cases} \dfrac{x - x_{i-1}}{x_i - x_{i-1}} & \text{for} \quad x_{i-1} \leq x \leq x_i \\ 1 - \dfrac{x - x_i}{x_{i+1} - x_i} & \text{for} \quad x_i \leq x \leq x_{i+1} \\ 0 & \text{elsewhere.} \end{cases} \quad (1.14)$$

That is, $f_i(x)$ is a linear function over elements $i-1$ ($x_{i-1} \leq x \leq x_i$) and i ($x_i \leq x \leq x_{i+1}$) and vanishes in the region outside these two elements. In this case $f_i(x)$ is continuous, but only piecewise differentiable. The representation of the function u in the form (1.10) is equivalent to

$$u = \sum_i [u_i f_i(x) + \alpha_i g_i(x)],$$

where u_i and α_i are unknown parameters, $f_i(x)$ is the same as that defined in (1.14), and

$$g_i(x) = \begin{cases} \dfrac{4(x_{i+1} - x)(x - x_i)}{(x_{i+1} - x_i)^2} & \text{for} \quad x_i \leq x \leq x_{i+1}. \\ 0 & \text{elsewhere.} \end{cases}$$

This function is shown in fig. 1.6.

For two-dimensional domains, which require the use of expressions such

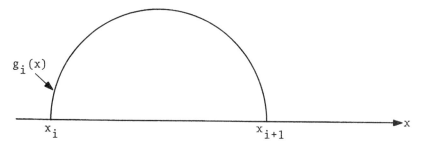

Fig. 1.6 A piecewise-smooth representation of u.

as (1.12), a typical interpolation function $f_i(x, y)$ is the pyramid function shown in Fig. 1.7; that is,

$$f_i(x, y) = \begin{cases} 1 & \text{at node } i \\ \text{a linear function} & \text{over each of the triangles surrounding node } i \\ 0 & \text{elsewhere} \end{cases} \quad (1.15)$$

Since $f_i(\mathbf{x})$ is nonzero only over a few elements around i, the summation in (1.13) will actually involve only a few terms for a given element. For a one-dimensional problem, with the interpolation function defined in (1.14), eq. (1.13) can be written as

$$u = u_i f_i(x) + u_{i+1} f_{i+1}(x) \quad (1.16)$$

for ith element. For a two-dimensional problem with the interpolation

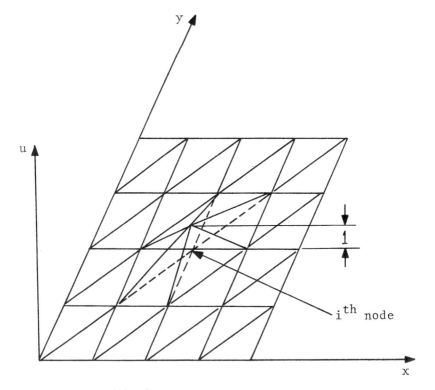

Fig. 1.7 The pyramid function.

function defined by (1.15), u of (1.13) is actually in the form (1.12) over the triangle. The functions in (1.12) or (1.16) are linear polynomials within the elements.

In almost all finite-element applications, one chooses an interpolation function such that within an element the summation in (1.13) involves only the generalized coordinates associated with nodes of the element; for example, in (1.16) within the ith element u depends only on u_i and u_{i+1}, which are the generalized coordinates of the two end nodes, and in (1.12) u also depends only on the nodal values of the element. This procedure is sometimes called *Hermite interpolation*.

In general, when using Hermite interpolation, we may say that the function u is actually interpolated through either the nodal values of u or a derivative of u within the element. However, the interpolation functions are so constructed that u is a smooth function which, depending on the application, possesses a certain order of differentiability over the entire domain. In other words, the interpolation functions are smooth functions within an element and can be constructed independently from those of the rest of the element; but they must be so constructed that when the functions at the nodes along the interelement boundary of two neighboring elements are compatible, the functions all along the corresponding interelement boundary are compatible, that is, the interpolated functions are at least differentiable at the interelement boundaries up to a certain order. In (1.16), when u is compatible at all the nodes, it is a continuous function over the entire domain, as shown in fig. 1.4. In (1.12), if u and v are compatible at nodes in common with their neighboring elements, and if u and v of the neighboring elements also vary linearly along their boundaries as (1.11), then u and v are continuous at the interelement boundaries. As a matter of fact, continuity is sufficient for a second-order differential equation. Since the unknown nodal values are to be determined from the stationary value of a functional such as (1.5) or (1.7), only continuity is required in evaluating the functional.

1.7 Selection of Polynomials for Displacement-Model Interpolation Functions

Polynomials are the almost exclusive choice for the interpolation functions in finite-element applications because they are easy to handle and because local sections of all smooth functions look like polynomials, that is, like truncated Taylor series. The order of the approximation can be measured by the degree of polynomials used for interpolation. In selecting the order and the terms of the polynomials, the most important consideration—aside from the compatibility constraints discussed in the last section—is therefore the *completeness* of the polynomial representation, as discussed in the next paragraph.

Let us write the field variable u within an element in the form

$$u(\mathbf{x}) = \sum_{i=1}^{M} f_i(\mathbf{x}) q_i, \tag{1.17}$$

where q_i is one of the M selected generalized coordinates associated with some discrete nodal points of the element and $\mathbf{x}$ is the spatial coordinate of the domain. By choosing the values of the q_i, so that u represents a complete polynomial of pth degree, then the interpolation is said to be pth-order complete. The error involved in approximating a smooth function u which possesses the first $p + 1$ derivatives within an element will be of order $\varepsilon^{p+1} D^{p+1} u$, where ε is the maximum size of the element and $D^{p+1} u$ is some mean value of the $(p + 1)$th partial derivative of u. In (1.9), for example, u_i corresponds to q_i here and denotes the nodal value of the function at the ith node. If we choose

$$u_i = \alpha_1 + \alpha_2 x_i, \qquad u_{i+1} = \alpha_1 + \alpha_2 x_{i+1},$$

where α_1 and α_2 are constants, then

$$u = \alpha_1 + \alpha_2 x,$$

which is a complete linear polynomial in one-dimensional space (i.e., $p = 1$). The error of using (1.9) in $x_i < x < x_{i+1}$ to approximate a function which possesses continuous second derivatives and has values u_i and u_{i+1} at x_i and x_{i+1}, respectively, will be of order $\varepsilon^2(d^2u/dx^2)$, where $\varepsilon = x_{i+1} - x_i$. This is so because, in general, u can be written as

$$u = u_i + \left(\frac{du}{dx}\right)_{x=x_i}(x - x_i) + \frac{1}{2}(x - x_i)^2\left(\frac{d^2u}{dx^2}\right)_{x=\zeta},$$

where $x_i \leq \zeta \leq x_{i+1}$. If u in (1.17) can be exactly represented by a polynomial

$$u = \alpha_1 + \alpha_2 x + \alpha_3 y$$

in a two-dimensional problem, then it is also first-order complete. If it can be represented by a polynomial

$$u = \alpha_1 + \alpha_2 x + \alpha_3 y + \alpha_4 xy + \alpha_5 x^2 + \alpha_6 y^2$$

exactly, where the α_i are arbitrary constants, then it is second-order complete. Therefore, purely from the point of view of the accuracy of the local representation, we should choose the polynomials for $f_i(\mathbf{x})$ in such a way that the interpolation u is complete to the highest possible value of p, once the number and the type of the generalized coordinates are determined.

In selecting the terms of the polynomial for $f_i(\mathbf{x})$, however, we must also observe the compatibility conditions. For example, suppose continuity is the compatibility requirement, the nodal values of the function are the generalized coordinates, and there are only two common nodes (e.g., nodes 1 and 2 shown in fig. 1.8, on the interelement boundary of the two adjacent elements I and II); then $f_i(x, y)$ for the two elements should be so chosen that the us vary linearly along all the boundaries. When the us are compatible at nodes 1 and 2, then the us are compatible all along the line between nodes 1 and 2; that is, u is continuous at the interelement boundary. If point A is also a node, $f_i(x, y)$ can be chosen such that u is a quadratic function along the boundaries; then u will be compatible all along the interelement boundaries when it is compatible at nodes 1, 2, and A.

It is also desirable (but not necessary) to select the terms of the polynomials in such a way that the interpolation is independent of the orientation of the local coordinate system, a condition known as geometric invariance.

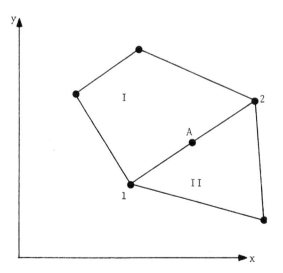

Fig. 1.8 Nodes on an interelement boundary.

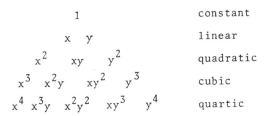

Fig. 1.9 Pascal triangle.

The purpose of such a choice is to avoid having a preferred or biased coordinate direction. One way to envision the achievement of such a property in the two-dimensional problem is to consider the Pascal triangle for the various terms of the polynomial (fig. 1.9).

For geometric invariance, the terms included in the polynomial should be symmetric about the central axis; in the quadratic terms, for example, if x^2 is included, y^2 should also be included.

In the selection of the polynomial, the most important considerations are completeness and compatibility. It is preferable to have geometric invariance here also, but it is not essential. In fact, if we use triangular elements other than isosceles, or rectangular elements of aspect ratio not equal to one, a preferable direction already exists.

To construct the interpolation, the function is assumed to be of the form (for a two-dimensional case)

$$u = \alpha_1 + \alpha_2 x + \alpha_3 y + \alpha_4 xy + \alpha_5 x^2 + \alpha_6 y^2,$$

and the αs are to be expressed in terms of the generalized coordinates. The total number of terms chosen in this equation is the same as the number of generalized coordinates. If the approximate solution is to converge, there must be a minimum value for the order of completeness, which will depend on the order of the differential equations. One should therefore have enough generalized coordinates for an element to ensure that such an order of completeness can be achieved. For a continuum of dimension greater than one, the highest order of complete polynomial which can be selected is usually fixed by the number of the generalized coordinates over the element boundaries, this because of compatibility considerations. Whether one should use interior nodes or not should therefore be judged by whether the excess coordinates associated with the interior nodes can make the interpolation achieve the highest order of completeness.

We shall discuss the construction of general interpolation functions in ch. 6.

1.8 Convergence

We can examine the convergence of the finite-element solution to the exact solution from two points of view: either the number of degrees of freedom per element can be increased to infinity with the element size fixed or else, by division of the domain into smaller and smaller elements, the size of each element can be allowed to shrink to zero with the number of degrees of freedom per element (and hence the form of the local approximate solution) fixed.

The former procedure is not commonly used because increasing the number of degrees of freedom of an element means reformulating the local, approximate solution, which can be tedious and uneconomical. In the latter procedure we only have to subdivide further the domain being considered, which is relatively straightforward. We shall therefore discuss convergence only from the standpoint of reducing the element size to zero.

As noted already, the finite-element method is a procedure for constructing the solution of the entire domain from the local, approximate solutions. There are various methods for accomplishing this. The most common practice is to obtain the finite-element solution from the extremum of a certain functional with respect to all the unknown generalized coordinates, just as in the Rayleigh-Ritz method, except that here the integrand of the functional is evaluated by using the local, approximate solution and its derivatives in each element.

If the finite-element solution is to converge to the exact solution, the local solution must be so constructed that the functional itself converges as the size of the element diminishes. In proving convergence, one only has to show that the convergence of the functional will imply the convergence of the function itself. The proof is beyond the scope of this book, and the reader is referred to the literature (e.g., Tong and Pian 1967; Strang and Fix 1973).

In practice, there are two convergence criteria:

1. The functional itself must be mathematically defined over the entire domain. Now, the local, approximate solutions have only low-order differentiability across the interelement boundaries. In order to be certain that the

functional converges, one must make sure that the discontinuity of local solutions and their derivatives across the interelement boundaries do not make the functional itself mathematically undefined over the entire domain. Consider, for example, a functional of the form

$$\Pi = \int_a^b \left(\frac{du}{dx}\right)^2 dx.$$

If u itself is not continuous within the range $a \leq x \leq b$, Π is not mathematically defined. This problem does not exist if the imaginary boundary conditions of each element have been properly assigned and compatibility along the interelement boundaries has been ensured. If the functional is not defined over the entire domain due to some incompatibility over the interelement boundaries, one cannot say for sure that the solution will converge. Experience with calculations shows that some cases will converge to correct solutions and some will converge to incorrect solutions.

2. The highest derivatives of the dependent variables in the integrand of the functional *in terms of the generalized coordinates* must be able to represent any *constant* within an element as the element size approaches zero.

1.9 Examples

Some simple examples are presented in this section to illustrate several aspects of the finite-element procedure. We will here look at boundary-value problems involving ordinary differential equations. In ch. 2, application to a partial differential equation—Poisson's equation—will provide a means to expand further on the ideas presented here. We start with the following boundary-value problem in the region $a \leq x \leq b$:

$$\frac{d^2 u}{dx^2} = p(x), \qquad u(a) = u(b) = 0. \tag{1.18}$$

The variational statement equivalent to (1.18) is given by $\delta \Pi = 0$, where

$$\Pi = \int_a^b \left[\frac{1}{2}\left(\frac{du}{dx}\right)^2 + p(x)u\right] dx. \tag{1.19}$$

The functional given by (1.19) can be derived directly from (1.18) by multiplying (1.18) by the variation δu and integrating by parts, as shown in app. D.

Equation (1.18) is actually a necessary and sufficient condition for a continuous function u with piecewise-continuous first derivatives to render Π stationary (i.e., $\delta\Pi = 0$), and this can easily be shown by forming $\delta\Pi$ as

$$\delta\Pi = \int_a^b [u'\delta u' + p\delta u]dx = u'\delta u \Big|_b^a - \int_a^b (u'' - p)\delta u\, dx = 0, \qquad (1.20)$$

where the prime indicates differentiation with respect to x. The integrated term vanishes since $\delta u(a) = \delta u(b) = 0$; therefore, for $\delta\Pi$ to vanish for arbitrary δu in $a \leq x \leq b$, the quantity in parentheses must vanish. This yields the differential equation $u'' - p = 0$. We now proceed to solve the boundary-value problem by making use of the variational statement. A description of the Rayleigh-Ritz method is given in app. D; there one assumes an expression for u over the entire region or interval $a \leq x \leq b$ in terms of a finite number of unknown parameters which are to be determined so that $\delta\Pi = 0$. In the finite-element method, $a \leq x \leq b$ is divided into a finite number of elements, and an expression for u is assumed for each element. As will be seen, the unknown values of u (nodal values) at the ends of each element (the nodes) are, in fact, parameters which play a similar role as the parameters that are sought for in the Rayleigh-Ritz procedure.

For the problem defined by (1.18), we make use of (1.19) by first dividing the $a \leq x \leq b$ into $n - 1$ elements (fig. 1.10a). Within an element such as $x_i \leq x \leq x_{i+1}$ (fig. 1.10b), assume that u can be represented by a linear interpolation,

$$u = u_i(1 - \xi) + u_{i+1}\xi, \qquad (1.21)$$

where

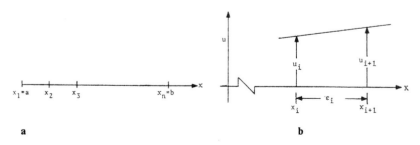

Fig. 1.10 (a) Division of an interval into elements. (b) A typical element, showing nodal values of the function u.

$$\xi = \frac{x - x_i}{x_{i+1} - x_i} = \frac{x - x_i}{\varepsilon_i}, \qquad dx = \varepsilon_i d\xi.$$

The value of u at $x = x_i$ is denoted by u_i, and this is the nodal value. The element size is $\varepsilon_i = x_{i+1} - x_i$. Note that by using linear interpolation, we ensure that the first derivative of u is a constant in each element—an important condition for convergence. There are $n - 1$ elements and n nodes, and by using (1.21), we can write the functional in (1.19) as a sum of the contributions of integrals over each element:

$$\Pi = \sum_{i=1}^{n-1} \int_{x_i}^{x_{i+1}} \left\{ \frac{1}{2} \left(\frac{u_{i+1} - u_i}{\varepsilon_i} \right)^2 + p(x)[u_i(1 - \xi) + u_{i+1}\xi] \right\} dx, \qquad (1.22)$$

or

$$\Pi = \sum_{i=1}^{n-1} \left[\frac{1}{2} \frac{(u_{i+1} - u_i)^2}{\varepsilon_i} + (F_1)_i u_i + (F_2)_i u_{i+1} \right], \qquad (1.23)$$

where

$$(F_1)_i = \int_{x_i}^{x_{i+1}} p(x)(1 - \xi) dx = \varepsilon_i \int_0^1 p(\xi)(1 - \xi) d\xi,$$
$$(F_2)_i = \int_{x_i}^{x_{i+1}} p(x)\xi dx = \varepsilon_i \int_0^1 p(\xi)\xi d\xi. \qquad (1.24)$$

The quantities $(F_1)_i$ and $(F_2)_i$ are therefore weighted integrals of $p(x)$ over the element interval with the interpolation functions $1 - \xi$ and ξ used as weighting functions. If the boundary-value problem in (1.18) is one in solid mechanics where $p(x)$ might be a loading function, for example, then $(F_1)_i$ and $(F_2)_i$ are often referred to as generalized forces at nodes 1 and 2 of element i; in general, however, they need not have this physical meaning.

It is clear from (1.23) that Π is a function of the nodal values $u_1, u_2, \cdots, u_n$. As shown in app. D, Π is rendered stationary with respect to these values when the u_i satisfy the relations

$$\frac{\partial \Pi}{\partial u_i} = 0, \qquad i = 1, 2, \cdots, n. \qquad (1.25)$$

This yields n equations in the n unknowns $u_1, u_2, \cdots, u_n$. Consider a two-element case. For $n = 3$, (1.23) gives

$$\Pi = \frac{1}{2}\frac{(u_2 - u_1)^2}{\varepsilon_1} + (F_1)_1 u_1 + (F_2)_1 u_2 + \frac{1}{2}\frac{(u_3 - u_2)^2}{\varepsilon_2} \\ + (F_1)_2 u_2 + (F_2)_2 u_3. \tag{1.26}$$

The boundary conditions require that $u_1 = u_3 = 0$, so that (1.26) becomes

$$\Pi = \frac{1}{2}\frac{u_2^2}{\varepsilon_1} + [(F_2)_1 + (F_1)_2] u_2 + \frac{1}{2}\frac{u_2^2}{\varepsilon_2}. \tag{1.27}$$

If (1.27) is to be stationary with respect to u_2, then

$$\frac{\partial \Pi}{\partial u_2} = 0 \quad \rightarrow \quad u_2\left(\frac{1}{\varepsilon_1} + \frac{1}{\varepsilon_2}\right) + (F_2)_1 + (F_1)_2 = 0$$

or

$$u_2 = -[(F_2)_1 + (F_1)_2]\frac{\varepsilon_1 \varepsilon_2}{\varepsilon_1 + \varepsilon_2}. \tag{1.28}$$

Consider the following special cases. Suppose, first, that in (1.18) $p(x) = 1$, $a = 0$, and $b = 1$. The continuous solution to (1.18) in this case is $u = x(x - 1)/2$. The finite-element solution for two elements is given by (1.28). From (1.24) we have

$$(F_2)_1 = \varepsilon_1 \int_0^1 \xi \, d\xi = \varepsilon_1/2, \\ (F_1)_2 = \varepsilon_2 \int_0^1 (1 - \xi) \, d\xi = \varepsilon_2/2. \tag{1.29}$$

Then, since $\varepsilon_1 = x_2$ and $\varepsilon_2 = 1 - x_2$ here, (1.28) and (1.29) yield

$$u_2 = -x_2(1 - x_2)\left[\frac{x_2}{2} + \frac{1 - x_2}{2}\right] = \frac{x_2(x_2 - 1)}{2}. \tag{1.30}$$

In this case, (1.30) agrees exactly at x_2 with the continuous solution.

As another special case, consider $p(x) = x$, $a = 0$, and $b = 1$. The continuous solution is $u = x(x^2 - 1)/6$, (1.24) becomes

$$(F_2)_1 = \int_{x_1}^{x_2} x\left(\frac{x - x_1}{x_2 - x_1}\right) dx = x_2^2/3, \\ (F_1)_2 = \int_{x_2}^1 x\left(1 - \frac{x - x_2}{x_3 - x_2}\right) dx = \frac{x_2(1 - x_2)}{2} + \frac{1}{6}(1 - x_2)^2, \tag{1.31}$$

and the corresponding finite-element solution is

$$u_2 = -\left[\frac{x_2^2}{3} + \frac{x_2(1-x_2)}{2} + \frac{(1-x_2)^3}{3}\right]x_2(1-x_2) = \frac{x_2(x_2^2-1)}{6}. \quad (1.32)$$

Equation (1.18) represents a boundary-value problem with homogeneous boundary conditions. We now turn to a problem with nonhomogeneous boundary conditions:

$$\begin{aligned} d^2u/dx^2 &= p(x), \quad \text{with} \quad u = u_a \quad \text{at} \quad x = a \\ \text{and} & \qquad\qquad\qquad du/dx = A \quad \text{at} \quad x = b. \end{aligned} \quad (1.33)$$

This problem is equivalent to $\delta \Pi = 0$, where

$$\Pi = \int_a^b \left[\frac{1}{2}\left(\frac{du}{dx}\right)^2 + pu\right]dx - Au\bigg|_{x=b}. \quad (1.34)$$

The last term in (1.34) states that Au is to be evaluated at $x = b$. That Π as given by (1.34) is correct can be established by the method indicated by eqs. (D.28–D.33) in app. D. The variation of Π is

$$\delta\Pi = (u' - A)\delta u\bigg|_{x=b} - \int_a^b (u'' - p)\delta u\, dx = 0. \quad (1.35)$$

Since $u = u_a$ at $x = a$, $\delta u(a) = 0$. However, $\delta u(b)$ is arbitrary since $u(b)$ is not known, and we must therefore have $u' - A = 0$. This yields the natural boundary condition, as discussed in app. D.

Another example is the Sturm-Liouville equation,

$$\frac{d}{dx}\left[f(x)\frac{du}{dx}\right] - g(x)u = p(x), \quad f, g > 0, \quad (1.36)$$

with boundary conditions $u = u_a$ at $x = a$ and $u' = A$ at $x = b$. The appropriate functional in this case is

$$\Pi = \int_a^b \left[\tfrac{1}{2}f(x)u'^2 + \tfrac{1}{2}g(x)u^2 + p(x)u\right]dx - f(x)Au\bigg|_{x=b}. \quad (1.37)$$

The method used in conjunction with (1.18), involving the linear interpolation

(1.21), can also be used to formulate a finite-element solution to this boundary-value problem. By summing the contributions of all elements, we obtain

$$\Pi = \sum_{i=1}^{n-1} \int_0^1 \left\{ \frac{f(\xi)}{2} \left(\frac{u_{i+1} - u_i}{\varepsilon_i} \right)^2 + \frac{g(\xi)}{2} [u_i(1 - \xi) + u_{i+1}\xi]^2 \right. \\ \left. + p(\xi)[u_i(1 - \xi) + u_{i+1}\xi] \right\} \varepsilon_i d\xi - f(b) A u_n. \tag{1.38}$$

The solution for the u_i can be obtained from the equations $\partial \Pi / \partial u_i = 0$, $i = 1, 2, \cdots, n$.

Now, if we wish to use *quadratic* interpolation for an element with only two nodes, we can proceed by assuming an expression for u such as

$$u = u_i(1 - \xi) + u_{i+1}\xi + \alpha(1 - \xi)\xi. \tag{1.39}$$

After substitution in Π, an additional equation to solve for the parameter α is obtained from the condition $\partial \Pi / \partial \alpha = 0$.

As indicated in (1.25), we must in general deal with a system of equations for the unknown nodal values. If there are a large number of elements, then the approach just discussed must be adapted for use on a computer. The idea is to develop the proper information once for a typical element of the field, and then to assemble the information from all elements to yield an appropriate system of equations. The motivation for such a procedure is as follows.

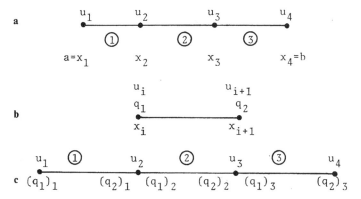

Fig. 1.11 (a) Three elements with nodal values $u_1, \ldots, u_4$. (b, c) Local values q_1, q_2 for a typical element. Note the relationship of q_1, q_2 to the global values $u_1, \ldots, u_4$.

Consider the boundary-value problem consisting of (1.36) and the boundary conditions $u = u_a$ at $x = a$ and $u = u_b$ at $x = b$.

Consider also that the interval $a \leq x \leq b$ is divided into three elements (fig. 1.11a).

The expression for Π as the sum of integrals over each element is given by

$$\Pi = \int_{x_1}^{x_2} \left\{ \frac{f}{2}\left(\frac{u_2 - u_1}{\varepsilon_1}\right)^2 + \frac{g}{2}[u_1(1 - \xi) + u_2\xi]^2 + p[u_1(1 - \xi) + u_2\xi] \right\} dx$$

$$+ \int_{x_2}^{x_3} \{\text{from element 2}\}\, dx + \int_{x_3}^{x_4} \{\text{from element 3}\}\, dx,$$

or

$$\Pi = \left[\tfrac{1}{2}(k_{11}u_1^2 + k_{12}u_1u_2 + k_{21}u_2u_1 + k_{22}u_2^2) - F_1u_1 - F_2u_2 \right]_1$$

$$+ \left[\tfrac{1}{2}(k_{11}u_2^2 + k_{12}u_2u_3 + k_{21}u_3u_2 + k_{22}u_3^2) - F_1u_2 - F_2u_3 \right]_2 \quad (1.40)$$

$$+ \left[\tfrac{1}{2}(k_{11}u_3^2 + k_{12}u_3u_4 + k_{21}u_4u_3 + k_{22}u_4^2) - F_1u_3 - F_2u_4 \right]_3.$$

Here subscripts outside brackets or parentheses denote the element number. For element 1, we then have

$$(k_{11})_1 \equiv \int_{x_1}^{x_2} \left[\frac{f}{\varepsilon_1^2} + g(1 - \xi)^2 \right] dx = \int_0^1 \left[\frac{f}{\varepsilon_1^2} + g(1 - \xi)^2 \right] \varepsilon_1 d\xi,$$

$$(k_{12})_1 \equiv \int_{x_1}^{x_2} \left[-\frac{f}{\varepsilon_1^2} + g(1 - \xi)\xi \right] dx,$$

etc.; for element 3,

$$(k_{12})_3 \equiv \int_{x_3}^{x_4} \left[-\frac{f}{\varepsilon_3^2} + g(1 - \xi)\xi \right] dx = \int_0^1 \left[-\frac{f}{\varepsilon_3^2} + g(1 - \xi)\xi \right] \varepsilon_3 d\xi,$$

etc. The quantities F_1 and F_2 for element 1 are

$$(F_1)_1 = -\int_{x_1}^{x_2} p(1 - \xi) dx, \qquad (F_2)_1 = -\int_{x_1}^{x_2} p\xi\, dx;$$

for element 2,

$$(F_1)_2 = -\int_{x_2}^{x_3} p(1-\xi)dx, \qquad (F_2)_2 = -\int_{x_2}^{x_3} p\xi dx.$$

One should note the similarities in the expressions within the brackets of (1.40) and in the way the quantities k_{ij} for different elements are being computed. The generalized coordinates of our problem—namely, the global values u_1, u_2, u_3, u_4—are related to the local coordinates q_1 and q_2 of fig. 1.11b by

$(q_1)_1 = u_1, \qquad (q_2)_1 = u_2,$
$(q_1)_2 = u_2, \qquad (q_2)_2 = u_3,$
$(q_1)_3 = u_3, \qquad (q_2)_3 = u_4.$

Now, for the ith element,

$$\pi_i = \left[\tfrac{1}{2}(k_{11}q_1^2 + k_{12}q_1q_2 + k_{21}q_2q_1 + k_{22}q_2^2) - F_1q_1 - F_2q_2\right]_i. \tag{1.41}$$

If we require $\delta\pi_i = 0$ with respect to q_1 and q_2 at the element level, the conditions $\partial\pi_i/\partial q_j = 0$ ($j = 1, 2$) lead to a system of equations which hold for the element:

$$\begin{aligned} k_{11}q_1 + k_{12}q_2 &= F_1, \\ k_{21}q_1 + k_{22}q_2 &= F_2. \end{aligned} \tag{1.42}$$

Equations (1.42), which define the relation between the qs and Fs for the element, can be represented in matrix form as

$$\mathbf{kq} = \mathbf{F},$$

where

$$\mathbf{k} = \begin{bmatrix} k_{11} & k_{12} \\ k_{21} & k_{22} \end{bmatrix}, \quad \mathbf{q} = \begin{Bmatrix} q_1 \\ q_2 \end{Bmatrix}, \quad \mathbf{F} = \begin{Bmatrix} F_1 \\ F_2 \end{Bmatrix}. \tag{1.43}$$

The quantities $\mathbf{k}$ and $\mathbf{F}$ are often referred to as the *element-stiffness matrix* and the *element-force vector*, respectively, Use of the words "stiffness" and "force" are carry-overs from structural mechanics, and since the physical meaning of the matrices depends on the problem, the connotations following from these names should not be taken literally. In general, one can think of

k as the transfer matrix, **q** as the input vector, and **F** as the output vector for the element.

The requirement that $\delta \Pi = 0$ for the entire system of elements leads to the notion of the *system* or *master* matrix **K**.

We must show next how each element matrix **k** contributes to **K**, so that a systematic procedure can be developed for forming **K** by first finding **k** for each element of the domain and then adding its contribution to the appropriate location (row and column) of the master matrix **K**. In our three-element example,

$$\Pi = \sum_{i=1}^{3} \left[\frac{1}{2}(k_{11}q_1^2 + k_{12}q_1q_2 + k_{21}q_2q_1 + k_{22}q_2^2) - F_1q_1 - F_2q_2 \right]_i. \qquad (1.44)$$

It should be clear, for example, that the local values q_1 and q_2 in the brackets of element 2 ($i = 2$) of (1.44) do in fact refer to the global values u_2 and u_3 as shown in (1.40), and that $(k_{12})_1$ is not, in general, the same as $(k_{12})_2$. In terms of global values, (1.44) is equivalent to

$$\Pi = \tfrac{1}{2}(K_{11}u_1^2 + K_{12}u_1u_2 + K_{21}u_2u_1 + K_{22}u_2^2 + K_{13}u_1u_3 + K_{23}u_2u_3 \\ + \cdots + K_{34}u_3u_4 + K_{44}u_4^2) - Q_1u_1 - Q_2u_2 - Q_3u_3 - Q_4u_4. \qquad (1.45)$$

The requirement that $\partial \Pi / \partial u_i = 0$ ($i = 1, 2, 3, 4$) leads to a system of four equations which are written in matrix form as

Ku = Q,

where

$$\mathbf{K} = \begin{bmatrix} K_{11} & K_{12} & K_{13} & K_{14} \\ K_{21} & K_{22} & K_{23} & K_{24} \\ K_{31} & K_{32} & K_{33} & K_{34} \\ K_{41} & K_{42} & K_{43} & K_{44} \end{bmatrix}, \quad \mathbf{u} = \begin{Bmatrix} u_1 \\ u_2 \\ u_3 \\ u_4 \end{Bmatrix}, \quad \mathbf{Q} = \begin{Bmatrix} Q_1 \\ Q_2 \\ Q_3 \\ Q_4 \end{Bmatrix}. \qquad (1.46)$$

The master matrix **K** refers to the overall system. The elements of **K** in terms of element-matrix quantities are

$$K_{11} = (k_{11})_1, \; K_{12} = (k_{12})_1, \; \cdots, \; K_{21} = (k_{21})_1, \; K_{22} = (k_{22})_1 + (k_{11})_2, \\ K_{23} = (k_{12})_2, \; K_{33} = (k_{22})_2 + (k_{11})_3, \; K_{34} = (k_{12})_3, \; \cdots, \qquad (1.47) \\ Q_1 = (F_1)_1, \; Q_2 = (F_2)_1 + (F_1)_2, \; \cdots.$$

Note, for example, that K_{22} consists of k_{22} from element 1 and k_{11} from element 2. Similarly, Q_2 consists of element forces from elements 1 and 2. Examination of these relations reveals that for the general case of many elements, one can determine the master-matrix component K_{IJ} by adding up the appropriate components k_{lm} of the element matrices. This is done in a straightforward manner by developing a transformation relating local subscripts l and m ($l, m = 1, 2$) of an element to global subscripts I and J ($I, J = 1, 2, 3, 4$); we shall use the $l \to I$, $m \to J$, to mean that k_{lm} is a component to be added to make up the final value of K_{IJ} and that q_l and q_m are the same as u_I and u_J, respectively. We shall also use the notation $q_l \to u_I$, $q_m \to u_J$, and $k_{lm} \to K_{IJ}$. Thus, for the first element,

$q_1 \to u_1 \quad q_2 \to u_2 \quad k_{11} \to K_{11}$
$1 \to 1 \quad\quad 2 \to 2 \quad\quad k_{12} \to K_{12}$
$ k_{22} \to K_{22}.$

For the second element,

$q_1 \to u_2 \quad q_2 \to u_3 \quad k_{11} \to K_{22}$
$1 \to 2 \quad\quad 2 \to 3 \quad\quad k_{12} \to K_{23}$
$ k_{22} \to K_{33}$

For the ith element the local element subscripts 1 and 2 transform as $1 \to i$ and $2 \to i + 1$, so that

$$(k_{11})_i \to K_{ii}, \quad (k_{12})_i \to K_{i,\,i+1}, \quad (k_{21})_i \to K_{i+1,\,i}, \quad (k_{22})_i \to K_{i+1,\,i+1}. \tag{1.48}$$

Once the relations involving such a transformation of subscripts are defined for the computer, it can do all the work of adding the element k_{ij} in the master matrix **K**. Note that the transformation in (1.48) holds for the case of a two-node element with a single degree of freedom per node. In later chapters, corresponding transformations will be established for the cases of three or more nodes per element and more than one degree of freedom per node.

It is instructive at this point to write several of the previous relations in more compact form. Equation (1.45) can be written as

$$\Pi = \sum_{i=1}^{n} \sum_{j=1}^{n} \frac{1}{2} K_{ij} u_i u_j - \sum_{i=1}^{n} u_i Q_i, \tag{1.49}$$

or in matrix form as

$$\Pi = \tfrac{1}{2} \mathbf{u}^T \mathbf{K} \mathbf{u} - \mathbf{u}^T \mathbf{Q}. \tag{1.50}$$

For each individual element we have

$$\pi_i = \tfrac{1}{2}\mathbf{q}^T\mathbf{k}\mathbf{q} - \mathbf{q}^T\mathbf{F}, \tag{1.51}$$

so that, for the overall system, we can write

$$\Pi = \sum_{i=1}^{n}\left[\frac{1}{2}\mathbf{q}^T\mathbf{k}\mathbf{q} - \mathbf{q}^T\mathbf{F}\right] = \frac{1}{2}\mathbf{u}^T\mathbf{K}\mathbf{u} - \mathbf{u}^T\mathbf{Q}. \tag{1.52}$$

With the notation of (1.52), the condition $\delta\Pi = 0$ with respect to $\mathbf{u}$ leads to the system of equations $\mathbf{K}\mathbf{u} = \mathbf{Q}$.

Once the system of algebraic equations for the unknown nodal values (or generalized coordinates) $\mathbf{u}$ is determined, appropriate boundary conditions or constraints must be applied. For the problem described by equations (1.46), the boundary conditions in terms of nodal values are given by $u_1 = u_a$ and $u_4 = u_b$ (see fig. 1.10a). Now, eqs. (1.46) are a consequence of arbitrary variations on u_i ($i = 1, 2, 3, 4$), which in turn has led to the system of equations

$$\begin{aligned}
\frac{\partial \Pi}{\partial u_1} &= 0 \to K_{11}u_1 + K_{12}u_2 + K_{13}u_3 + K_{14}u_4 = Q_1, \\
\frac{\partial \Pi}{\partial u_2} &= 0 \to K_{21}u_1 + K_{22}u_2 + K_{23}u_3 + K_{24}u_4 = Q_2, \\
\frac{\partial \Pi}{\partial u_3} &= 0 \to K_{31}u_1 + K_{32}u_2 + K_{33}u_3 + K_{34}u_4 = Q_3, \\
\frac{\partial \Pi}{\partial u_4} &= 0 \to K_{41}u_1 + K_{42}u_2 + K_{43}u_3 + K_{44}u_4 = Q_4.
\end{aligned} \tag{1.53}$$

Since u_1 and u_4 are, in fact, prescribed, they cannot have arbitrary variations (i.e., $\delta u_1 = \delta u_4 = 0$), and so the first and fourth of eqs. (1.53) do not hold. More precisely, u_1 and u_4 should be replaced by u_a and u_b, respectively, and only the second and third equations are correct and are to be used for evaluation of the unknowns u_2 and u_3. We can, however, replace the first equation by $u_1 = u_a$ and the fourth equation by $u_4 = u_b$. By applying this procedure, we obtain a set of four equations for the four unknowns u_1, u_2, u_3, u_4:

$$\begin{aligned}
u_1 &= u_a, \\
K_{21}u_1 &+ K_{22}u_2 + K_{23}u_3 + K_{24}u_4 = Q_2, \\
K_{31}u_1 &+ K_{32}u_2 + K_{33}u_3 + K_{34}u_4 = Q_3, \\
u_4 &= u_b.
\end{aligned} \tag{1.54}$$

The coefficients of the unknown u_i in (1.54) no longer form a symmetric matrix, a fact which can create difficulties in large-scale computation. This situation can be remedied by moving the first and the fourth terms of the second and the third equations to the right-hand sides and using u_a and u_b for u_1 and u_4 in these two equations. We then have

$$
\begin{aligned}
u_1 + 0 + 0 + 0 &= u_a, \\
0 + K_{22}u_2 + K_{23}u_3 + 0 &= Q_2 - K_{21}u_a - K_{24}u_b, \\
0 + K_{32}u_2 + K_{33}u_3 + 0 &= Q_3 - K_{31}u_a - K_{34}u_b, \\
0 + 0 + 0 + u_4 &= u_b.
\end{aligned}
\quad (1.55)
$$

This is called the *constrained set of equations*. The reason for introducing the constrained set of equations is its convenience in large-scale computation, a factor that will be examined in detail in ch. 3.

In general, one will have n nodes and $n-1$ elements, and the boundary conditions will be $u_1 = u_a$ and $u_n = u_b$. A constrained set of equations is then obtained by replacing appropriate terms in the first and last equations by zeroes and adjusting their right-hand sides so that these equations are equivalent to $u_1 = u_a$ and $u_n = u_b$, respectively. For the remaining equations, one places the known quantities ($u_1 = u_a$ and $u_n = u_b$ in this case) on the right-hand sides. This process leads to

$$
\begin{aligned}
u_1 + 0 + 0 + \cdots + 0 + 0 &= u_a, \\
0 + K_{22}u_2 + K_{23}u_3 + \cdots + K_{2,n-1}u_{n-1} + 0 &= Q_2 - K_{21}u_a - K_{2n}u_b, \\
0 + K_{32}u_2 + K_{33}u_3 + \cdots + K_{3,n-1}u_{n-1} + 0 &= Q_3 - K_{31}u_a - K_{3n}u_b, \\
&\quad \vdots \\
0 + K_{n-1,2}u_2 + K_{n-1,3}u_3 + \cdots + K_{n-1,n-1}u_{n-1} + 0 &= Q_{n-1} - K_{n-1,1}u_a \\
&\qquad - K_{n-1,n}u_b, \\
0 + 0 + 0 + \cdots + 0 + u_n &= u_b.
\end{aligned}
\quad (1.56)
$$

It should be clear now how this is to be put in a form suited to a computer. First, it is assumed that the system matrix **K** and the vector **Q** have been obtained. For this problem all elements in the first and last rows of the **K** matrix are set equal to zero except for the K_{11} and K_{nn} terms, which are set equal to unity. In this regard the first and nth columns are also set equal to zero except for diagonal terms. In addition, the first and nth elements of the **Q** vector are set equal to u_a and u_b, respectively, and, as shown in (1.56), the indicated terms are subtracted from the other elements of the **Q** vector. The problem is thereby reduced to the solution of what we shall refer to as a constrained set of equations, which in this case may be written as

$$\mathbf{K}^*\mathbf{u} = \mathbf{Q}^*, \tag{1.57}$$

where

$$\mathbf{K}^* = \begin{bmatrix} 1 & 0 & 0 & \cdots & 0 & 0 \\ 0 & K_{22} & K_{23} & \cdots & K_{2,n-1} & 0 \\ 0 & K_{32} & K_{33} & \cdots & K_{3,n-1} & 0 \\ \vdots & \vdots & \vdots & \vdots & \vdots & \vdots \\ 0 & K_{n-1,2} & K_{n-1,3} & \cdots & K_{n-1,n-1} & 0 \\ 0 & 0 & 0 & \cdots & 0 & 1 \end{bmatrix},$$

$$\mathbf{Q}^* = \begin{Bmatrix} u_a \\ Q_2 - K_{21}u_a - K_{2n}u_b \\ Q_3 - K_{31}u_a - K_{3n}u_b \\ \vdots \\ Q_{n-1} - K_{n-1,1}u_a - K_{n-1,n}u_b \\ u_b \end{Bmatrix}. \tag{1.58}$$

The same procedure is applicable if we had decided to constrain u_2 instead of u_1, in which case the second row and column are affected instead of the first. Further discussion with additional examples involving constrained sets of equations will be given in ch. 3.

In summary, the finite-element procedure consists of:

1 discretization of the physical system

2 characterization of the element or elements (i.e., determination of the k_{ij} from variational statements and proper interpolation procedures)

3 assembly of the local element k_{ij} and F_i to obtain the global or master K_{IJ} and Q_I

4 application of appropriate constraints to yield a solvable system of algebraic equations.

After all this has been done, the subsequent steps fall into the category of well-known numerical techniques.

Problems

In several of the following problems, familiarity with the material in app. D is assumed.

1 It is known that the problem of solving the Sturm-Liouville equation in $0 \le x \le L$—that is,

$$\frac{d}{dx}\left[p(x)\frac{du}{dx}\right] - r(x)u = f(x)$$

with boundary conditions such as $u = \bar{u}_1$ at $x = 0$ and $u = \bar{u}_L$ at $x = L$—is mathematically equivalent to the problem of finding the extremum of the functional

$$\Pi = \frac{1}{2}\int_0^L \left[p(x)\left(\frac{du}{dx}\right)^2 + r(x)u^2\right]dx + \int_0^L f(x)u\,dx$$

for all smooth functions u satisfying the boundary conditions. Here $p(x)$, $r(x)$, and $f(x)$ are piecewise-continuous functions. Divide the interval $0 \le x \le L$ into n equal segments at points $x_{i+1} = iL/n$ ($i = 0, 1, 2, \cdots, n$) and show that Π can be written as

$$\Pi = \sum_{i=1}^n \left\{\frac{1}{2}\int_{x_i}^{x_{i+1}}\left[p(x)\left(\frac{du}{dx}\right)^2 + r(x)u^2\right]dx + \int_{x_i}^{x_{i+1}} f(x)u\,dx\right\}$$

if u is continuous at all points $x = x_i$.

2 (a) Consider the following boundary conditions for the segment bounded by x_i and x_{i+1}: $u = q_i$ at $x = x_i$ and $u = q_{i+1}$ at $x = x_{i+1}$. This segment is called the ith element. Assuming $(x_{i+1} - x_i)/L \ll 1$, construct a simple approximate solution of u for this segment. (b) One of the simplest approximate solutions (representations) for u in the ith element is

$$u = q_i(1 - y) + q_{i+1}y,$$

where $y = (x - x_i)/h$ and $h = x_{i+1} - x_i$. Assuming that the Taylor series for u exists, and that the values q_i and q_{i+1} are exact, estimate the error for u within the element in terms of h. What is the error in du/dx?

3 (a) Using the linear representation for u given in prob. 2(b), and assuming that $p(x) = p_i$, $r(x) = r_i$, and $f(x) = f_i$ are constants within the ith element, find the following quantities, which are associated with problem 1:

$$U_i = \frac{1}{2}\int_{x_i}^{x_{i+1}}\left[p(x)\left(\frac{du}{dx}\right)^2 + r(x)u^2\right]dx \quad \text{and} \quad F_i = \int_{x_i}^{x_{i+1}} f(x)u\,dx.$$

Estimate the error in U_i in terms of h, assuming that the values of q_i, q_{i+1}, p_i, r_i, and f_i are exact. (b) In conjunction with prob. 1, consider the case where the domain is divided into two elements. With the linear representation for u, and constants p_i, r_i, and f_i in each element, find Π of prob. 1 in terms of the q_i. (c) Since u at $x = 0 (= x_1)$ and $L (= x_3)$ are given by $\bar{u}_1$ and $\bar{u}_L$, respectively, in the notation of prob. 1, we must have $q_1 = \bar{u}_1$ and $q_3 = \bar{u}_L$. Find q_2 from the extremum of Π obtained in part (b). (d) Compare q_2 with the exact solution of prob. 1 at $x = x_2 (= L/2)$ for the case where $p(x) = 1$, $r(x) = 0$, and $f(x) = x$.

4 Consider the following boundary-value problem for the region $0 \le x \le 1$:

$$\frac{d^2 u}{dx^2} - u = 5, \quad u(0) = 5, \quad u(1) = 6.$$

Use linear interpolation together with an appropriate variational statement to obtain a finite-element solution. Divide the region into 2, 3, and 4 elements and compare successive results with the exact solution.

5 Redo prob. 4, with the new boundary conditions $u(0) = 5$ and $du(1)/dx = 2$.

6 Heat conduction in a particular circular fin of uniform thickness can be described by the normalized equation

$$\frac{d^2 u}{dr^2} + \frac{1}{r}\frac{du}{dr} - 4u = 0, \quad \frac{1}{2} \le r \le 1.$$

Normalized boundary conditions involving a prescribed temperature at the inner radius and insulation at the outer boundary are given by $u(\frac{1}{2}) = 1$ and $du(1)/dr = 0$. (a) Obtain an equivalent variational statement for this problem. (b) Develop the finite-element set of equations, using matrix representation, for the case of n elements. (c) Use linear interpolation to obtain finite-element solutions for two and three elements. (d) Compare with the exact solution, which involves modified Bessel functions

$$\left(u_{ex} = \frac{K_1(2) I_0(2r) + I_1(2) K_0(2r)}{I_0(1) K_1(2) + K_0(1) I_1(2)} \right).$$

7 Redo prob. 6 using quadratic interpolation and compare the results with those of linear interpolation with three elements.

8 (a) Determine the associated variational statement for the problem

$$\frac{d^2y}{dx^2} + y = 4x, \quad y(0) = y(1) = 0.$$

(b) Find solutions by the Rayleigh-Ritz method using both quadratic and cubic polynomials. (c) Obtain a finite-element solution using linear interpolation and four elements. Compare with the results of (b) and the exact solution.

9 The deflection of a beam on an elastic foundation can be expressed in nondimensional form by the following equation and boundary conditions

$$\frac{d^4u}{dx^4} + u = 1,$$

$$u(0) = \frac{d^2u(0)}{dx^2} = 0,$$

$$u(1) = \frac{d^2u(1)}{dx^2} = 0.$$

Derive the associated variational problem and natural boundary conditions.

10 (a) Use the Rayleigh-Ritz method to obtain an approximate solution of the problem

$$\frac{d}{dx}\left(x\frac{dy}{dx}\right) + y = x, \quad y(0) = 0, \quad y(1) = 1$$

in the form

$$y \approx x + x(1 - x)(c_1 + c_2 x),$$

where c_1 and c_2 are constants and where the variational problem is expressed in the form

$$\int_0^1 \left[\frac{d}{dx}\left(x\frac{dy}{dx}\right) + y - x\right]\delta y \, dx = 0.$$

(b) Obtain a finite-element solution using linear interpolation with four elements, and compare with the Rayleigh-Ritz solution.

2 The Finite-Element Method for Poisson's Equation

2.1 Introduction

Poisson's equation, and its special case, the Laplace equation, are encountered in many parts of mathematical physics. The temperature over surfaces or volumes, potential fields, torsion of rods, deflection of membranes, etc., all can be described in terms of Poisson's equation. The reader should apply the techniques introduced in this chapter to a field problem he is familiar with. In this way he will be able to appreciate the physical reasoning behind the mathematical procedure.

Solving the two-dimensional Poisson equation by the finite-element method is one of the simplest problems involving most of the features of this method. In this chapter we shall consider this problem in detail, using both triangular and rectangular elements. The basic steps in the setting up of the finite-element equations are, however, applicable to more general cases.

2.2 Mathematical Formulation

In a two-dimensional domain A, the equation we wish to solve is

$$\nabla^2 \phi = f(x, y). \tag{2.1}$$

We take the boundary conditions to be

$$\phi(x, y) = \bar{\phi} \quad \text{on} \quad \partial A_\phi, \tag{2.2}$$

$$\frac{\partial \phi(x, y)}{\partial \nu} = \bar{\phi}_\nu \quad \text{on} \quad \partial A_\nu, \tag{2.3}$$

where $\partial A = \partial A_\phi + \partial A_\nu$ denotes the boundary of A; ∂A_ϕ and ∂A_ν are the parts of the boundary where ϕ and $\partial \phi / \partial \nu$ are prescribed, respectively. The term $\partial \phi / \partial \nu$ denotes the normal derivative along the boundaries; that is,

$$\frac{\partial \phi}{\partial \nu} = \boldsymbol{\nu} \cdot \nabla \phi, \tag{2.4}$$

where $\boldsymbol{\nu}$ is a unit normal vector pointing outward at the boundaries and $\nabla \phi$ is the gradient of ϕ. Both $\bar{\phi}$ and $\bar{\phi}_\nu$ are given functions along the boundaries.

Variational formulations can also be used to express Poisson's equation. We shall consider the one that corresponds to a minimum principle, namely,

$$\pi(\phi) = \tfrac{1}{2}\int_A (\nabla\phi)^2\, dA + \int_A f(x,y)\phi\, dA - \int_{\partial A_\nu} \bar{\phi}_\nu \phi\, ds, \tag{2.5}$$

where

$$(\nabla\phi)^2 = \left(\frac{\partial\phi}{\partial x}\right)^2 + \left(\frac{\partial\phi}{\partial y}\right)^2 \quad \text{and} \quad dA = dxdy, \tag{2.6}$$

and ds is a line segment along the boundary of A. The function π is called the potential energy.

Theorem: Of all continuous functions ϕ satisfying (2.2), the exact solution of (2.1)–(2.3) is distinguished by the minimum of the potential energy. A continuous function that satisfies (2.2) is called an admissible function.

Proof: Let ϕ be the function satisfying (2.2) and associated with the minimum of π in (2.5). Consider another function

$$\psi = \phi(x,y) + \varepsilon\eta(x,y)$$

that also satisfies (2.2); that is, the function η satisfies

$$\eta(x,y) = 0 \quad \text{on} \quad \partial A_\phi \tag{2.7}$$

and ε is an arbitrary constant. We can now evaluate the difference

$$\pi(\psi) - \pi(\phi) = \delta\pi(\phi, \varepsilon\eta) + \delta^2\pi(\varepsilon\eta), \tag{2.8}$$

where

$$\delta\pi(\phi, \varepsilon\eta) = \varepsilon\left\{\int_A \left[\frac{\partial\phi}{\partial x}\frac{\partial\eta}{\partial x} + \frac{\partial\phi}{\partial y}\frac{\partial\eta}{\partial y}\right] dA \right. \tag{2.9}$$

$$\left. + \int_A f(x,y)\eta\, dA - \int_{\partial A_\nu} \bar{\phi}_\nu \eta\, ds\right\},$$

$$\delta^2\pi(\varepsilon\eta) = \frac{\varepsilon^2}{2}\int_A \left[\left(\frac{\partial\eta}{\partial x}\right)^2 + \left(\frac{\partial\eta}{\partial y}\right)^2\right] dA. \tag{2.10}$$

Since $\pi(\phi)$ is the minimum, we must have $\pi(\psi) \geq \pi(\phi)$, and thus

$$\delta\pi = \delta^2\pi \geq 0 \tag{2.11}$$

for any continuous function η satisfying (2.7). This will require $\delta\pi = 0$. Now, $\delta^2\pi$ is always greater than or equal to zero and is proportional to ε^2, while $\delta\pi$ is only proportional to ε. If $\delta\pi \neq 0$ for arbitrary $\varepsilon\eta$, we can hold the function η fixed and choose ε so small that the absolute value of $\delta\pi$ is much greater than ε^2; that is, the sign of $\delta\pi + \delta^2\pi$ depends only on the sign of $\delta\pi$. Obviously (2.11) cannot be satisfied, because the sign of $\delta\pi$ depends on the sign of ε, which can be positive or negative. ∎

Consider the consequence of $\delta\pi = 0$. For convenience, we denote the function $\varepsilon\eta$ by $\delta\phi$, and (2.7) becomes

$$\delta\phi(x, y) = 0 \quad \text{on} \quad \partial A_\phi. \tag{2.12}$$

The first integral in (2.9) can be written in a slightly different form by the use of the divergence theorem

$$\int_A \left(\frac{\partial \phi}{\partial x} \frac{\partial \delta\phi}{\partial x} + \frac{\partial \phi}{\partial y} \frac{\partial \delta\phi}{\partial y} \right) dA = \int_{\partial A} \frac{\partial \phi}{\partial \nu} \delta\phi \, ds - \int_A \left(\frac{\partial^2 \phi}{\partial x^2} + \frac{\partial^2 \phi}{\partial y^2} \right) \delta\phi \, dA.$$

Substituting into (2.9), we obtain

$$\begin{aligned}\delta\pi(\phi, \delta\phi) = &-\int_A \left(\frac{\partial^2 \phi}{\partial x^2} + \frac{\partial^2 \phi}{\partial y^2} - f(x, y) \right) \delta\phi \, dA \\ &+ \int_{\partial A_\nu} \left(\frac{\partial \phi}{\partial \nu} - \bar{\phi}_\nu \right) \delta\phi \, ds \\ &+ \int_{\partial A_\phi} \frac{\partial \phi}{\partial \nu} \delta\phi \, ds = 0,\end{aligned} \tag{2.13}$$

where $\partial A = \partial A_\phi + \partial A_\nu$. The restriction of $\delta\phi$ given in (2.12) implies that the last integral in (2.13) vanishes identically. If (2.13) is to be satisfied for arbitrary $\delta\phi$ over A and ∂A_ν, however, the integrands of both the first and second integral must vanish separately, and this leads to the governing equation and natural boundary conditions.

2.3 Rigid and Natural Boundary Conditions

We usually call $\delta\pi(\phi, \delta\phi)$ the first variation of π, and $\delta\phi$ the variational quantity. The vanishing of the boundary integral over ∂A_ϕ in (2.13) is due to vanishing of the variational quantity. This is equivalent to requiring that $\phi = \bar{\phi}$ on ∂A_ϕ. This type of boundary condition is called a *rigid boundary condition*. It can be shown that if a variational approach is used to formulate the problem, this condition must be satisfied in order to obtain a unique solution. The boundary integral over ∂A_ν vanishes because the coefficient of the variational quantity is zero, that is, because the boundary condition (2.3) is satisfied. This is called a *natural boundary condition*, and it is a consequence of $\delta\pi = 0$ for arbitrary $\delta\phi$ on ∂A_ν. Similarly the vanishing of the first integral, which is the differential equation (2.1), is also implied by $\delta\pi = 0$ for arbitrary $\delta\phi$ over A.

In constructing a solution based on the extremum of a functional, we are seeking the solution based on $\delta\pi = 0$ for arbitrary admissible $\delta\phi$. We must therefore require that the rigid boundary conditions be satisfied, while the natural boundary conditions as well as the differential equations will result from $\delta\pi = 0$.

It should be noted that the rigid boundary conditions, the natural boundary conditions, and the form of the governing equation are not intrinsic properties of a physical problem. We can formulate a problem in different ways and have different forms of governing differential equations. Whether a boundary condition is rigid or natural depends upon how we formulate the functional for the variational principle, because the rigid conditions are associated with the variational quantities themselves, and the natural conditions are associated with the coefficients of the variational quantities over the boundary integrals. It makes no difference what form of the governing differential equation we use and what the types of boundary conditions are if we are seeking the exact solution, because all forms of the governing equations are equivalent and all the boundary conditions are to be satisfied exactly. In constructing the approximate solution based on a variational principle, however, we have to make such a distinction. The rigid conditions have to be enforced, while the differential equations and the natural boundary conditions will be implied from the vanishing of the first variation. Of course, in this case both the differential equations and the natural boundary conditions will only be satisfied approximately.

In using the finite-element method to solve this problem, we subdivide the domain A into a finite number of discrete elements, construct a local, approximate representation of ϕ for each element, and obtain the final solution based on the minimization of π by using the local, approximate ϕ. The local, approximate ϕ of an element satisfies the rigid boundary conditions if the element has a part of its boundary coincident with ∂A_ϕ. The following sec-

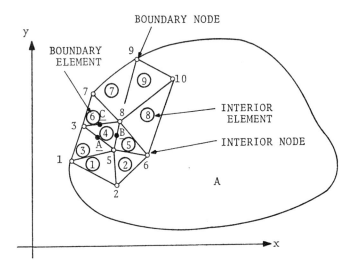

Fig. 2.1 Subdivision of a domain into triangular elements.

tions present a sequence of steps involved in constructing a finite-element solution.

2.4 Subdivision of the Domain

The domain A can be divided into a finite number of discrete polygons such as triangles and/or quadrilaterals, that are called *elements* (fig. 2.1). The vertices of the elements are chosen as the nodal points. The nodes lying on the boundary of the domain are called *boundary nodes*. An element with at least two boundary nodes is called a *boundary element*, while the elements having only one or no boundary nodes are called *interior elements*. It is common practice to approximate the curved boundary of the domain by straight-line segments between the boundary nodes.

All the nodes and all the elements are numbered with positive consecutive integers for purposes of bookkeeping and identification. The number assigned to each node is called the *global nodal number*. In fig. 2.1, the circled numbers are the element numbers and the others are the nodal numbers. Each element is associated with a fixed set of nodes. For instance, the first element has nodes 1, 2, and 5; the second element has nodes 2, 6, and 5, etc. A typical triangular element with nodes I, J, and L is shown in fig. 2.2. The coordinates of the nodes are (x_I, y_I), (x_J, y_J), and (x_L, y_L).

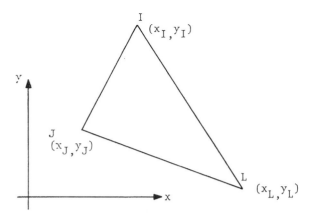

Fig. 2.2 A typical triangular element.

2.5 Local Approximate Representation ϕ

As noted in the previous chapter, the most general approximate representation of a smooth function over a small area of an element is a polynomial of finite degree. We may therefore represent ϕ in the form

$$\phi(x, y) = \alpha_1 + \alpha_2 x + \alpha_3 y + \alpha_4 xy + \alpha_5 x^2 + \cdots. \tag{2.14}$$

Terms will be selected for completeness and compatibility. For different element shapes and different types of nodal generalized coordinates, the terms to be selected will be different. In the variational formulation, ϕ must be at least continuous because π of (2.5) will not be mathematically defined where ϕ is discontinuous. For an element having the shape of a polygon, and having the values of ϕ at the corner points as generalized coordinates (i.e., ϕ will only be matched with two nodes on each side), we should choose ϕ to vary linearly along all the sides. If we also have the value of ϕ at a point between two adjacent corners of an element as a generalized coordinate (i.e., if ϕ is matched at three nodes on this side), we should choose ϕ to vary quadratically on this side. For example, for element 4 shown in fig. 2.1, if ϕ is to match only at nodes 3, 5, and 8 with its neighboring elements, we construct it so that it varies linearly between nodes 3 and 5, 5 and 8, and 8 and 3. If, in addition, ϕ is matched at points *A*, *B*, and *C*, we construct it such that it is a quadratic function between nodes 3 and 5, 5 and 8, and 8 and 3 for elements 3, 4, 5, and 6.

The number of terms used in the polynomial (2.14) will depend on the number of matching conditions with neighboring elements that ϕ must fulfill. For a triangular element with matching only at the three vertices, three terms are required in (2.14). If ϕ is also to be matched at the three midpoints of the sides, six terms are required. For a rectangular element with matching at the four nodes, four terms are required in (2.14), and these are chosen to lead to a linear variation along the sides, so that compatibility between elements is not violated. The highest order of approximation within an element that can be achieved by a polynomial is usually fixed by the requirement of compatibility. Therefore, unless the selection of more terms in (2.14) is to achieve a higher order of approximation, taking more terms than the number of assigned boundary generalized coordinates may not lead to much improvement in the approximation.

2.5.1 Triangular Element. Some Details of Notation

Let us consider the typical triangular element shown in fig. 2.2. For convenience, we shall replace the nodal labels I, J, and L of the figure by 1, 2, and 3, respectively. The number 1, 2 or 3 is then called the *local nodal number* for the element. The corresponding coordinates are (x_1, y_1), (x_2, y_2), and (x_3, y_3).

We shall construct, within the triangle, an approximate representation of ϕ which will be matched with the ϕs of the neighboring elements at the nodes. It will be convenient to use the values of ϕ at the nodes— ϕ_I, ϕ_J, and ϕ_L — as the unknown parameters; these are called the *generalized coordinates* or the *degrees of freedom*, ϕ_I being the Ith degree of freedom. For convenience, we shall also relabel ϕ_I, ϕ_J, and ϕ_L as $(\phi_1)_n$, $(\phi_2)_n$, and $(\phi_3)_n$, respectively. The subscripts I, J, and L are the global numbers while 1, 2, and 3 are the local numbers of the generalized coordinates. The subscript n indicates the nth element. It is used only to avoid confusion in notation; when the sense is clear, it will be dropped. Capital letters and lowercase letters will be used in the subscript to distinguish the global and the local numbering systems, respectively. There is a one-to-one relationship between the two numbering systems in each element. For instance, the Ith degree of freedom, ϕ_I, is the first degree of freedom $(\phi_1)_n$ of the nth element now under consideration. For each generalized coordinate, there will be a corresponding generalized load. In the present example there is only one degree of freedom at each node, and the nodal number and the degree number are therefore the same; that is, the Ith (ith) node is associated with the Ith (ith) degree of freedom and the Ith (ith) generalized load in the global (local) numbering system.

To represent ϕ within the triangle in terms of the three nodal values, we choose the simplest form—a linear polynomial,

$$\phi = \alpha_1 + \alpha_2 x + \alpha_3 y. \tag{2.15}$$

Equation (2.15) can be regarded as a truncated Taylor series and is a complete first-order polynomial. Given this representation, we can determine the coefficients in (2.15) uniquely in terms of the nodal values of ϕ, without having to refer to the original governing equations.

Let us determine the αs by evaluating (2.15) at the nodes, that is, by solving

$$\phi_i = \alpha_1 + \alpha_2 x_i + \alpha_3 y_i \qquad i = 1, 2, 3, \tag{2.16a}$$

or

$$\begin{aligned} \phi_1 &= \alpha_1 + \alpha_2 x_1 + \alpha_3 y_1, \\ \phi_2 &= \alpha_1 + \alpha_2 x_2 + \alpha_3 y_2, \\ \phi_3 &= \alpha_1 + \alpha_2 x_3 + \alpha_3 y_3, \end{aligned} \tag{2.16b}$$

where the subscripts on the ϕs are the local numbers. The solution to (2.16) is

$$\begin{aligned} \alpha_1 &= (a_1 \phi_1 + a_2 \phi_2 + a_3 \phi_3)/2\Delta, \\ \alpha_2 &= (b_1 \phi_1 + b_2 \phi_2 + b_3 \phi_3)/2\Delta, \\ \alpha_3 &= (c_1 \phi_1 + c_2 \phi_2 + c_3 \phi_3)/2\Delta, \end{aligned} \tag{2.17}$$

where

$$\begin{aligned} \Delta &= \tfrac{1}{2}(x_2 y_3 - x_3 y_2 + x_3 y_1 - x_1 y_3 + x_1 y_2 - x_2 y_1), \\ a_1 &= x_2 y_3 - x_3 y_2, \\ b_1 &= y_2 - y_3, \\ c_1 &= x_3 - x_2, \end{aligned} \tag{2.18}$$

with the other as, bs, and cs obtainable by cyclic permutation of the subscript 1, 2, 3. The quantity Δ is simply the area of the triangle.

A substitution of (2.17) into (2.15) yields

$$\phi = \phi_1 f_1(x, y) + \phi_2 f_2(x, y) + \phi_3 f_3(x, y), \tag{2.19}$$

where

$$f_i(x, y) = (a_i + b_i x + c_i y)/2\Delta, \qquad i = 1, 2, 3. \tag{2.19a}$$

The function $f_i(x, y)$, is called an *interpolation function*; it has the value 1 at the ith node and the value 0 at the other two nodes. Since $f_i(x, y)$ is linear in x and y, it is identically zero on the side between these two nodes. For instance, $f_1(x, y)$ is 1 at node 1 and 0 between nodes 2 and 3.

The expression for ϕ in (2.19) can be looked upon from a more general point of view (see sec. 1.2). This function is an approximate solution within the element for the given boundary conditions that ϕ equals ϕ_1, ϕ_2, and ϕ_3 at the three vertices and varies linearly along the element boundaries. Of course, if it is desired, an approximate solution within the element other than that of (2.19) can be constructed for the same boundary value of ϕ.

2.5.2 Rectangular Element

We shall consider a special case of the quadrilateral element—namely, the rectangular element shown in fig. 2.3—and we shall construct a ϕ that will be matched to the ϕs of the adjacent elements at the four corners. Let the value of ϕ at the four nodes be denoted by ϕ_1, ϕ_2, ϕ_3, and ϕ_4. To represent ϕ within the element in terms of these nodal values, we choose

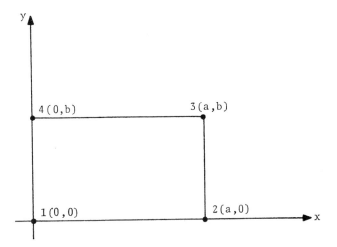

Fig. 2.3 A rectangular element.

$$\phi(x, y) = \alpha_1 + \alpha_2 x + \alpha_3 y + \alpha_4 xy. \tag{2.20}$$

This form has all the linear terms in x and y and one quadratic term, and it varies linearly along the four sides. The ϕ of two adjacent elements will be identically the same along their entire common boundary, as long as the ϕs are the same at the two common nodes. A function in the form (2.20) is called a *bilinear function*.

There are only four αs, and these can be uniquely expressed in terms of the four nodal values of ϕ. We evaluate (2.20) at the nodes:

$$\begin{aligned} \alpha_1 &= \phi_1, \\ \alpha_1 + a\alpha_2 &= \phi_2, \\ \alpha_1 + a\alpha_2 + b\alpha_3 + ab\alpha_4 &= \phi_3, \\ \alpha_1 + b\alpha_3 &= \phi_4. \end{aligned} \tag{2.21}$$

Solving the above four algebraic equations, we have

$$\begin{aligned} \alpha_1 &= \phi_1, \\ \alpha_2 &= (\phi_2 - \phi_1)/a, \\ \alpha_3 &= (\phi_4 - \phi_1)/b, \\ \alpha_4 &= (\phi_1 + \phi_2 + \phi_3 - \phi_4)/ab. \end{aligned} \tag{2.22}$$

Substitution of (2.22) into (2.20) yields

$$\phi(x, y) = \phi_1 f_1(x, y) + \phi_2 f_2(x, y) + \phi_3 f_3(x, y) + \phi_4 f_4(x, y), \tag{2.23}$$

where

$$\begin{aligned} f_1(x, y) &= (a - x)(b - y)/ab, \\ f_2(x, y) &= x(b - y)/ab, \\ f_3(x, y) &= xy/ab, \\ f_4(x, y) &= (a - x)y/ab. \end{aligned} \tag{2.24}$$

The function $f_i(x, y)$ is also called an interpolation function; it has the value 1 at the ith node and the value 0 at the other three nodes.

If the ϕ for each element is chosen to vary linearly along all of its sides, then the ϕs for adjacent elements will be identically the same along their entire common boundaries, because the ϕs have the same value at the common nodes for different elements. In other words, the approximate ϕ is continuous over the entire domain if the same value of ϕ is used for all the elements at the common nodes. This is true no matter what approximation is used within the element.

One may attempt to improve the approximate solution by using

$$\phi(x, y) = \sum_i \phi_i f_i(x, y) + \alpha g(x, y)$$

within an element, where the f_i are defined in (2.19) for a triangular element and in (2.24) for a rectangular element, $g(x, y)$ is some known function that is zero on all sides, and α is an additional unknown parameter. It can be shown that if the three-node triangular elements of fig. 2.2 are used, the final solution for ϕ_i is independent of α and the form of $g(x, y)$.* That is, improving the representation of ϕ within the element without improving the matching conditions with the adjacent elements may not lead to an improved finite-element solution at all. The effort of introducing the additional unknown parameter α is clearly not worthwhile in this case. We should emphasize, however, that this statement cannot be blindly generalized and should be examined from case to case.

2.6 Finite-Element Equations for the Entire Domain

In the finite-element method, instead of using the expression (2.5) for the potential energy, we write

$$\Pi = \sum_{n=1}^{N} \pi_n, \tag{2.25}$$

where N is the total number of elements in the domain, and where

$$\pi_n = \tfrac{1}{2} \int_{A_n} (\nabla \phi)^2 \, dA + \int_{A_n} f(x, y) \phi \, dA - \int_{(\partial A_\nu)_n} \bar{\phi}_\nu \phi \, ds, \tag{2.26}$$

with

A_n = area of the nth element,

$(\partial A_\nu)_n$ = portion of the boundary of the nth element over which the normal gradient $\partial \phi / \partial \nu$ is prescribed.

It should be noted that $(\partial A_\nu)_n = 0$ for all interior elements and for all elements having no boundaries on ∂A_ν. In this case π_n consists of only the first two integrals in (2.26). It is clear that π_n is the potential energy of the

---

*See prob. 8. This assertion is not true for the more general case discussed in sec. 2.9.

nth element. Equation (2.25) states that the potential energy of the entire domain is equal to the sum of the potential energies of all the elements. (Prove that this is true if ϕ is continuous over the entire domain. In other words, prove that Π as defined in (2.25) is the same as that of (2.5) only if ϕ is a continuous function.)

The finite-element equations are set up according to the following procedure:

Step 1 π_n is evaluated by using ϕ in the form (2.19) for a triangular element and (2.23) for a rectangular element.

Step 2 Π is obtained by summing the π_n for all the elements, with the nodal value of ϕ being the same for different elements at a common node.

Step 3 The nodal values of ϕ are set equal to those of $\bar{\phi}$ for those nodes on the boundary ∂A_ϕ, so that these nodal values are known quantities in the functional Π.

Step 4 The solution of ϕ is obtained from the minimum of Π with respect to all unknown nodal values of ϕ. This is equivalent to having $\delta \Pi = 0$ with

$$\delta\phi = \sum_{i=1}^{j} [f_i(x, y)\delta\phi_i]_n \tag{2.27}$$

for the nth element. In (2.27) $j = 3$ if the element is a triangle, and $j = 4$ if the element is a rectangle.

For each element, where $\delta\phi_1$, $\delta\phi_2$, $\delta\phi_3$, and $\delta\phi_4$ are arbitrary variations of the nodal values, except for the nodes where the nodal value is known from the foregoing step 3, the corresponding variational quantity is zero. Obviously this last step 4 is equivalent to having (2.13) satisfied with the choice of $\delta\phi$ in the particular form of (2.27).

We shall now consider each step in detail.

Step 1 Evaluation of the Element Matrices

Triangular Element To evaluate π_n we first differentiate ϕ of (2.19):

$$\frac{\partial \phi}{\partial x} = \sum_{i=1}^{3} \phi_i \frac{\partial f_i}{\partial x} = (\phi_1 b_1 + \phi_2 b_2 + \phi_3 b_3)/2\Delta,$$

$$\frac{\partial \phi}{\partial y} = \sum_{i=1}^{3} \phi_i \frac{\partial f_i}{\partial y} = (\phi_1 c_1 + \phi_2 c_2 + \phi_3 c_3)/2\Delta.$$

These are both constants. (Note that we are still using the local notations 1, 2, 3 for the subscripts.) Substitution into (2.26) yields

$$\pi_n = \sum_{i=1}^{3} \sum_{j=1}^{3} \frac{1}{2} \phi_i \phi_j k_{ij} - \sum_{i=1}^{3} \phi_i (Q_i)_n, \tag{2.28}$$

where

$$\begin{aligned} k_{ij} &= \int_{A_n} \left(\frac{\partial f_i}{\partial x} \frac{\partial f_j}{\partial x} + \frac{\partial f_i}{\partial y} \frac{\partial f_j}{\partial y} \right) dA, \\ (Q_i)_n &= -\int_{A_n} f(x, y) f_i(x, y) \, dA + \int_{(\partial A_\nu)_n} \bar{\phi}_\nu f_i(x, y) \, ds. \end{aligned} \tag{2.29}$$

(Note that $\partial \phi/\partial x$ and $\partial \phi/\partial y$ are constant within the elements.) For the interior elements, or the elements having no boundaries coincident with ∂A_ν, the second integral vanishes. In matrix form we can write

$$\pi_n = \frac{1}{2} \boldsymbol{\phi}_n^T \mathbf{k}_n \boldsymbol{\phi}_n - \boldsymbol{\phi}_n^T \mathbf{Q}_n, \tag{2.30}$$

where

$$\begin{aligned} \mathbf{k}_n \atop {3 \times 3} &\equiv \frac{1}{4\Delta} \begin{bmatrix} b_1^2 + c_1^2 & & \text{sym} \\ b_2 b_1 + c_2 c_1 & b_2^2 + c_2^2 & \\ b_3 b_1 + c_3 c_1 & b_3 b_2 + c_3 c_2 & b_3^2 + c_3^2 \end{bmatrix}_n \equiv (k_{ij})_n, \\ \boldsymbol{\phi}_n &\equiv \begin{Bmatrix} \phi_I \\ \phi_J \\ \phi_L \end{Bmatrix} = \begin{Bmatrix} \phi_1 \\ \phi_2 \\ \phi_3 \end{Bmatrix}_n, \quad \mathbf{Q}_n \equiv \begin{Bmatrix} Q_1 \\ Q_2 \\ Q_3 \end{Bmatrix}_n. \end{aligned} \tag{2.31}$$

We can express k_{ij} in a different form:

$$\begin{aligned} k_{11} &= l_{23}^2 / 4\Delta = l_{23}/2h_1, \\ k_{12} &= -\frac{l_{13}^2 + l_{23}^2 - l_{12}^2}{8\Delta} = -\frac{1}{4} \left(\frac{l_{13}}{h_2} + \frac{l_{23}}{h_1} - \frac{l_{12}}{h_3} \right), \end{aligned} \tag{2.32}$$

where $l_{23} = l_{32}$ is the distance between nodes 2 and 3 and h_1 is the height of

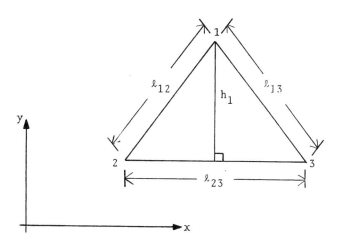

Fig. 2.4 A triangular element [see (2.32)].

the triangle measured from node 1 (fig. 2.4). The other k_{ij} are obtainable by cyclic permutation of the subscripts. It is clear from (2.32) that the direct-influence coefficient* is linearly proportional to the width of the base and inversely proportional to the height of the triangle.

Rectangular Element We first differentiate (2.23):

$$\frac{\partial \phi}{\partial x} = \sum_{i=1}^{4} \frac{\partial f_i}{\partial x} \phi_i = \frac{b-y}{ab}(\phi_2 - \phi_1) + \frac{y}{ab}(\phi_3 - \phi_4),$$
$$\frac{\partial \phi}{\partial y} = \sum_{i=1}^{4} \frac{\partial f_i}{\partial y} \phi_i = \frac{a-x}{ab}(\phi_4 - \phi_1) + \frac{x}{ab}(\phi_3 - \phi_2).$$
(2.33)

Substitution of (2.23) and (2.33) into (2.26) yields.

$$\pi_n = \sum_{i=1}^{4} \sum_{j=1}^{4} \frac{1}{2} \phi_i \phi_j k_{ij} - \sum_{i=1}^{4} \phi_i (Q_i)_n = \frac{1}{2} \boldsymbol{\phi}_n^T \mathbf{k}_n \boldsymbol{\phi}_n - \boldsymbol{\phi}_n^T \mathbf{Q}_n,$$

where k_{ij} and $(Q_i)_n$ are defined in (2.29) and the $f_i(x, y)$ are defined in (2.24).

This can be written in the same matrix form as (2.30), but with

---

*The *direct-influence coefficient* $k_{ij} \equiv i$th generalized load due to a unit value of the jth generalized coordinate.

$$\mathbf{k}_n \atop {4 \times 4} = \frac{1}{3ab} \begin{bmatrix} a^2 + b^2 & -b^2 + \dfrac{a^2}{2} & -\dfrac{b^2}{2} - \dfrac{a^2}{2} & \dfrac{b^2}{2} - a^2 \\ & a^2 + b^2 & \dfrac{b^2}{2} - a^2 & -\dfrac{b^2}{2} - \dfrac{a^2}{2} \\ & \text{sym} & a^2 + b^2 & -b^2 + \dfrac{a^2}{2} \\ & & & a^2 + b^2 \end{bmatrix}. \qquad (2.34)$$

Using the terminology of solid mechanics, we call $\mathbf{k}_n$ the *element-stiffness matrix*, $\boldsymbol{\phi}_n$ the *element generalized-coordinate vector*, and $\mathbf{Q}_n$ the *applied* (or *external*) *element generalized-load vector* (this is the equivalent nodal load defined from the element potential energy; it is consistent with the assumed form of ϕ). It can be said that the element-stiffness matrix characterizes the behavior of the element, because when there is only one isolated element the equation

$$\mathbf{k}_n \boldsymbol{\phi}_n = \mathbf{Q}_n \qquad (2.35)$$

is an approximate relation between the generalized coordinate vector $\boldsymbol{\phi}_n$ and the applied generalized load vector $\mathbf{Q}_n$.

It should be noted that the stiffness matrix $\mathbf{k}_n$ is symmetric and positive semidefinite (why?) and that its elements depend only on differences between the spatial coordinates and the area of the elements.

Step 2 Assembling the Element Matrices into the Master Matrix.

Substitution of (2.30) into (2.25) yields

$$\Pi = \sum_{n=1}^{N} \left(\frac{1}{2} \boldsymbol{\phi}_n^{\mathrm{T}} \mathbf{k}_n \boldsymbol{\phi}_n - \boldsymbol{\phi}_n^{\mathrm{T}} \mathbf{Q}_n \right). \qquad (2.36)$$

With some algebraic manipulation, this can be written as

$$\Pi = \tfrac{1}{2} \boldsymbol{\phi}^{\mathrm{T}} \mathbf{K} \boldsymbol{\phi} - \boldsymbol{\phi}^{\mathrm{T}} \mathbf{Q}, \qquad (2.37)$$

where

$$\boldsymbol{\phi} = \begin{Bmatrix} \phi_1 \\ \phi_2 \\ \vdots \\ \phi_M \end{Bmatrix},$$

M is the number of total nodal points, and **K** and **Q** are, respectively, the master stiffness matrix of the entire domain and the master column matrix of the total applied generalized nodal loading. The actual assembly procedure implied by (2.36) and (2.37) will be discussed in ch. 3; it involves a procedure similar to that introduced in sec. 1.8.*

Step 3 Constraining Master Matrices to Satisfy Rigid Boundary Conditions.

The matrix **K** is symmetric and positive semidefinite since it is assembled from the element matrices $\mathbf{k}_n$, which are symmetric and positive semidefinite. Some of the components of ϕ are known since the nodal values of ϕ for nodes on ∂A_ϕ are equal to $\bar{\phi}$ at the corresponding points. In principle, the components of ϕ can be so ordered that we can partition the matrices and express (2.37) in the form

$$\Pi = \frac{1}{2}[\phi_\alpha \bar{\phi}_\beta]\begin{bmatrix} \mathbf{K}_{\alpha\alpha} & \mathbf{K}_{\alpha\beta} \\ \mathbf{K}_{\alpha\beta}^T & \mathbf{K}_{\beta\beta} \end{bmatrix}\begin{Bmatrix} \phi_\alpha \\ \bar{\phi}_\beta \end{Bmatrix} - [\phi_\alpha \bar{\phi}_\beta]\begin{Bmatrix} \mathbf{Q}_\alpha \\ \mathbf{Q}_\beta \end{Bmatrix}, \tag{2.38}$$

where ϕ_α is unknown, $\bar{\phi}_\beta$ is the known vector whose components are the prescribed nodal values of ϕ on ∂A_ϕ, and $\mathbf{Q}_\alpha$ and $\mathbf{Q}_\beta$ are the known prescribed load vectors.† The components of $\bar{\phi}_\beta$ are called the *constrained degrees of freedom*. Then the application of the variational principle $\delta \Pi = 0$ with respect to ϕ_α will yield the system of equations

$$\mathbf{K}_{\alpha\alpha}\phi_\alpha = \mathbf{Q}_\alpha - \mathbf{K}_{\alpha\beta}\bar{\phi}_\beta, \tag{2.39}$$

which can be used to evaluate the unknown nodal ϕ_α (see prob. 1).

Step 4 Solving for the Unknown Generalized Coordinates

Solving (2.39) is an algebraic process. We may express the solution ϕ_α in the form

$$\phi_\alpha = \mathbf{K}_{\alpha\alpha}^{-1}(\mathbf{Q}_\alpha - \mathbf{K}_{\alpha\beta}\bar{\phi}_\beta).$$

We must emphasize that to obtain the solution of a large system of equations by inverting $\mathbf{K}_{\alpha\alpha}$ is a very time-consuming, inaccurate, and impractical pro-

---

*In the next two chapters we shall discuss how, in practice, **K** and **Q** are assembled for use in high-speed digital computation. It is a good exercise to compute **K** for $N = 2$ or 3.
†The practical way to carry this out for digital computation will be discussed in the next chapter.

2.7 An Example, Using Triangular Elements

To illustrate the procedure we consider the simple example shown in fig. 2.5. The governing equation is

$$\frac{\partial^2 \phi}{\partial x^2} + \frac{\partial^2 \phi}{\partial y^2} = 2. \tag{2.40}$$

The rigid boundary condition is

$$\phi = 5 - 1.5y + 2.5y^2 \tag{2.41}$$

between nodes 1 and 4, and the natural boundary conditions are

$$\begin{aligned}\frac{\partial \phi}{\partial \nu} &= -\frac{\partial \phi}{\partial y} = -x \quad \text{at} \quad y = 0, \\ \frac{\partial \phi}{\partial \nu} &= \frac{\partial \phi}{\partial x} = y \quad \text{at} \quad x = 0.\end{aligned} \tag{2.42}$$

The particular finite-element discretization and the numbering of nodes

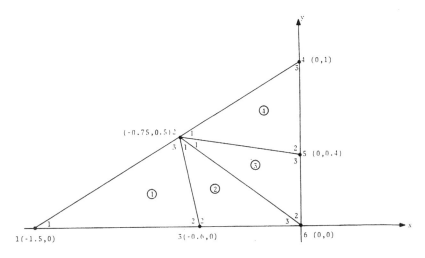

Fig. 2.5 Example. Node numbers outside the triangles are global; those inside are local.

2.7 An Example, Using Triangular Elements

and elements given in fig 2.5 were chosen for purposes of illustration; others could have been used.

We shall first use (2.31) to evaluate all the element matrices. For element 1, we designate the global node numbers 1, 3 and 2 as the local node numbers 1, 2, and 3, respectively. The corresponding nodal coordinates are $(-1.5, 0.0)$ $(-0.6, 0)$, and $(-0.75, 0.5)$. From (2.18) we have

$\Delta = 0.225$.
$b_1 = y_2 - y_3 = -0.5, \qquad b_2 = y_3 - y_1 = 0.5, \qquad b_3 = y_1 - y_2 = 0$,
$c_1 = x_3 - x_2 = -0.15, \qquad c_2 = x_1 - x_3 = -0.75$,
$c_3 = x_2 - x_1 = 0.90$.

In (2.29), the boundary $(\partial A_\nu)_1$ consists of the segment $y = 0, -1.5 \le x \le -0.6$, and $\bar{\phi}_\nu = -x$; thus

$$(Q_i)_1 = -\int_{A_1} 2f_i(x, y)dxdy + \int_{-1.5}^{-0.6} (-x)f_i(x, 0)dx,$$

where $f_i(x, y)$ is as given by (2.19a). From (2.31), we have

$$\mathbf{k}_1 = \begin{bmatrix} 0.30278 & & \text{sym} \\ -0.15278 & 0.90278 & \\ -0.15000 & -0.75000 & 0.90000 \end{bmatrix},$$

$$\boldsymbol{\phi}_1 = \begin{Bmatrix} \phi_1 \\ \phi_2 \\ \phi_3 \end{Bmatrix}_1 = \begin{Bmatrix} \phi_1 \\ \phi_3 \\ \phi_2 \end{Bmatrix}, \qquad \mathbf{Q}_1 = \begin{Bmatrix} 0.3900 \\ 0.2550 \\ -0.1500 \end{Bmatrix}.$$

For element 2, the local nodal numbers 1, 2, and 3 correspond to nodes 2, 3, and 6, respectively. The corresponding nodal coordinates are $(-0.75, 0.5)$, $(-0.6, 0.0)$, and $(0.0, 0.0)$. From (2.17), we have

$\Delta = 0.15$,
$b_1 = 0.00, \qquad b_2 = -0.5, \qquad b_3 = 0.50$,
$c_1 = 0.60, \qquad c_2 = -0.75, \qquad c_3 = 0.15$.

The boundary $(\partial A_\nu)_2$ consists of the line segments $y = 0, -0.6 \le x \le 0$, and $\bar{\phi}_\nu = -x$; thus

$$(Q_i)_2 = -\int_{A_2} 2f_i(x, y)dxdy + \int_{-0.6}^{0} (-x)f_i(x, 0)dx.$$

From (2.31), we have

$$\mathbf{k}_2 = \begin{bmatrix} 0.6000 & & \text{sym} \\ -0.7500 & 1.35417 & \\ 0.1500 & -0.60417 & 0.45417 \end{bmatrix},$$

$$\boldsymbol{\phi}_2 = \begin{Bmatrix} \phi_1 \\ \phi_2 \\ \phi_3 \end{Bmatrix}_2 = \begin{Bmatrix} \phi_2 \\ \phi_3 \\ \phi_6 \end{Bmatrix}, \qquad \mathbf{Q}_2 = \begin{Bmatrix} -0.10 \\ 0.02 \\ 0.04 \end{Bmatrix}.$$

Similarly, for element 3, we find the matrix and vectors

$$\mathbf{k}_3 = \begin{bmatrix} 0.26667 & & \text{sym} \\ 0.06667 & 0.95417 & \\ -0.33333 & -1.02083 & 1.35417 \end{bmatrix},$$

$$\boldsymbol{\phi}_3 = \begin{Bmatrix} \phi_1 \\ \phi_2 \\ \phi_3 \end{Bmatrix}_3 = \begin{Bmatrix} \phi_2 \\ \phi_6 \\ \phi_5 \end{Bmatrix}, \qquad \mathbf{Q}_3 = \begin{Bmatrix} -0.1000 \\ -0.0730 \\ -0.0470 \end{Bmatrix};$$

and for element 4,

$$\mathbf{k}_4 = \begin{bmatrix} 0.40000 & & \text{sym} \\ -0.33333 & 0.90278 & \\ -0.06667 & -0.56944 & 0.63611 \end{bmatrix},$$

$$\boldsymbol{\phi}_4 = \begin{Bmatrix} \phi_1 \\ \phi_2 \\ \phi_3 \end{Bmatrix}_4 = \begin{Bmatrix} \phi_2 \\ \phi_5 \\ \phi_4 \end{Bmatrix}, \qquad \mathbf{Q}_4 = \begin{Bmatrix} -0.15 \\ 0.03 \\ 0.09 \end{Bmatrix}.$$

Substituting into (2.36), we obtain

$$\Pi = \sum_{n=1}^{4} \left\{ \frac{1}{2} \boldsymbol{\phi}_n^\mathrm{T} \mathbf{k}_n \boldsymbol{\phi}_n - \boldsymbol{\phi}_n^\mathrm{T} \mathbf{Q}_n \right\} = \frac{1}{2} \boldsymbol{\phi}^\mathrm{T} \mathbf{K} \boldsymbol{\phi} - \boldsymbol{\phi}^\mathrm{T} \mathbf{Q}, \tag{2.43}$$

where

$$\mathbf{K} = \begin{bmatrix} 0.30278 & & & & & \text{sym} \\ -0.15000 & 2.1667 & & & & \\ -0.15278 & -1.5000 & 2.25695 & & & \\ 0.0 & -0.0667 & 0.0 & 0.63611 & & \\ 0.0 & -0.6667 & 0.0 & -0.56944 & 2.25695 & \\ 0.0 & 0.21667 & -0.60417 & 0 & -1.02083 & 1.40833 \end{bmatrix},$$

$$\boldsymbol{\phi} = \begin{Bmatrix} \phi_1 \\ \phi_2 \\ \vdots \\ \phi_6 \end{Bmatrix}, \qquad \mathbf{Q} = \begin{Bmatrix} 0.39000 \\ -0.50000 \\ 0.27200 \\ 0.09000 \\ -0.01700 \\ -0.11300 \end{Bmatrix}. \tag{2.44}$$

The values of ϕ on nodes 1, 2, and 4 are given by (2.41), so we have

$$\boldsymbol{\phi} = \begin{Bmatrix} 5.000 \\ 4.875 \\ \phi_3 \\ 6.000 \\ \phi_5 \\ \phi_6 \end{Bmatrix}.$$

That is, there are only three unknowns in (2.43)—ϕ_3, ϕ_5, and ϕ_6. Minimizing π with respect to these unknowns, we obtain

$$2.2569\phi_3 + 0.0\phi_5 - 0.60417\phi_6$$
$$= 0.272 + 0.15278\phi_1 + 1.5\phi_2 + 0.0\phi_4 = 8.350,$$
$$0.0\phi_3 + 2.2569\phi_5 - 1.02083\phi_6$$
$$= -0.117 + 0.0\phi_1 + 0.6667\phi_2 + 0.56944\phi_4 = 6.647,$$
$$-0.60417\phi_3 - 1.02083\phi_5 + 1.40833\phi_6$$
$$= -0.113 + 0.0\phi_1 - 0.21667\phi_2 + 0.0\phi_4 = -1.170,$$

and the finite-element solution for the ϕs is

$$\phi_3 = 5.09, \qquad \phi_5 = 5.29, \qquad \phi_6 = 5.19.$$

It can be shown that the exact solution of the problem is

$$\phi_{\text{ex}} = 5 + xy + y^2$$

therefore,

$$\phi_{3,\text{ex}} = \phi_{ex}(-0.6, 0) = 5, \qquad \phi_{5,\text{ex}} = \phi_{ex}(0, 0.4) = 5.16$$
$$\phi_{6,\text{ex}} = \phi_{ex}(0, 0) = 5.$$

Another Way To Evaluate the Element Matrices In sec. 2.6 the element matrix $\mathbf{k}_n$ and vector $\mathbf{Q}_n$ were evaluated from the integral of π_n, with

ϕ expressed in the form (2.19) or (2.23). Actually, the integration can be carried out in a slightly different sequence. For a triangular element, we first use the expression (2.15) for ϕ and substitute into (2.26). With

$$\frac{\partial \phi}{\partial x} = \alpha_2 \quad \text{and} \quad \frac{\partial \phi}{\partial y} = \alpha_3,$$

integration in (2.26) yields

$$\pi = \frac{1}{2} \boldsymbol{\alpha}^T \mathbf{k}'_n \boldsymbol{\alpha} - \boldsymbol{\alpha}^T \mathbf{Q}'_n, \tag{2.45}$$

where

$$\boldsymbol{\alpha} = \begin{Bmatrix} \alpha_1 \\ \alpha_2 \\ \alpha_3 \end{Bmatrix}, \qquad \mathbf{k}'_n = \Delta \begin{bmatrix} 0 & 0 & 0 \\ 0 & 1 & 0 \\ 0 & 0 & 1 \end{bmatrix}, \qquad \mathbf{Q}'_n = \begin{Bmatrix} Q'_1 \\ Q'_2 \\ Q'_3 \end{Bmatrix}_n, \tag{2.46}$$

and

$$\begin{aligned}
Q'_1 &= -\int_{A_n} f(x, y)\, dA + \int_{(\partial A_\nu)_n} \bar{\phi}_\nu ds, \\
Q'_2 &= -\int_{A_n} x f(x, y)\, dA + \int_{(\partial A_\nu)_n} x \bar{\phi}_\nu ds, \\
Q'_3 &= -\int_{A_n} y f(x, y)\, dA + \int_{(\partial A_\nu)_n} y \bar{\phi}_\nu ds.
\end{aligned} \tag{2.47}$$

The vector $\boldsymbol{\alpha}$ is related to $\boldsymbol{\phi}_n$ of (2.31) by (2.16). In matrix form we have

$$\boldsymbol{\alpha} = \mathbf{T}_n \boldsymbol{\phi}_n, \tag{2.48}$$

where

$$\mathbf{T}_n = \frac{1}{2\Delta} \begin{bmatrix} a_1 & a_2 & a_3 \\ b_1 & b_2 & b_3 \\ c_1 & c_2 & c_3 \end{bmatrix}.$$

A substitution of (2.48) into (2.45) yields

$$\pi_n = \tfrac{1}{2} \boldsymbol{\phi}_n^T \mathbf{T}_n^T \mathbf{k}'_n \mathbf{T}_n \boldsymbol{\phi}_n - \boldsymbol{\phi}_n^T \mathbf{T}_n^T \mathbf{Q}'_n. \tag{2.49}$$

Since
$$\pi_n = \frac{1}{2} \boldsymbol{\phi}_n^T \mathbf{k}_n \boldsymbol{\phi}_n - \boldsymbol{\phi}_n^T \mathbf{Q}_n,$$
it follows that
$$\mathbf{k}_n = \mathbf{T}_n^T \mathbf{k}_n' \mathbf{T}_n,$$
$$\mathbf{Q}_n = \mathbf{T}_n^T \mathbf{Q}_n'. \tag{2.50}$$

For a rectangle, we may first use the expression (2.20) for ϕ; we therefore have
$$\frac{\partial \phi}{\partial x} = \alpha_2 + \alpha_4 y, \qquad \frac{\partial \phi}{\partial y} = \alpha_3 + \alpha_4 x. \tag{2.51}$$

We substitute these into (2.26), carry out the integration, and have
$$\pi_n = \tfrac{1}{2} \boldsymbol{\alpha}^T \mathbf{k}_n' \boldsymbol{\alpha} - \boldsymbol{\alpha}^T \mathbf{Q}_n', \tag{2.52}$$
where
$$\boldsymbol{\alpha}^T = \{\alpha_1 \quad \alpha_2 \quad \alpha_3 \quad \alpha_4\},$$
$$\mathbf{Q}_n'^T = \{Q_1' \quad Q_2' \quad Q_3' \quad Q_4'\}_n$$
$$\mathbf{k}_n' = \begin{bmatrix} 0 & 0 & 0 & 0 \\ 0 & 1 & 0 & b/2 \\ 0 & 0 & 1 & a/2 \\ 0 & \dfrac{b}{2} & \dfrac{a}{2} & \dfrac{a^2+b^2}{3} \end{bmatrix} ab. \tag{2.53}$$

Here Q_1', $Q_{2'}'$, and Q_3' are the quantities defined by (2.47) and
$$Q_4' = -\int_{A_n} xy f(x,y) dx dy + \int_{(\partial A_\nu)_n} xy \bar{\phi}_\nu ds. \tag{2.54}$$

The vector $\boldsymbol{\alpha}$ is related to ϕ by (2.22). In matrix form,
$$\boldsymbol{\alpha} = \mathbf{T}_n \boldsymbol{\phi}_n, \tag{2.55}$$
where
$$\mathbf{T}_n = \begin{bmatrix} 1 & 0 & 0 & 0 \\ -1/a & 1/a & 0 & 0 \\ -1/b & 0 & 0 & 1/b \\ 1/ab & -1/ab & 1/ab & -1/ab \end{bmatrix}. \tag{2.56}$$

Note that if the origin is at the center of the element, $\mathbf{k}'_n$ and $\mathbf{T}_n$ in (2.53) and (2.56) become

$$\mathbf{k}'_n = \begin{bmatrix} 0 & 0 & 0 & 0 \\ 0 & 1 & 0 & 0 \\ 0 & 0 & 1 & 0 \\ 0 & 0 & 0 & \dfrac{a^2+b^2}{12} \end{bmatrix} ab, \qquad \mathbf{T}_n = \begin{bmatrix} \dfrac{1}{4} & \dfrac{1}{4} & \dfrac{1}{4} & \dfrac{1}{4} \\ -\dfrac{1}{2a} & \dfrac{1}{2a} & \dfrac{1}{2a} & -\dfrac{1}{2a} \\ -\dfrac{1}{2b} & -\dfrac{1}{2b} & \dfrac{1}{2b} & \dfrac{1}{2b} \\ \dfrac{1}{ab} & -\dfrac{1}{ab} & \dfrac{1}{ab} & -\dfrac{1}{ab} \end{bmatrix}.$$

Substituting (2.55) into (2.52), we find $\mathbf{k}_n$ and $\mathbf{Q}_n$ in the same form as those of (2.50), with $\mathbf{k}'_n$ and $\mathbf{T}_n$ defined in (2.53) and (2.56).

In the present problem the two approaches to the evaluation of $\mathbf{k}_n$ and $\mathbf{Q}_n$ are comparable in terms of convenience. Since the coefficients of $\partial\phi/\partial x$ and $\partial\phi/\partial y$ are constant, the integration for $\mathbf{k}_n$ is trivial. In problems with more complicated expressions for π_n, however, the present approach is sometimes simpler because the expressions for $\partial\phi/\partial x$ and $\partial\phi/\partial y$ are simple.

2.8 Mixed Boundary Conditions

If, instead of having the boundary conditions in the form (2.3), we have

$$\frac{\partial \phi}{\partial \nu} + \theta(x,y)\phi = \bar{\phi}_\nu \quad \text{on} \quad \partial A_\nu, \tag{2.57}$$

we can construct the finite-element solution in the same manner as before except that in evaluating the element matrices the expression for π_n in (2.26) is modified by

$$\pi_n = \tfrac{1}{2}\int_{A_n} (\nabla\phi)^2 dA + \tfrac{1}{2}\int_{(\partial A_\nu)_n} \theta(x\,y)\phi^2\,ds + \int_{A_n} f(x,y)\phi\,dA - \int_{(\partial A_\nu)_n} \bar{\phi}_\nu\phi\,ds, \tag{2.58}$$

which is the potential energy of the element. The second integral accounts for the additional term $\theta\phi$ in (2.57). It can be shown that (2.57) is still a natural boundary condition in this formulation. Once the element function ϕ has been chosen, $\mathbf{k}_n$ and $\mathbf{Q}_n$ can be evaluated from (2.58) in the same manner as in secs. 2.6. and 2.7. Thus

$$\mathbf{k}_n = (k_{ij})_n,$$

where

$$k_{ij} = \int_{A_n} \left(\frac{\partial f_i}{\partial x} \frac{\partial f_j}{\partial x} + \frac{\partial f_i}{\partial y} \frac{\partial f_j}{\partial y} \right) dx\, dy + \int_{(\partial A_\nu)_n} \theta(x,y) f_i f_j\, ds. \qquad (2.59)$$

(Where i and j range from 1 to 3 for a triangular element and from 1 to 4 for a rectangular element, and f_i is the interpolation function defined in (2.19a) or (2.23).) The vector $\mathbf{Q}_n$ is the same as that in (2.29). For the interior elements and for the elements having no boundaries on ∂A_ν, the second integral in (2.59) vanishes, so its element-stiffness matrices are the same as in the case given in (2.29).

2.9 Equation with Variable Coefficients

Poisson's equation (2.1) is a special form of the second-order elliptic equation. The more general form can be written as

$$\sum_{\alpha=1}^{2} \sum_{\beta=1}^{2} \frac{\partial}{\partial x_\alpha} a_{\alpha\beta} \frac{\partial \phi}{\partial x_\beta} - b(x,y)\phi = f(x,y), \qquad (2.60)$$

where $x_1 = x$, $x_2 = y$, and $a_{\alpha\beta} = a_{\beta\alpha}$ ($\alpha = 1, 2$; $\beta = 1, 2$) are functions of x and y such that all the $a_{\alpha\beta}$ are nonnegative and $a_{11}a_{22} > a_{12}^2$. The boundary conditions are (2.2) and (2.3) or (2.57). The variational principle corresponding to the set of equations is

$$\begin{aligned} \Pi = \frac{1}{2} \int_A \left[a_{11}\left(\frac{\partial \phi}{\partial x}\right)^2 + 2a_{12} \frac{\partial \phi}{\partial x} \frac{\partial \phi}{\partial y} + a_{22}\left(\frac{\partial \phi}{\partial y}\right)^2 + b\phi^2 \right] dA \\ + \frac{1}{2} \int_{\partial A_\nu} \theta(x,y) \phi^2\, ds + \int_A f(x,y)\phi\, dA - \int_{\partial A_\nu} \bar{\phi}_\nu \phi\, ds. \end{aligned} \qquad (2.61)$$

In the finite-element method, the step for setting up the algebraic equations is identically the same as that of sec. 2.6. Of course, for the present case, π_n should be evaluated according to the new expression

$$\begin{aligned} \pi_n = \frac{1}{2} \int_{A_n} \left[a_{11}\left(\frac{\partial \phi}{\partial x}\right)^2 + 2a_{12} \frac{\partial \phi}{\partial x} \frac{\partial \phi}{\partial y} + a_{22}\left(\frac{\partial \phi}{\partial y}\right)^2 + b\phi^2 \right] dA \\ + \frac{1}{2} \int_{(\partial A_\nu)_n} \theta \phi^2\, ds + \int_{A_n} f(x,y)\phi\, dA - \int_{(\partial A_\nu)_n} \bar{\phi}_\nu \phi\, ds. \end{aligned} \qquad (2.62)$$

In other words, the element matrix $\mathbf{k}_n$ and the vector $\mathbf{Q}_n$ are now evaluated according to (2.62), with ϕ in the form (2.15) or (2.19) for a triangular element and in the form (2.20) or (2.23) for a rectangular element. The expression for $\mathbf{Q}_n$ is the same as that of (2.29) because the terms of (2.62), which are linearly proportional to ϕ, are the same as those of (2.26). In particular, when we assume

$$\phi = \sum_i \phi_i f_i(x, y),$$

with the f_i given by (2.19a) or (2.24), we find that

$$\mathbf{k}_n = (k_{ij})_n, \tag{2.63}$$

where

$$k_{ij} = \int_{A_n} \left[a_{11} \frac{\partial f_i}{\partial x} \frac{\partial f_j}{\partial x} + a_{12}\left(\frac{\partial f_i}{\partial x} \frac{\partial f_j}{\partial y} + \frac{\partial f_i}{\partial y} \frac{\partial f_j}{\partial x} \right) \right.$$
$$\left. + a_{22} \frac{\partial f_i}{\partial y} \frac{\partial f_j}{\partial y} + b f_i f_j \right] dA + \int_{(\partial A_\nu)_n} \theta f_i f_j \, ds.$$

We can see that the difference in the setting up of the finite-element equations for different problems is in the evaluation of element matrices by integration, especially the element-stiffness matrix, such as in (2.29) and (2.63). In (2.29) the integration is trivial because the coefficients of $\partial \phi/\partial x$ and $\partial \phi/\partial y$ are constants. In (2.63) the integration is a bit more complicated because a_{11}, a_{12}, a_{22}, b, and θ can, in general, be functions of x and y. Nevertheless, the integration can always be evaluated numerically. In practice, we may approximate $a_{11}, \cdots, \theta$ within an element by constants, which we shall call $\bar{a}_{11}, \bar{a}_{12}, \bar{a}_{22}, \bar{b}$ and $\bar{\theta}$. Usually the constants are the values of the corresponding function at the centroid of the element or the midpoint of the line segment. Then the integration in (2.63) can be carried out explicitly.

For example, for a triangular element,

$$k_{ij} = \frac{1}{4\triangle}\left[\bar{a}_{11} b_i b_j + \bar{a}_{12}(b_i c_j + b_j c_i) + \bar{a}_{22} c_i c_j \right] + \frac{\bar{b}}{12} \triangle (1 + \delta_{ij}), \tag{2.64}$$

where δ_{ij} is the Kronecker delta, for an interior element or the boundary element having no boundaries on ∂A_ν [i.e., $(\partial A_\nu)_n = 0$], and

$$k_{ij} = \frac{1}{4\Delta}\left[\bar{a}_{11}b_ib_j + \bar{a}_{12}(b_ic_j + b_jc_i) + \bar{a}_{22}c_ic_j\right]$$
$$+ \frac{\bar{b}}{12}\Delta(1 + \delta_{ij}) + c_{ij}, \tag{2.65}$$

where c_{ij} is the contribution from the second integral of (2.63), for elements with $(\partial A_\nu)_n \neq 0$. The expression for c_{ij} depends on the side or sides of the element that are on ∂A_{ν_n}. For example, if the side between nodes 1 and 2 coincides with ∂A_ν, then $f_1(x, y)$ and $f_2(x, y)$ are linear functions between the two nodes; that is,

$$f_1 = 1 - \frac{s}{l_{12}},$$
$$f_2 = \frac{s}{l_{12}}, \tag{2.66}$$
$$f_3(x, y) = 0,$$

where s is the distance along the side from node 1 to 2 and l_{12} is the distance between the two nodes. If we carrying out the integration

$$c_{ij} = \int_{\text{node1}}^{\text{node2}} \theta f_i f_j \, ds$$

by approximating θ as a constant $\bar{\theta}$, we find that

$$\begin{aligned} c_{i3} &= c_{3i} = 0, \quad i = 1, 2, 3, \\ c_{11} &= c_{22} = l_{12}\bar{\theta}/3, \\ c_{12} &= c_{21} = l_{12}\bar{\theta}/6. \end{aligned} \tag{2.67}$$

If the other side is on the boundary ∂A_ν, the c_{ij} can be obtained by cyclic permutation of 1, 2, 3.

2.10 A Three-Dimensional Problem

To generalize the finite-element method to three dimensions is, in principle, trivial. The appropriate simplest element is a tetrahedron. We can choose the four vertices of the tetrahedron as the nodes and represent ϕ within the element in the form

$$\phi = \alpha_1 + \alpha_2 x + \alpha_3 y + \alpha_4 z. \tag{2.68}$$

As in (2.19), we can express the αs in terms of the four nodal ϕ and thereby put ϕ in the form

$$\phi = \sum_{i=1}^{4} \phi_i f_i(x, y, z). \tag{2.69}$$

A slightly more complicated three-dimensional element is the hexahedron, for which ϕ can be expressed in the form

$$\phi = \alpha_1 + \alpha_2 x + \alpha_3 y + \alpha_4 z + \alpha_5 xy + \alpha_6 xz + \alpha_7 yz + \alpha_8 xyz \tag{2.70}$$

or in the form

$$\phi = \sum_{i=1}^{8} \phi_i f_i(x, y, z) \tag{2.71}$$

by expressing the αs in terms of the nodal values of ϕ, where the $f_i(x, y, z)$ are three-dimensional interpolation functions.

The steps in setting up the finite-element equations are the same as those of sec. 2.5 except that, to evaluate the element matrices, we use

$$\pi_n = \tfrac{1}{2} \int_{V_n} (\nabla \phi)^2 dV + \int_{V_n} f(x, y, z) \phi dV - \int_{\partial V_n} \bar{\phi}_\nu \phi dS, \tag{2.72}$$

where

V_n = volume of the nth element,
∂V_n = portion of the boundary of the nth element over which the normal gradient $\partial \phi / \partial \nu$ is prescribed.

Therefore, the evaluation of the element matrices can be carried out in a straightforward manner. The practical complication in the three-dimensional problem is the assembling of the element matrices $\mathbf{k}_n$ into the master matrix $\mathbf{K}$, because of the complexity of a three-dimensional body itself. Another difficulty is the solution of the final algebraic equations, which generally comprise a very large system. Due to the finite core storage of the computer, and the execution time required for large systems, economic considerations can become an important factor.

Problems

1 This problem illustrates the application of the variational principle to

a system with a finite number of degrees of freedom. Let Π be a quadratic functional with a finite number of degrees of freedom q_i; that is

$$\Pi = \frac{1}{2} \sum_{i=1}^{n} \sum_{j=1}^{n} a_{ij} q_i q_j - \sum_{i=1}^{n} q_i Q_i;$$

or, in matrix notation,

$$\Pi = \tfrac{1}{2} \mathbf{q}^T \mathbf{K} \mathbf{q} - \mathbf{q}^T \mathbf{Q},$$

where $\mathbf{q}$ and $\mathbf{Q}$ are column vectors with components q_i and Q_i, respectively, and $\mathbf{K}$ is a symmetric square matrix with entries

$$K_{ij} = \tfrac{1}{2}(a_{ij} + a_{ji}).$$

The extremum of Π is found from

$$\frac{\partial \Pi}{\partial q_i} = K_{ij} q_j - Q_i = 0$$

or, in matrix form,

$$\mathbf{K}\mathbf{q} - \mathbf{Q} = 0.$$

Show that if $\mathbf{K}$ is positive definite, the extremum is a minimum. Show further that if $\mathbf{K}$ is nonsingular, the solution for $\mathbf{q}$ in the equation above is unique.

2 Find the element-stiffness matrices for the triangles 1 and 2 in fig. 2.6 using the functional defined in (2.26).

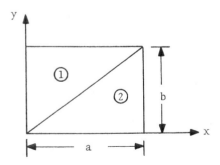

Fig. 2.6

3 In fig. 2.6, let

$$\Pi = \int_0^b \int_0^a \tfrac{1}{2} (\nabla \phi)^2 dx dy = \tfrac{1}{2} \boldsymbol{\phi}_1^T \mathbf{k}_1 \boldsymbol{\phi}_1 + \tfrac{1}{2} \boldsymbol{\phi}_2^T \mathbf{k}_2 \boldsymbol{\phi}_2 = \tfrac{1}{2} \boldsymbol{\phi}^T \mathbf{K} \boldsymbol{\phi},$$

where

$$\boldsymbol{\phi} = \begin{Bmatrix} \phi_1 \\ \phi_2 \\ \phi_3 \\ \phi_4 \end{Bmatrix}.$$

Write out **K** explicitly. Compare **K** with **k** of (2.34) and show that

$$K_{ii} - k_{ii} > 0, \qquad i = 1, 2, 3, 4.$$

4 Derive finite-element equations using the mesh shown in fig. 2.7. The governing differential equation is $\nabla^2 \phi = 2$. and the boundary conditions are

$\phi = 5 + xy + x^2$ on $y - x = 1$,
$\phi = 5 + x^2$ on $y = 0$,
$\partial \phi / \partial x = y$ on $x = 0$.

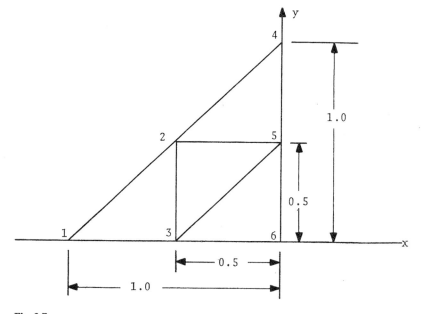

Fig. 2.7

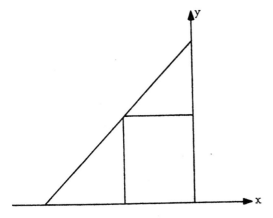

Fig. 2.8

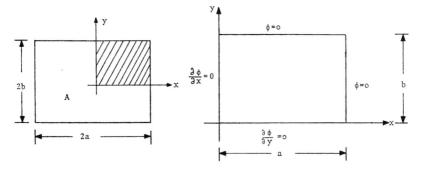

Fig. 2.9

5 Redo prob. 4 with boundary conditions

$\phi = 5 + xy + x^2$ on $y - x = 1$,
$\partial\phi/\partial x = y$ on $x = 0$,
$\partial\phi/\partial y = x$ on $y = 0$.

It should be noted here the exact solution is the same as that of prob. 4. Compare the two solutions.

6 Redo prob. 4 using the mesh shown in fig. 2.8.

7 In St-Venant's theory of torsion (Courant 1943, Kellogg 1929), the stress function satisfies Poisson's equation

$$\frac{\partial^2\phi}{\partial x^2} + \frac{\partial^2\phi}{\partial y^2} = -2$$

over the cross section A of a cylindrical shaft. The boundary condition is $\phi = 0$. The torsional rigidity of the cylinder is given by

$$J = \int_A \phi\, dA.$$

Consider the case in which A is a rectangular area. Because of the symmetry, one need only treat one-quarter of the area—for example, the region $x, y \geq 0$, with boundary conditions as shown in fig. 2.9.

Find the torsional rigidity by the finite-element method using each of the meshes in fig. 2.10 for the shaded quarter of fig. 2.9.

8 In solving Poisson's equation (2.1) by the finite-element method using triangular and/or rectangular meshes, one can assume ϕ in the form

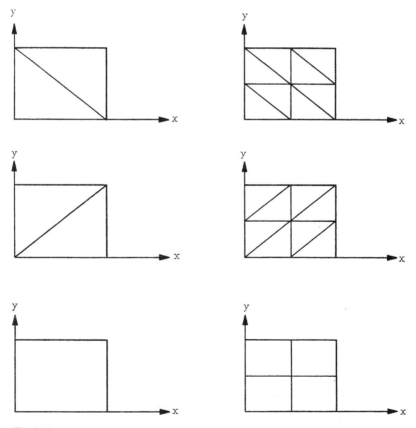

Fig. 2.10

$$\phi = \sum_{i=1}^{j} \phi_i f_i(x, y) + \alpha g(x, y)$$

within an element, where $j = 3$ and the f_i are defined in (2.19a) for a triangular element, and $j = 4$ and the f_i are defined in (2.23) for a rectangular element; $g(x, y)$ is a known function which is 0 on all the sides of the element, and α is an additional unknown parameter. Show that the finite-element solution for the ϕ_i is independent of α and the form of $g(x, y)$ so that the additional term αg does not improve the solution. (Hint: α of the nth element can be determined from (2.26) by minimizing π_n with respect to α.

3 Assembly and Solution for Large Systems

3.1 Introduction

In ch. 2 we saw that there is a great deal of algebraic manipulation involved in the finite-element method. Besides the evaluation of the element-stiffness matrices and the element-loading vector, there is the assembling of all the element matrices to obtain a master stiffness matrix and a total applied-load vector, as in (2.37), the imposing of constrained boundary conditions to obtain (2.39) for the unknowns, and finally the procedure of obtaining the solution for the unknowns in an efficient manner. Efficiency is the key to the successful application of the finite-element method. To describe most problems with sufficient accuracy usually requires a large number of degrees of freedom, so that (2.39) will often in practice be a system containing thousands of equations. The amount of information and the number of algebraic operations needed to obtain values of the unknowns is huge, and all these manipulations must be performed by high-speed computers. Thus the efficient use of the computer is essential to making the application of the finite-element method economically feasible.

In this chapter we shall discuss the basic ideas of assembly, the constraints and the method of solution. In the next chapter we shall discuss how these ideas are actually carried out by the computer.

3.2 Assembling the Master Matrices

In sec. 2.7 we showed that the assembling of the master matrices $\mathbf{K}$ and $\mathbf{Q}$ from the element matrices $\mathbf{k}_m$ and $\mathbf{Q}_m$ is nothing but a summing procedure. The component K_{IJ} of the master stiffness matrix $\mathbf{K}$ is a sort of influence coefficient which gives the Ith generalized load due to a unit value of Jth generalized coordinate, where I and J are global degree numbers. The component k_{ij} of $\mathbf{k}_m$ is the influence coefficient relating the ith generalized load and the jth generalized coordinate of the mth element, where i and j are local degree numbers. The total value of K_{IJ} is the sum of the contributions of all the elements. That is, if the ith and jth generalized coordinates of the mth element are the Ith and Jth coordinates in the global system, the contribution from the mth element to the Ith generalized load due to a unit value of the Jth generalized coordinate is k_{ij}. If the mth element is not associated with the Ith or the Jth degree of freedom, then there is no contribution to K_{IJ} from this element. The component Q_I of the applied-load vector $\mathbf{Q}$ is the total known external Ith generalized load, and $(Q_i)_m$ of $\mathbf{Q}_m$ is the known external ith generalized load applied to the mth element. The total value of Q_I is the sum of the Ith generalized loads from all the elements. In other words, the contribution to Q_I from the mth element is $(Q_i)_m$, if its ith generalized load is the Ith generalized load in the global system. If an element is not associated with the Ith generalized coordinate at all, there is no contribution to Q_I from the element.

In the actual evaluation of the master matrices, instead of determining K_{IJ} and Q_I for a fixed I and J by adding the contributions from each of the elements at one time, we perform the summation element by element; that is, we assemble one element after another, and we add the contribution of each element to the proper location in the master matrix. The element is the basic unit in the finite-element method. It is best to generate all the element matrices for one element at the same time. The most convenient procedure is, of course, to add their contributions to the master matrices as soon as they have been generated, that is, to add all the k_{ij} and $(Q_i)_m$ of the mth element to the corresponding K_{IJ} and Q_I, where I and J are the global degree numbers of the ith and jth degrees of freedom of the element, respectively, and then to proceed to the next element.

To express the assembling procedure in mathematical form, let the master matrices be denoted by $\mathbf{K}^{(0)}$ ($=0$) and $\mathbf{Q}^{(0)}$ ($=0$) before any contributions are added from an element and by $\mathbf{K}^{(m)}$ and $\mathbf{Q}^{(m)}$ after the mth element has been assembled. Let $i_1, i_2, \cdots, i_p$ be the global degree numbers and p be the number of degrees of freedom of the element. Then

$$K_{i_r i_s}^{(m)} = K_{i_r i_s}^{(m-1)} + k_{rs}, \quad r, s = 1, 2, \cdots, p,$$
$$Q_{i_r}^{(m)} = Q_{i_r}^{(m-1)} + (Q_r)_m, \quad r = 1, 2, \cdots, p, \qquad (3.1)$$

and the rest of the $K_{ij}^{(m)}$ are the same as $K_{ij}^{(m-1)}$. The array $i_1, i_2, \cdots, i_p$ represents the transformation from local degree numbers to global degree numbers.

In many practical problems, $\mathbf{k}$ and $\mathbf{K}$ are symmetric, and we therefore only have to assemble the lower (or the upper) triangle of $\mathbf{K}$ using the lower (or upper) triangle of $\mathbf{k}$. In this case (3.1) will have to be modified slightly; we first define

$$M \equiv \max(i_r, i_s),$$
$$N \equiv \min(i_r, i_s);$$

then we have

$$K_{MN}^{(m)} = K_{MN}^{(m-1)} + k_{rs}, \qquad r = 1, 2, \cdots, p,$$
$$\qquad\qquad\qquad\qquad s = 1, 2, \cdots, r, \qquad (3.2)$$
$$Q_{ir}^{(m)} = Q_{ir}^{(m-1)} + (Q_r)_m, \qquad r = 1, 2, \cdots, p.$$

After we proceed through all the elements, we have the final master matrices **K** and **Q**.

3.3 An Example of Assembling

We can make the assembly procedure clearer by going through the example considered in sec. 2.7. Figure 2.5 is reproduced in fig. 3.1 with the local number of each element indicated at the corners within each element. Since there is only one degree of freedom for each node, no distinction is needed for the nodal numbers and the degree numbers.

Table 3.1 emphasizes that the local number is just a way of indicating the ordering of the degrees of freedom in an element.

There are six degrees of freedom for the entire domain. The order of **K** and **Q** will be 6 × 6 and 6 × 1. Before we assemble any of the element matrices, we initialize all the components of **K** and **Q** to zero. Then we start from the first element. Since its first, second, and third degrees of freedom are the

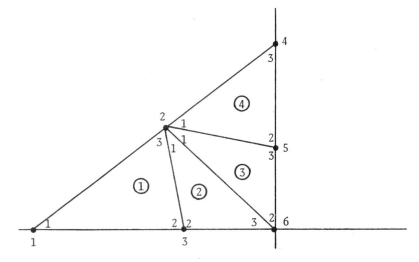

Fig. 3.1 Local and global numbers in a finite-element division of a domain.

3.3 An Example of Assembling

Table 3.1 Global and local numbers for the elements of fig. 3.1.

Element	Global Numbers			Local Numbers		
1	1	3	2	1	2	3
2	2	3	6	1	2	3
3	2	6	5	1	2	3
4	2	5	4	1	2	3

first, third, and second degrees for the entire domain, respectively, that is, since the transformation of the local degree number to the global degree number is

$i_1 = 1,$
$i_2 = 3,$
$i_3 = 2,$

we add the contribution of the first element to the master matrices as follows:

add $k_{11} (= 0.3028)$ to $K_{11}(= 0)$,

add $(Q_1)_1 (= 0.3900)$ to $Q_1(= 0)$,

add $k_{21} (= -0.1528)$ to $K_{31} (= 0)$,

add $(Q_2)_1 (= 0.2550)$ to $Q_3(= 0)$,

add $k_{31} (-0.1500)$ to $K_{21} (= 0)$,

add $(Q_3)_1 (= -0.1500)$ to $Q_2(= 0)$,

add $k_{22} (= 0.9028)$ to $K_{33} (= 0)$,
add $k_{23} (= -0.7500)$ to $K_{32} (= 0)$,
add $k_{33} (= 0.9000)$ to $K_{22} (= 0)$.

The values of the k_{ij} and $(Q_i)_1$ are those that we obtained in sec. 2.7. We have taken advantage of the fact that **k** and **K** are symmetric; we only have to assemble the lower (or upper) triangle of **K** from the lower (or upper) triangle of **k**. The resulting master matrices are now (i.e., after the first element is assembled)

$$\mathbf{K}^{(1)} = \begin{bmatrix} 0.3028 & & & & & \text{sym} & \\ -0.1500 & 0.9000 & & & & & \\ -0.1528 & -0.7500 & 0.9028 & & & & \\ 0 & 0 & 0 & 0 & & & \\ 0 & 0 & 0 & 0 & 0 & & \\ 0 & 0 & 0 & 0 & 0 & 0 \end{bmatrix},$$

$$\mathbf{Q}^{(1)} = \left\{ \begin{array}{c} 0.3900 \\ -0.1500 \\ 0.2550 \\ 0 \\ 0 \\ 0 \end{array} \right\}.$$

We now assemble the second element. Its first, second, and third degrees of freedom are the second, third, and sixth for the entire domain, respectively. The transformation of the local to the global number is

$i_1 = 2,$
$i_2 = 3,$
$i_3 = 6.$

Therefore, we

add $k_{11}(= 0.6000)$ to $K_{22}(= 0.9000)$,
          add $(Q_1)_2(= -0.1000)$ to $Q_2(= -0.1500)$,
add $k_{21}(= -0.7500)$ to $K_{32}(= -0.7500)$,
          add $(Q_2)_2(= 0.0200)$ to $Q_3(= 0.2550)$,
add $k_{31}(= -0.1500)$ to $K_{62}(= 0.0000)$,
          add $(Q_3)_2(= -0.0400)$ to $Q_6(=0)$,
add $k_{22}(= 1.3542)$ to $K_{33}(= 0.9028)$,
add $k_{32}(= -0.6042)$ to $K_{63}(= 0.0000)$,
add $k_{33}(= 0.4517)$ to $K_{66}(= 0.0000)$.

The resulting master matrices are now

$$\mathbf{K}^{(2)} = \begin{bmatrix} 0.3028 & & & & & & \text{sym} \\ -0.1500 & 1.5000 & & & & & \\ -0.1528 & -1.5000 & 2.2570 & & & & \\ 0 & 0 & 0 & 0 & & & \\ 0 & 0 & 0 & 0 & 0 & & \\ 0 & 0.1500 & -0.6042 & 0 & 0 & 0.4517 \end{bmatrix},$$

$$\mathbf{Q}^{(2)} = \begin{Bmatrix} 0.3900 \\ -0.2500 \\ 0.2750 \\ 0 \\ 0 \\ -0.0400 \end{Bmatrix}.$$

For the third element, we have

$i_1 = 2,$
$i_2 = 6,$
$i_3 = 5.$

Therefore, we

add $k_{11}(= 0.2667)$ to $K_{22}(= 1.5000)$,
                              add $(Q_1)_3(= -0.1000)$ to $Q_2(= -0.2500)$,
add $k_{21}(= 0.0667)$ to $K_{62}(= 0.1500)$,
                              add $(Q_2)_3(= -0.073)$ to $Q_6(= -0.0400)$,
add $k_{31}(= -0.3333)$ to $K_{52}(= 0.0000)$,
                              add $(Q_3)_3(= -0.047)$ to $Q_5(= 0)$,
add $k_{22}(= 0.9542)$ to $K_{66}(= 0.4517)$,
add $k_{32}(= -1.021)$ to $K_{65}(= 0)$,
add $k_{33}(= 1.354)$ to $K_{55}(= 0)$.

It should be noted that we have added k_{32} to K_{65} instead of K_{56}. This is because we are only assembling the lower traingle of **k** to that of **K**.

Finally, for the fourth element, we have

$i_1 = 2,$
$i_2 = 5,$
$i_3 = 4.$

Therefore, we

add $k_{11}(= 0.4000)$ to $K_{22}(= 1.767)$,
                              add $(Q_1)_4(= -0.1500)$ to $Q_2(= -0.3500)$,
add $k_{21}(= -0.3333)$ to $K_{52}(= -0.3333)$,

add $(Q_2)_4(= 0.0300)$ to $Q_5(= -0.0470)$,
add $k_{31}(= -0.0667)$ to $K_{42}(= 0)$,
add $(Q_3)_4(= 0.0900)$ to $Q_4(= 0.0000)$,
add $k_{22}(= 0.9028)$ to $K_{55}(= 1.354)$,
add $k_{32}(= -0.5694)$ to $K_{54}(= 0)$,
add $k_{33}(= 0.6361)$ to $K_{44}(= 0)$.

Note that k_{32} is added to K_{54}. The resulting master matrices **K** and **Q** are

$$\mathbf{K} = \begin{bmatrix} 0.3028 & & & & \text{sym} & \\ -0.1500 & 2.167 & & & & \\ -0.1528 & -1.500 & 2.257 & & & \\ 0 & -0.0667 & 0 & 0.6361 & & \\ 0 & -0.6667 & 0 & 0.5694 & 2.257 & \\ 0 & 0.2167 & -0.6042 & 0 & -1.021 & 1.406 \end{bmatrix},$$

$$\mathbf{Q} = \begin{Bmatrix} 0.390 \\ -0.500 \\ 0.272 \\ 0.090 \\ -0.017 \\ -0.113 \end{Bmatrix},$$

(3.3)

which are, of course, the same as (2.44).

3.4 Constraining

When the rigid boundary conditions are imposed in (2.38), the generalized-coordinate vector ϕ is partitioned into ϕ_α and $\bar{\phi}_\beta$, where ϕ_α is unknown and $\bar{\phi}_\beta$ is a known vector having its components equal to the nodal values ϕ on ∂A_ϕ. The master matrices **K** and **Q** are also partitioned accordingly. Then we carry out the variation of Π with respect to ϕ_α only and obtain (2.39) for the unknowns. In the form (2.38), the vectors ϕ and **Q** and the matrix **K** generally have to be reordered almost completely before the partitioning can be carried out,* and the number of equations in (2.39) is different from the total number of degrees of freedom. In practice, it is very time-consuming and cumbersome to reorder completely the matrices of a large system. We shall therefore look

---

*This is because the degrees of freedom of ϕ are usually ordered in such a way that **K** will have a small bandwidth.

into a means of avoiding such a step. Let us make the following modification in (2.38): replace

$\bar{\boldsymbol{\phi}}_\beta$ by $\boldsymbol{\phi}_\beta$,
$\mathbf{Q}_\alpha$ by $\mathbf{Q}_\alpha - K_{\alpha\beta}\bar{\boldsymbol{\phi}}_\beta$,
$\mathbf{Q}_\beta$ by $\bar{\boldsymbol{\phi}}_\beta$,
$\mathbf{K}_{\alpha\beta}$ by 0,
$\mathbf{K}_{\beta\beta}$ by $\mathbf{I}$;

we then have

$$\Pi = \frac{1}{2}\{\boldsymbol{\phi}_\alpha^T \boldsymbol{\phi}_\beta^T\}\mathbf{K}'\begin{Bmatrix}\boldsymbol{\phi}_\alpha \\ \boldsymbol{\phi}_\beta\end{Bmatrix} - \{\boldsymbol{\phi}_\alpha^T \boldsymbol{\phi}_\beta^T\}\mathbf{Q}', \tag{3.4}$$

where

$$\mathbf{K}' = \begin{bmatrix}\mathbf{K}_{\alpha\alpha} & 0 \\ 0 & \mathbf{I}\end{bmatrix}, \quad \mathbf{Q}' = \begin{Bmatrix}\mathbf{Q}_\alpha - \mathbf{K}_{\alpha\beta}\bar{\boldsymbol{\phi}}_\beta \\ \bar{\boldsymbol{\phi}}_\beta\end{Bmatrix}. \tag{3.5}$$

The variation of Π with respect to $\boldsymbol{\phi}_\alpha$ and $\boldsymbol{\phi}_\beta$ yields

$$\mathbf{K}'\begin{Bmatrix}\boldsymbol{\phi}_\alpha \\ \boldsymbol{\phi}_\beta\end{Bmatrix} = \begin{bmatrix}\mathbf{K}_{\alpha\alpha} & 0 \\ 0 & \mathbf{I}\end{bmatrix}\begin{Bmatrix}\boldsymbol{\phi}_\alpha \\ \boldsymbol{\phi}_\beta\end{Bmatrix} = \mathbf{Q}'. \tag{3.6}$$

It is obvious that the solution of (3.6) is the same as that for $\boldsymbol{\phi}_\alpha$ in (2.39), and also,

$$\boldsymbol{\phi}_\beta = \bar{\boldsymbol{\phi}}_\beta,$$

which satisfies the rigid boundary conditions. Equation (3.6) can be reordered so that the combination of $\boldsymbol{\phi}_\alpha$ and $\boldsymbol{\phi}_\beta$ has the same order as the original $\boldsymbol{\phi}$. Reorder the matrices of $\mathbf{K}'$ and $\mathbf{Q}'$ accordingly, and denote them by $\mathbf{K}^*$ and $\mathbf{Q}^*$, respectively; then (3.6) can be written as

$$\mathbf{K}^*\boldsymbol{\phi} = \mathbf{Q}^*. \tag{3.7}$$

The solution of (3.7) is, of course, identical to that of (3.6) so that the boundary constraints are automatically satisfied.

It should be noted that we do not directly form and use as such the matrices

$\mathbf{K}_{\alpha\alpha}$, $\mathbf{K}_{\alpha\beta}$, $\mathbf{K}_{\beta\beta}$, $\mathbf{K}'$, and $\mathbf{Q}'$. In (3.6) the right-hand side for the unknown generalized coordinates is equal to the known applied generalized load minus $\mathbf{K}_{\alpha\beta}\bar{\phi}_\beta$, which is the sum of all the known generalized coordinates multiplied by their corresponding column of the original matrix $\mathbf{K}$: the right-hand side for the known generalized coordinates is the known value itself. In the left-hand side of (3.6) all the rows and the column of $\mathbf{K}'$ corresponding to the constrained degree of freedom are 0 except for the diagonal terms, which are 1s. This is also true for (3.7) because it is the same as (3.6) except that the order of the equations is different. We can thus obtain (3.7) directly from the original $\mathbf{K}$ and $\mathbf{Q}$ in the following way: The constrained conditions are imposed one at a time by modifying $\mathbf{K}$ and $\mathbf{Q}$. If the jth component of ϕ is a known degree of freedom and is equal to $\bar{\phi}_j$, we define $\mathbf{Q}^*$ by multiplying the jth column of $\mathbf{K}$ ($\mathbf{K}_j$) by $\bar{\phi}_j$ and subtracting it from the load $\mathbf{Q}$, except for the jth component, which is set equal to $\bar{\phi}_j$. Then we define $\mathbf{K}^*$ by setting the jth column and jth row of $\mathbf{K}$ to zero, except for the diagonal term K_{jj}, which is set equal to 1. We have

$$Q_i^* = Q_i - K_{ij}\bar{\phi}_j,$$
$$K_{ij}^* = K_{ji}^* = 0, \qquad (3.8a)$$

for all $i \leq j - 1$ and $i \geq j + 1$, and

$$Q_i^* = \bar{\phi}_j,$$
$$K_{ii}^* = 1, \qquad (3.8b)$$

for $i = j$. The rest of the components of $\mathbf{K}^*$ are the same as those of $\mathbf{K}$. The matrices $\mathbf{K}^*$ and $\mathbf{Q}^*$ are the constrained master stiffness matrix and the constrained applied-load vector if ϕ_j is the only constrained degree of freedom. Otherwise, by letting $\mathbf{K}^*$ be $\mathbf{K}$ and $\mathbf{Q}^*$ be $\mathbf{Q}$, respectively, this process can be repeated until we have all the constrained conditions imposed. In each cycle the load vector $\mathbf{Q}^*$ is altered according to (3.8). It should be noted that those components of $\mathbf{K}$ and $\mathbf{Q}$ that are associated with an already constrained degree of freedom will not be changed in this process, because the rows of the degrees of freedom that have already been constrained are already set equal to zero.

We illustrate the constraining procedure by using the example presented in sec. 2.7. The constrained degrees of freedom are 1, 2, and 4, with

$$\phi_1 = 5.000, \qquad \phi_2 = 4.875, \qquad \phi_4 = 6.000.$$

3.4 Constraining

To impose the condition that $\phi_1 = 5.00$, we multiply the first column of **K** by 5.00; subtract it from **Q**; and then set $Q_1 = 5.00$: thus

$$\mathbf{Q}^* = \begin{Bmatrix} 5 \\ -0.5000 - (-0.15) \times 5.00 \\ 0.272 - (-0.1528) \times 5.00 \\ 0.090 - 0.0 \times 5.00 \\ -0.017 - 0.0 \times 5.00 \\ -0.113 - 0.0 \times 5.00 \end{Bmatrix} = \begin{Bmatrix} 5 \\ 0.2500 \\ 1.0370 \\ 0.090 \\ -0.017 \\ -0.113 \end{Bmatrix}. \tag{3.9a}$$

We then set the first row and the first column of **K** equal to 0, except for K_{11} which is set equal to 1; thus

$$\mathbf{K}^* = \begin{bmatrix} 1 & & & & & \\ 0 & 2.1607 & & & \text{sym} & \\ 0 & -1.500 & 2.257 & & & \\ 0 & -0.0667 & -0.000 & 0.6361 & & \\ 0 & -0.6667 & 0 & 0.5694 & 2.257 & \\ 0 & 0.2167 & -0.6042 & 0 & -1.021 & 1.408 \end{bmatrix}. \tag{3.9b}$$

To impose the condition $\phi_2 = 4.875$, we multiply the second column of $\mathbf{K}^*$ in (3.9b) by 4.875; subtract it from $\mathbf{Q}^*$ of (3.9a); and then set $Q_2^* = 4.875$: thus

$$\mathbf{Q}^* = \begin{bmatrix} 5.000 - 0.0 \times 4.875 \\ 4.875 \\ 1.037 - (-1.5) \times 4.875 \\ 0.090 - (-0.0667) \times 4.875 \\ -0.017 - (-0.6667) \times 4.875 \\ -0.113 - 0.2167 \times 4.875 \end{bmatrix} = \begin{Bmatrix} 5.000 \\ 4.875 \\ 8.350 \\ 3.340 \\ 3.223 \\ -1.170 \end{Bmatrix}. \tag{3.10}$$

We then set the second row and second column to 0 and the second diagonal term of $\mathbf{K}^*$ to 1; thus

$$\mathbf{K}^* = \begin{bmatrix} 1 & & & & & \\ 0 & 1 & & & \text{sym} & \\ 0 & 0 & 2.257 & & & \\ 0 & 0 & 0 & 0.6361 & & \\ 0 & 0 & 0 & -0.5694 & 2.257 & \\ 0 & 0 & -0.6042 & 0 & -1.021 & 1.408 \end{bmatrix}. \tag{3.11}$$

Finally, we can impose the condition $\phi_4 = 6.00$ in a similar manner and obtain the constrained stiffness matrix $\mathbf{K}^*$ and the constrained load vector $\mathbf{Q}^*$. The final equation for the vector $\boldsymbol{\phi}$ is

$$\begin{bmatrix} 1 & & & & & \\ 0 & 1 & & & \text{sym} & \\ 0 & 0 & 2.257 & & & \\ 0 & 0 & 0 & 1 & & \\ 0 & 0 & 0 & 0 & 2.257 & \\ 0 & 0 & -0.6042 & 0 & -1.021 & 1.408 \end{bmatrix} \begin{Bmatrix} \phi_1 \\ \phi_2 \\ \phi_3 \\ \phi_4 \\ \phi_5 \\ \phi_6 \end{Bmatrix} = \begin{Bmatrix} 5.000 \\ 4.875 \\ 8.35 \\ 6.00 \\ 6.647 \\ -1.17 \end{Bmatrix}. \quad (3.12)$$

3.5 Solution

For our convenience in later chapters, we shall use $\mathbf{q}$ to denote the generalized-coordinate vector, and $\mathbf{K}$ and $\mathbf{Q}$ without an asterisk to denote the constrained master stiffness matrix and the constrained generalized-load vector. Then (3.7) becomes

$$\mathbf{Kq} = \mathbf{Q}. \quad (3.13)$$

The use of the finite-element method for a linear problem inevitably involves the solution of such a set of n linear algebraic equations. Usually the order of the equations is large and $\mathbf{K}$ is nonsingular and sparse, with most of its nonzero entries lying within a narrow band of the diagonal. (In many cases $\mathbf{K}$ is symmetric and positive definite.)

The feasibility of the application of the finite-element method hinges on the answers to two questions:

1 How fast can the equations be solved?
2 How accurate is the computed solution?

The first question can be answered by counting the operations involved in the solution algorithm used. The second question is more fundamental since its answer determines whether it is meaningful to solve the equations numerically; the major consideration here is roundoff error, which will be discussed in the next chapter.

There are usually two kinds of methods used for solving a system of equations of large order: (1) direct methods, and (2) iterative methods. A direct method is one which, after a finite number of operations, if all computations were carried out without roundoff, would lead to the true solution of the algebraic system. An iterative method usually requires an infinite number of iterations to converge to the true solution. Within a tolerable error, there is no clear-cut answer as to which of these methods is best for a system such as (3.13). The general rule is that the iterative methods are better if **K** is of the order of tens of thousands and highly sparse, and if the diagonal terms are dominant. The direct methods are usually faster, if they can be used at all, when **K** is of the order of thousands or less. In practice, the roundoff error is usually the controlling factor in determining whether a direct method of solution can be used. Other obvious considerations are which of the two methods one is more familiar with and what sorts of computer programs are available.

We shall now discuss the basic procedures of the most commonly used methods.

3.5.1 Gaussian Elimination (A Direct Method)

The idea underlying Gaussian elimination is simple. For a system of n algebraic equations, we use the first equation to eliminate the "first" variable from the last $n - 1$ equations, then use the second equation to eliminate the second variable from the last $n - 2$ equations, and so on. After $n - 1$ such eliminations have been performed, the resulting matrix equation is triangular and can be solved easily.

The details of the method are as follows. Starting from the given system (3.13) we write

$$
\begin{aligned}
K_{11}q_1 + K_{12}q_2 + \cdots + K_{1n}q_n &= Q_1, \\
K_{21}q_1 + K_{22}q_2 + \cdots + K_{2n}q_n &= Q_2, \\
&\vdots \\
K_{n1}q_1 + K_{n2}q_2 + \cdots + K_{nn}q_n &= Q_n.
\end{aligned}
\tag{3.14}
$$

To eliminate q_1 from the second through the last equations, we subtract the second equation from the first equation multiplied by K_{21}/K_{11}, subtract the third equation from the first equation multiplied by K_{31}/K_{11}, and so on. The reduced system of equations is

$$
\begin{aligned}
K_{11}q_1 + K_{12}q_2 + \cdots + K_{1n}q_n &= Q_1, \\
\left(K_{22} - \frac{K_{21}K_{12}}{K_{11}}\right)q_2 + \cdots + \left(K_{2n} - \frac{K_{21}K_{1n}}{K_{11}}\right)q_n &= Q_2 - \frac{K_{21}}{K_{11}}Q_1, \\
&\vdots
\end{aligned}
\tag{3.15}
$$

$$\left(K_{n2} - \frac{K_{n1}K_{12}}{K_{11}}\right)q_2 + \cdots + \left(K_{nn} - \frac{K_{n1}K_{1n}}{K_{11}}\right)q_n = Q_n - \frac{K_{n1}}{K_{11}}Q_1.$$

The first equation of (3.15) will be used later to evaluate q_1.

Now we can proceed to eliminate q_2 from the last $n-1$ equations of (3.15). We shall denote the reduced system, in which $q_1, q_2, \cdots, q_{m-1}$ have already been eliminated, by

$$\mathbf{K}^{(m)}\mathbf{q} = \mathbf{Q}^{(m)}, \tag{3.16}$$

where

$$\mathbf{K}^{(m)} = [K_{ij}^{(m)}] \quad \text{and} \quad \mathbf{Q}^{(m)} = \begin{Bmatrix} Q_1^{(m)} \\ \vdots \\ Q_n^{(m)} \end{Bmatrix}.$$

It can be easily shown that $K_{ij}^{(m)}$ and $Q_i^{(m)}$ can be computed recursively by

$$K_{ij}^{(m)} = \begin{cases} K_{ij}^{(m-1)} & \text{for } i \leq m-1, \\ 0 & \text{for } i \geq m \text{ and } j \leq m-1, \\ K_{ij}^{(m-1)} - \dfrac{K_{i,m-1}^{(m-1)}}{K_{m-1,m-1}^{(m-1)}} K_{m-1,j}^{(m-1)} & \text{for } i \geq m \text{ and } j \geq m, \end{cases} \tag{3.17a}$$

$$Q_i^{(m)} = \begin{cases} Q_i^{(m-1)} & \text{for } i \leq m-1, \\ Q_i^{(m-1)} - \dfrac{K_{i,m-1}^{(m-1)}}{K_{m-1,m-1}^{(m-1)}} Q_{m-1}^{(m-1)} & \text{for } i \geq m. \end{cases} \tag{3.17b}$$

After $n-1$ such steps the equations are reduced to the triangular form

$$\begin{aligned}
K_{11}^{(n)}q_1 + K_{12}^{(n)}q_2 + \cdots + K_{1n}^{(n)}q_n &= Q_1^{(n)}, \\
K_{22}^{(n)}q_2 + \cdots + K_{2n}^{(n)}q_n &= Q_2^{(n)}, \\
&\vdots \\
K_{n-1,n-1}^{(n)}q_{n-1} + K_{n-1,n}^{(n)}q_n &= Q_{n-1}^{(n)}, \\
K_{nn}^{(n)}q_n &= Q_n^{(n)}.
\end{aligned} \tag{3.18}$$

This is called *forward elimination*. Equation (3.18) can be solved easily for the qs because we can determine q_n from the last equation, then q_{n-1} from the $(n-1)$th equation, and so on; that is,

$$q_n = \frac{Q_n^{(n)}}{K_{nn}^{(n)}}, \tag{3.19}$$

$$q_i = \frac{1}{K_{ii}^{(n)}}\left[Q_i^{(n)} - \sum_{j=i+1}^{n} K_{ij}^{(n)} q_j\right], \quad i = n-1, n-2, \cdots, 1.$$

This procedure is called *back substitution*.

It is easy to count the number of operations required in (3.17) and (3.19). Since additions and subtractions can be performed much faster than multiplications and divisions on most computers, we shall only count multiplications and divisions. The number of such operations is

$$(n - m + 1)^2 + (n - m + 1)$$

for (3.17a), and

$$(n - m + 1)$$

for (3.17b) for a given m, where $m = 2, 3, \cdots, n$. The total number of operations required to reduce the system to (3.18) is

$$\sum_{m=2}^{n} [(n - m + 1)^2 + (n - m + 1)] + \sum_{m=2}^{n} (n - m + 1)$$
$$= \frac{n(n^2 - 1)}{3} + \frac{n(n - 1)}{2}. \qquad (3.20)$$

The total number of operations required for (3.19) is

$$\frac{n(n + 1)}{2}. \qquad (3.21)$$

Adding (3.20) and (3.21), we find that the total number of operations required to solve (3.13) is

$$\frac{n^3}{3} + n^2 - \frac{n}{3}. \qquad (3.22)$$

This count does not account for the sparseness of the matrix **K** and includes multiplication by zero terms, which need not be performed. Methods for taking advantage of matrix sparseness will be discussed in ch. 4.

3.5.2 Factorization (A Direct Method)

We shall assume that **K** is symmetric, because this is the most common case encountered in the finite-element method. (The procedure for handling a general nonsymmetric matrix can be found in Isaacson and Keller 1966.) The basic idea in the factorization method is to express **K** in the form

$$\mathbf{K} = \mathbf{LDL}^T, \qquad (3.23)$$

where **L** is a lower triangular matrix and **D** is a diagonal matrix:

$$\mathbf{L} = \begin{bmatrix} L_{11} & & & \\ L_{21} L_{22} & 0 & & \\ \vdots & & \ddots & \\ L_{n1} & \cdots & & L_{nn} \end{bmatrix}, \quad \mathbf{D} = \begin{bmatrix} D_1 & & & \\ & D_2 & & 0 \\ & & \ddots & \\ 0 & & & D_n \end{bmatrix}. \tag{3.24}$$

Expressing **K** as a product of such matrix factors is called *factorization*. Equation (3.13) now becomes

$$\mathbf{Kq} = \mathbf{LDL}^T\mathbf{q} = \mathbf{Q}. \tag{3.25}$$

The solution for **q** can be obtained easily in two steps; First, we solve **Lg** = **Q**, that is,

$$\begin{aligned}
L_{11}g_1 &= Q_1, \\
L_{21}g_1 + L_{22}g_2 &= Q_2, \\
&\vdots \\
L_{n1}g_1 + L_{n2}g_2 + \cdots + L_{nn}g_n &= Q_n,
\end{aligned} \tag{3.26}$$

by forward substitution. In the first equation of (3.26) the single unknown g_1 can be determined immediately; then g_2 can be determined from the second equation, and so on. Thus

$$\begin{aligned}
g_1 &= \frac{Q_1}{L_1}, \\
g_2 &= \frac{1}{L_{22}}(Q_2 - L_{21}g_1),
\end{aligned}$$

and, in general,

$$g_i = \frac{1}{L_{ii}}(Q_i - \sum_{j=1}^{i-1} L_{ij}g_j), \quad i = 2, 3, \cdots, n. \tag{3.27}$$

Next, we solve $\mathbf{L}^T\mathbf{q} = \mathbf{D}^{-1}\mathbf{g}$, that is,

$$\begin{aligned}
L_{11}q_1 + L_{21}q_2 + \cdots + L_{n1}q_n &= g_1/D_1, \\
L_{22}q_2 + \cdots + L_{n2}q_n &= g_2/D_2, \\
&\vdots \\
L_{nn}q_n &= g_n/D_n,
\end{aligned} \tag{3.28}$$

by back substitution. This procedure is similar to the solution (3.19), so that

$$q_n = \frac{(g_n/D_n)}{L_{nn}},$$

$$q_i = \frac{1}{L_{ii}} \left(\frac{g_i}{D_i} - \sum_{j=i+1}^{n} L_{ji} q_j \right), \quad i = n-1, n-2, \cdots, 1. \tag{3.29}$$

To determine L_{ij} and D_i, we shall carry out the matrix multiplication of (3.23) and equate the corresponding entries of the matrices of both sides. Thus

$$K_{11} = L_{11} D_1 L_{11},$$
$$K_{21} = L_{21} D_1 L_{11}.$$
$$\vdots$$
$$K_{n1} = L_{n1} D_1 L_{11}, \tag{3.30}$$
$$K_{22} = L_{21} D_1 L_{21} + L_{22} D_2 L_{22},$$
$$K_{32} = L_{31} D_1 L_{21} + L_{32} D_2 L_{22},$$
$$\vdots$$

and, in general,

$$K_{ij} = \sum_{l=1}^{j} L_{il} D_l L_{jl}, \quad j \leq i; i = 1, 2 \cdots, n. \tag{3.31}$$

Since $K_{ij} = K_{ji}$, there are at most $n(n+1)/2$ independent equations in (3.31); but there are $n(n+1)/2$ entries L_{ij} and n entries D_i to be evaluated. We therefore have at least n free choices in assigning the values of L_{ij} or D_i. Two methods are commonly used:

(a) Set all n entries D_i to one (Cholesky method). From (3.30), we then have

$$L_{11} = (K_{11})^{1/2},$$
$$L_{21} = K_{21}/L_{11},$$
$$\vdots \tag{3.32}$$
$$L_{n1} = K_{n1}/L_{11},$$
$$L_{22} = (K_{22} - L_{21}^2)^{1/2},$$
$$L_{32} = (K_{32} - L_{31} L_{21})/L_{22},$$
$$\vdots$$

and, in general,

$$L_{jj} = \left[K_{jj} - \sum_{l=1}^{j-1} L_{jl}L_{jl} \right]^{1/2},$$

$$L_{ij} = \frac{1}{L_{jj}} \left(K_{ij} - \sum_{l=1}^{j-1} L_{il}L_{jl} \right), \quad i = 2, 3, \cdots, n_j, \; j = 1, 2, \cdots, i-1. \quad (3.33)$$

(b) Set the n diagonal entries L_{ii} to one (triple-factoring method). From (3.30), we then have

$$D_1 = K_{11},$$
$$L_{21} = K_{21}/D_1,$$
$$\vdots$$
$$L_{n1} = K_{n1}/D_n, \quad (3.34)$$
$$D_2 = K_{22} - L_{21}^2 D_1$$
$$L_{32} = (K_{32} - L_{31}D_1L_{21})/D_2,$$
$$\vdots$$

and, in general,

$$D_j = K_{jj} - \sum_{l=1}^{j-1} L_{jl}D_l L_{jl},$$

$$L_{ij} = \frac{1}{D_j} \left(K_{ij} - \sum_{l=1}^{j-1} L_{il}D_l L_{jl} \right), \quad i = 1, 2, \cdots, n, \; j = 1, 2, \cdots, i-1. \quad (3.35)$$

The number of steps involved in the determination of a fully populated **K** by the two methods is shown in Table 3.2 This counting does not account for the sparseness of **K**. We can see that the first choice requires $n(n-1)/2$ fewer multiplications than that of the second choice (in the banded matrix of semiband $B \ll n$, this number changes to roughly Bn), but with the price of taking n additional square roots. The Cholesky method works efficiently only for positive-definite matrices, because otherwise some of the L_{ij} will be imaginary. Therefore, the triple-factoring method is generally recommended. As a matter of fact, since no square root is necessary in the latter approach, there is less chance of numerical error.

We shall use the triple-factoring method to solve (3.12). First, we factor **K** according to (3.24). We have

Table 3.2 Number of steps to determine **K** by the Cholesky and triple-factoring methods.

	Multiplication or Division	Square Root
Cholesky method		
Factorization [(3.32) and (3.33)]	$\dfrac{n^3}{6} + \dfrac{n^2}{2} - \dfrac{n}{2}$	n
Forward Substitution (3.27)	$\dfrac{(n+1)n}{2}$	0
Back Substitution [(3.29) with $D_i = 1$]	$\dfrac{n(n+1)}{2}$	0
Total	$\dfrac{n^3}{6} + \dfrac{3n^2}{2} + \dfrac{n}{2}$	n
Triple-Factoring method		
Factorization [(3.34) and (3.35)]	$\dfrac{n^3}{6} + n^2 + n$	0
Forward Substitution [(3.27) with $L_{ii}=1$]	$\dfrac{n(n-1)}{2}$	0
Back Substitution [(3.29) with $L_{ii}=1$]	$\dfrac{n(n-1)}{2}$	0
Total	$\dfrac{n^3}{6} + n^2$	0

[a] The assumption has been made that, in (3.35), $D_l L_{jl}$ ($j = 1, 2, \cdots, i - 1$) is stored during the factorization of the ith row.

$$D_1 = 1.0,$$
$$L_{21} = 0.0/1.0 = 0,$$
$$L_{31} = 0.0/1.0 = 0,$$
$$\vdots \qquad\qquad\qquad\qquad\qquad (3.36)$$
$$L_{61} = 0.0/1.0 = 0,$$
$$D_2 = 1.0 - 0.0^2 \times 1.0 = 1.0,$$
$$L_{32} = 0.0 - 0.0 \times 1.0 \times 0.0 = 0,$$

and, in general, using (3.35), we have

$$\mathbf{L} = \begin{bmatrix} 1.00 & & & & & \\ 0 & 1.00 & & \text{sym} & & \\ 0 & 0 & 1.00 & & & \\ 0 & 0 & 0 & 1.00 & & \\ 0 & 0 & 0 & 0 & 1.00 & \\ 0 & 0 & -0.267 & 0 & -0.4523 & 1.0 \end{bmatrix}, \quad \mathbf{D} = \begin{bmatrix} 1.0 & & & & & \\ & 1.0 & & & 0 & \\ & & 2.26 & & & \\ & & & 1.0 & & \\ & 0 & & & 2.26 & \\ & & & & & 0.785 \end{bmatrix}.$$

Performing the forward substitution according to (3.27), we find that

$$g_1 = Q_1 = 5.00,$$
$$g_2 = 4.88 - 0 \times 5.0 = 4.88,$$
$$g_3 = 8.35 - 0 \times 5.0 - 0 \times 4.88 = 8.35,$$
$$g_4 = 6.00 - 0 \times 5.0 - 0 \times 4.88 - 0 \times 8.35 = 6.00, \qquad (3.37)$$
$$g_5 = 6.65 - 0 \times 5.0 - 0 \times 4.88 - 0 \times 8.35 - 0 \times 6.0 = 6.65,$$
$$g_6 = -1.17 - 0 \times 5 - 0 \times 4.88 - (-0.267 \times 8.35) - 0 \times 6.0$$
$$- (-0.453) \times 6.65 = 4.07.$$

Performing the back substitution according to (3.29), we then have

$$\phi_6 = \frac{4.07}{0.758} = 5.19,$$
$$\phi_5 = \frac{6.65}{2.26} - (-0.452) \times 5.19 = 5.28,$$
$$\phi_4 = \frac{6.00}{1.0} - 0 \times 5.28 - 0 \times 5.19 = 6.00,$$
$$\phi_3 = \frac{8.35}{2.26} - 0 \times 6.0 - 0 \times 5.28 - (-0.267) \times 5.19 = 5.09, \qquad (3.38)$$
$$\phi_2 = \frac{4.88}{1.0} - 0 \times 5.09 - 0 \times 6.0 - 0 \times 5.28 - 0 \times 5.19 = 4.88,$$
$$\phi_1 = \frac{5.00}{1.0} - 0 \times 4.88 - 0 \times 5.09 - 0 \times 6.00 - 0 \times 5.28 - 0 \times 5.19$$
$$= 5.00.$$

3.5.3 Iterative Methods

These methods proceed from some initial "guess" $\mathbf{q}^{(0)}$ and define a sequence of successive approximations $\mathbf{q}^{(1)}$, $\mathbf{q}^{(2)}, \cdots$ which is required to converge to the exact solution. One advantage of such methods is the fact that roundoff errors are usually damped out as the procedure continues. However, this method must terminate after a finite number of iterations, and this will introduce a so-called truncation error.

A large class of iterative methods proceed in the following fashion: In order to solve (3.13), the matrix $\mathbf{K}$ is split into two parts, $\mathbf{N}$ and $\mathbf{P}$, such that

$$\mathbf{K} = \mathbf{N} - \mathbf{P}, \qquad (3.39)$$

with $\det \mathbf{N} \neq 0^*$. Starting with some guess $\mathbf{q}^{(0)}$, we then define a sequence of vectors

*det $\mathbf{N}$ denotes the determinant of the matrix $\mathbf{N}$.

$$\mathbf{N}\mathbf{q}^{(r)} = \mathbf{P}\mathbf{q}^{(r-1)} + \mathbf{Q}. \tag{3.40}$$

As a practical matter, $\mathbf{N}$ should be chosen so that (3.40) can be easily solved for $\mathbf{q}^{(r)}$.

A discussion of the convergence of this iteration process can be found in Isaacson and Keller (1966). The rate of convergence depends on the magnitude of the maximum eigenvalue of the matrix $\mathbf{M} = \mathbf{N}^{-1}\mathbf{P}$, which is usually denoted by $\rho(\mathbf{M})$. We must have $0 < \rho(\mathbf{M}) < 1$ for convergence.

Two of the most commonly used iterative methods are: *Jacobi's iteration*, in which

$$\mathbf{N} = \begin{bmatrix} K_{11} & & & \\ & K_{22} & & 0 \\ & 0 & \ddots & \\ & & & K_{nn} \end{bmatrix}, \quad \mathbf{P} = \mathbf{N} - \mathbf{K}, \tag{3.41}$$

with n^2 operations for each iteration, and *Gauss-Seidel* (or *successive*) *iteration*, in which

$$\mathbf{N} = \begin{bmatrix} K_{11} & & & \\ K_{21} & K_{22} & & 0 \\ \vdots & & \ddots & \\ K_{n1} & & & K_{nn} \end{bmatrix}, \quad \mathbf{P} = \mathbf{N} - \mathbf{K}. \tag{3.42}$$

It can be shown that the Gauss-Seidel method will converge if

$$\max_i \left(\sum_{\substack{j=1 \\ j \neq i}}^{n} \left[\frac{k_{ij}}{k_{ii}} \right] \right) < 1.$$

The rate of convergence of iterative methods may be improved by the method of *overrelaxation*. Let

$$\mathbf{K} = \mathbf{N}(\alpha) - \mathbf{P}(\alpha), \tag{3.43}$$

where α is some appropriate parameter. An ideal case would be to choose an α^* such that

$$\rho(\mathbf{M}(\alpha^*)) = \min_\alpha \left[\rho(\mathbf{M}(\alpha)) \right]. \tag{3.44}$$

For example, let

$$\mathbf{N}(\alpha) = (1 + \alpha)\,\mathbf{N}_0, \quad \mathbf{P}(\alpha) = (1 + \alpha)\,\mathbf{N}_0 - \mathbf{K} = \mathbf{P}_0 + \alpha\mathbf{N}_0.$$

Then

$$\alpha^* = -\frac{\lambda_1 + \lambda_n}{2}, \qquad (3.45)$$

where

$$\lambda_1 \leq \lambda_2 \leq \cdots \leq \lambda_n < 1$$

are the eigenvalues of $\mathbf{N}_0^{-1}\,\mathbf{P}_0$, and

$$\rho[\mathbf{M}(\alpha^*)] = \frac{\lambda_n - \lambda_1}{2 - \lambda_1 - \lambda_n} < 1. \qquad (3.46)$$

3.6 Minimization Techniques for the Solution of (3.13)

If $\mathbf{K}$ is positive definite, one may obtain the solution of (3.13) by direct minimization of

$$\Pi(\mathbf{q}) = \tfrac{1}{2}\,\mathbf{q}^\mathrm{T}\mathbf{K}\mathbf{q} - \mathbf{q}^\mathrm{T}\mathbf{Q}. \qquad (3.47)$$

We shall briefly discuss two minimization methods in this section.

3.6.1 Conjugate-Gradient Method
Let

$$\mathbf{g} = \nabla\Pi = \mathbf{K}\mathbf{q} - \mathbf{Q}. \qquad (3.48)$$

Then the algorithmic form of the minimization process is:

$\mathbf{q}_0 =$ initial guess vector,
$\mathbf{g}_0 = \mathbf{K}\mathbf{q}_0 - \mathbf{Q}$,
$\mathbf{s}_0 = -\mathbf{g}_0$,
$\mathbf{q}_{i+1} = \mathbf{q}_i + \alpha_i \mathbf{s}_i$,
$\mathbf{g}_{i+1} = \mathbf{K}\mathbf{q}_{i+1} - \mathbf{Q}$,
$\mathbf{s}_{i+1} = -\mathbf{g}_i + \dfrac{\mathbf{g}_{i+1}^\mathrm{T}\mathbf{g}_{i+1}}{\mathbf{g}_i^\mathrm{T}\mathbf{g}_i}\mathbf{s}_i$,

where the α_i are obtained from the equations $\partial\Pi(\mathbf{q}_i + \alpha_i\mathbf{s}_i)/\partial\alpha_i = 0$.

It can be shown that the solution converges to an exact solution after the nth iteration (Fox and Kapoor 1969). Another advantage of this method is that one does not have to directly form and use as such the matrix $\mathbf{K}$. The computation of the vector $\mathbf{Kq}_i$ for $\mathbf{g}_i$ can be accomplished by adding up the contributions from all of the elements. We therefore have to store only the element matrices and the vectors $\mathbf{g}_i$ and $\mathbf{s}_i$ in order to perform the iteration. The core storage required for this method is thus very small.

3.6.2 Variable-Metric Method

Let $\mathbf{q}_0$ be an initial guessed vector and $\mathbf{H}_0$ be an arbitrary positive-definite matrix. (The unit matrix is usually chosen for $\mathbf{H}_0$.) The algorithm of the variable-metric method, as described by Fletcher and Powell (1963) is then as follows:

1 Define

$$\mathbf{g}_i \equiv \mathbf{Kq}_i - \mathbf{Q}, \tag{3.49}$$
$$\mathbf{s}_i \equiv -\mathbf{H}_i\mathbf{g}_i.$$

2 Find

$$\mathbf{q}_{i+1} = \mathbf{q}_i + \alpha_i\mathbf{s}_i, \tag{3.50}$$

where α_i is chosen to minimize $\Pi(\mathbf{q}_i + \alpha_i\mathbf{s}_i)$. It can be shown that

$$\alpha_i = \frac{\mathbf{s}_i^T\mathbf{g}_i}{\mathbf{s}_i^T\mathbf{K}\mathbf{s}_i}. \tag{3.51}$$

3 Generate an improvement to $\mathbf{H}$:

$$\mathbf{H}_{i+1} = \mathbf{H}_i + \frac{\boldsymbol{\sigma}_i\boldsymbol{\sigma}_i^T}{\boldsymbol{\sigma}_i^T\mathbf{y}_i} - \frac{\mathbf{H}_i\mathbf{y}_i\mathbf{y}_i^T\mathbf{H}_i}{\mathbf{y}_i^T\mathbf{H}_i\mathbf{y}_i}, \tag{3.52}$$

where

$$\mathbf{y}_i \equiv \mathbf{g}_{i+1} - \mathbf{g}_i, \tag{3.53}$$
$$\boldsymbol{\sigma}_i \equiv \alpha_i\mathbf{s}_i. \tag{3.54}$$

4 Go back to step 1. Terminate when $\mathbf{g}_i$ goes to 0.

This method suffers from the necessity of storing the matrix $\mathbf{H}$. If $\Pi(\mathbf{q})$ is a quadratic form (as is usual for many structural problems), however, then at the minimum $\mathbf{H}$ is the inverse of $\mathbf{K}$.

There are many other iteration methods: the reader may refer to Isaacson and Keller (1966) and Ralston (1965).

3.7 Static-Condensation or Substructure Method

In practice, the assembly and solution procedures can be separated into several steps. First, the entire domain under consideration is divided into several parts which are connected to each other by a finite number of nodes and the interelement boundaries between these nodes. Each "part" may itself have many elements. The element matrices of each part are assembled into a "submaster" matrix for the part through the procedure described in sec. 3.2. These submaster matrices for each part are just like the element matrices for an element; each part can thus be viewed as a superelement and is, in fact, usually called a *substructure*. All these submaster matrices are to be assembled into the final master matrices $\mathbf{K}$ and $\mathbf{Q}$. Before the final assembling is performed, however, the generalized coordinates associated with the nodes in the interior of each substructure are expressed in terms of those on its boundaries, so that reduced matrices associated only with the generalized coordinates of the boundary nodes can be developed. This procedure is called *static condensation*. The reduced matrices are then assembled to produce a system of equations involving only the generalized coordinates of the boundary nodes of each substructure. The final equations can be solved by any of the methods described in secs. 3.4 and 3.5.

To illustrate this procedure, we consider an example in which a domain is divided into two parts by an imaginary boundary C (fig. 3.2a). Both parts are subdivided into many elements, and they have nodes in common along their boundary C (fig. 3.2b). Each of these two parts is a substructure. The element numbers are circled in fig 3.2b. The common boundary C has nodes 21 through 24. The rest of the boundaries of the substructures I and II are C_I and C_{II}, respectively. When we assemble all the element matrices of elements 1 through 9, we obtain the submaster matrices $\mathbf{K}_I$ and $\mathbf{Q}_I$ for substructure I, and when we assemble elements 10 through 15, we obtain $\mathbf{K}_{II}$ and $\mathbf{Q}_{II}$ for substructure II. The corresponding generalized coordinate vectors are denoted by $\mathbf{q}_I$ and $\mathbf{q}_{II}$.

We now partition both $\mathbf{q}_I$ and $\mathbf{q}_{II}$ into three parts:

3.7 Static-Condensation or Substructure Method

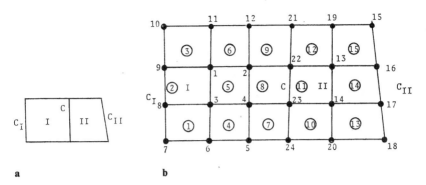

Fig. 3.2 Schematic example of a domain divided into two parts.

$$\mathbf{q}_I = \begin{Bmatrix} \mathbf{q}_1 \\ \mathbf{q}_2 \\ \mathbf{q}_3 \end{Bmatrix}_I, \qquad \mathbf{q}_{II} = \begin{Bmatrix} \mathbf{q}_1 \\ \mathbf{q}_2 \\ \mathbf{q}_3 \end{Bmatrix}_{II}, \qquad (3.55)$$

where $(\mathbf{q}_1)_I$ and $(\mathbf{q}_2)_I$ are, respectively, the unconstrained and the constrained generalized coordinates (or degrees of freedom) of substructure I which are not in common with the rest of the structure (i.e., substructure II in this case) and the $(\mathbf{q}_3)_I$ *are* generalized coordinates in common with the rest of the structure. Similar definitions are made for $(\mathbf{q}_1)_{II}$, $(\mathbf{q}_2)_{II}$ and $(\mathbf{q}_3)_{II}$. In fig. 3.2, if there is only one degree of freedom per node and if all the degrees of freedom associated with the boundary nodes are constrained, we have

$$(\mathbf{q}_1)_I = \begin{Bmatrix} q_1 \\ q_2 \\ q_3 \\ q_4 \end{Bmatrix}, \qquad (\mathbf{q}_2)_I = \begin{Bmatrix} q_5 \\ \vdots \\ q_{12} \end{Bmatrix}, \qquad (\mathbf{q}_3)_I = \begin{Bmatrix} q_{21} \\ \vdots \\ q_{24} \end{Bmatrix},$$

$$(\mathbf{q}_1)_{II} = \begin{Bmatrix} q_{13} \\ q_{14} \end{Bmatrix}, \qquad (\mathbf{q}_2)_{II} = \begin{Bmatrix} q_{15} \\ \vdots \\ q_{20} \end{Bmatrix}, \qquad (\mathbf{q}_3)_{II} = \begin{Bmatrix} q_{21} \\ \vdots \\ q_{24} \end{Bmatrix}.$$

Note that $(\mathbf{q}_3)_I$ is the same as $(\mathbf{q}_3)_{II}$.

Now let the submaster matrices also be partitioned:

$$\mathbf{K}_I = \begin{bmatrix} \mathbf{K}_{11} & \mathbf{K}_{12} & \mathbf{K}_{13} \\ \mathbf{K}_{12}^T & \mathbf{K}_{22} & \mathbf{K}_{23} \\ \mathbf{K}_{13}^T & \mathbf{K}_{23}^T & \mathbf{K}_{33} \end{bmatrix}_I, \qquad \mathbf{Q}_I = \begin{Bmatrix} \mathbf{Q}_1 \\ \mathbf{Q}_2 \\ \mathbf{Q}_3 \end{Bmatrix}_I. \qquad (3.56)$$

The substructures I and II can then assembled to form the master matrices

$$\mathbf{K} = \begin{bmatrix} (\mathbf{K}_{11})_\mathrm{I} & & & & \\ (\mathbf{K}_{12})_\mathrm{I}^T & (\mathbf{K}_{22})_\mathrm{I} & & \mathrm{sym} & \\ 0 & 0 & (\mathbf{K}_{11})_\mathrm{II} & & \\ 0 & 0 & (\mathbf{K}_{12})_\mathrm{II}^T & (\mathbf{K}_{22})_\mathrm{II} & \\ (\mathbf{K}_{13})_\mathrm{I}^T & (\mathbf{K}_{23})_\mathrm{I}^T & (\mathbf{K}_{13})_\mathrm{II}^T & (\mathbf{K}_{23})_\mathrm{II}^T & (\mathbf{K}_{33})_\mathrm{I} + (\mathbf{K}_{33})_\mathrm{II} \end{bmatrix},$$

$$\mathbf{Q} = \begin{Bmatrix} (\mathbf{Q}_1)_\mathrm{I} \\ (\mathbf{Q}_2)_\mathrm{I} \\ (\mathbf{Q}_1)_\mathrm{II} \\ (\mathbf{Q}_2)_\mathrm{II} \\ (\mathbf{Q}_3)_\mathrm{I} + (\mathbf{Q}_3)_\mathrm{II} \end{Bmatrix}. \tag{3.57a}$$

The corresponding generalized-coordinate vector of the entire domain is

$$\mathbf{q} = \begin{Bmatrix} (\mathbf{q}_1)_\mathrm{I} \\ (\mathbf{q}_2)_\mathrm{I} \\ (\mathbf{q}_1)_\mathrm{II} \\ (\mathbf{q}_2)_\mathrm{II} \\ (\mathbf{q}_3)_\mathrm{II} \end{Bmatrix} = \begin{Bmatrix} q_1 \\ q_2 \\ \vdots \\ q_{24} \end{Bmatrix}, \tag{3.57b}$$

and the corresponding functional is

$$\Pi = \tfrac{1}{2} \mathbf{q}^T \mathbf{K} \mathbf{q} - \mathbf{q}^T \mathbf{Q}. \tag{3.58}$$

Minimizing Π with respect to $(\mathbf{q}_1)_\mathrm{I}$, and $(\mathbf{q}_1)_\mathrm{II}$, we obtain

$$(\mathbf{K}_{11})_\mathrm{I}(\mathbf{q}_1)_\mathrm{I} + (\mathbf{K}_{12})_\mathrm{I}(\mathbf{q}_2)_\mathrm{I} + (\mathbf{K}_{13})_\mathrm{I}(\mathbf{q}_3)_\mathrm{I} = (\mathbf{Q}_1)_\mathrm{I}, \tag{3.59a}$$

$$(\mathbf{K}_{11})_\mathrm{II}(\mathbf{q}_1)_\mathrm{II} + (\mathbf{K}_{12})_\mathrm{II}(\mathbf{q}_2)_\mathrm{II} + (\mathbf{K}_{13})_\mathrm{II}(\mathbf{q}_3)_\mathrm{II} = (\mathbf{Q}_1)_\mathrm{II}. \tag{3.59b}$$

We can, in principle, express $(\mathbf{q}_1)_\mathrm{I}$ and $(\mathbf{q}_1)_\mathrm{II}$ in terms of the other generalized coordinates and thus completely eliminate them from Π in (3.58). Symbolically, we write

$$\begin{aligned}(\mathbf{q}_1)_\mathrm{I} &= (\mathbf{K}_{11})_\mathrm{I}^{-1}\{(\mathbf{Q}_1)_\mathrm{I} - (\mathbf{K}_{12})_\mathrm{I}(\mathbf{q}_2)_\mathrm{I} - (\mathbf{K}_{13})_\mathrm{I}(\mathbf{q}_3)_\mathrm{I}\}, \\ (\mathbf{q}_1)_\mathrm{II} &= (\mathbf{K}_{11})_\mathrm{II}^{-1}\{(\mathbf{Q}_1)_\mathrm{II} - (\mathbf{K}_{12})_\mathrm{II}(\mathbf{q}_2)_\mathrm{II} - (\mathbf{K}_{13})_\mathrm{II}(\mathbf{q}_3)_\mathrm{II}\}.\end{aligned} \tag{3.60}$$

Substitution into (3.58) yields

$$\Pi = \tfrac{1}{2} \mathbf{q}'^T \mathbf{K}' \mathbf{q}' - \mathbf{q}'^T \mathbf{Q}' - C, \tag{3.61}$$

where

$$\mathbf{q}' = \begin{Bmatrix} (\mathbf{q}_2)_\mathrm{I} \\ (\mathbf{q}_2)_\mathrm{II} \\ (\mathbf{q}_3)_\mathrm{II} \end{Bmatrix}, \qquad C = \tfrac{1}{2}(\mathbf{Q}_1^T \mathbf{K}_{11}^{-1} \mathbf{Q}_1)_\mathrm{I} + \tfrac{1}{2}(\mathbf{Q}_1^T \mathbf{K}_{11}^{-1} \mathbf{Q}_1)_\mathrm{II}, \qquad (3.62a)$$

and

$$\mathbf{K}' = \begin{bmatrix} (\mathbf{K}'_{22})_\mathrm{I} & & \mathrm{sym} \\ 0 & (\mathbf{K}'_{22})_\mathrm{II} & \\ (\mathbf{K}'_{23})_\mathrm{I}^T & (\mathbf{K}'_{23})_\mathrm{II}^T & (\mathbf{K}'_{33})_\mathrm{I} + (\mathbf{K}'_{33})_\mathrm{II} \end{bmatrix}, \qquad \mathbf{Q}' = \begin{Bmatrix} (\mathbf{Q}'_2)_\mathrm{I} \\ (\mathbf{Q}'_2)_\mathrm{II} \\ (\mathbf{Q}'_3)_\mathrm{I} + (\mathbf{Q}'_3)_\mathrm{II} \end{Bmatrix}. \qquad (3.62b)$$

Here

$$\begin{aligned}
(\mathbf{K}'_{22})_\mathrm{I} &= (\mathbf{K}_{22} - \mathbf{K}_{12}^T \mathbf{K}_{11}^{-1} \mathbf{K}_{12})_\mathrm{I}, \\
(\mathbf{K}'_{23})_\mathrm{I} &= (\mathbf{K}_{23} - \mathbf{K}_{12}^T \mathbf{K}_{11}^{-1} \mathbf{K}_{13})_\mathrm{I}, \\
(\mathbf{K}'_{33})_\mathrm{I} &= (\mathbf{K}_{33} - \mathbf{K}_{13}^T \mathbf{K}_{11}^{-1} \mathbf{K}_{13})_\mathrm{I}, \\
(\mathbf{Q}'_2)_\mathrm{I} &= (\mathbf{Q}_2 - \mathbf{K}_{12}^T \mathbf{K}_{11}^{-1} \mathbf{Q}_1)_\mathrm{I}, \\
(\mathbf{Q}'_3)_\mathrm{I} &= (\mathbf{Q}_3 - \mathbf{K}_{13}^T \mathbf{K}_{11}^{-1} \mathbf{Q}_1)_\mathrm{I},
\end{aligned} \qquad (3.63)$$

with similar expressions for quantities with subscript II. The matrices

$$\mathbf{K}'_\mathrm{I} = \begin{bmatrix} \mathbf{K}'_{22} & \mathbf{K}'_{23} \\ (\mathbf{K}'_{23})^T & \mathbf{K}'_{33} \end{bmatrix}_\mathrm{I}, \qquad \mathbf{Q}'_\mathrm{I} = \begin{Bmatrix} \mathbf{Q}'_2 \\ \mathbf{Q}'_3 \end{Bmatrix}_\mathrm{I}. \qquad (3.64)$$

are called the reduced submatrices for substructure I. Similar reduced matrices can be obtained for substructure II. The constraint conditions are imposed by altering $\mathbf{K}'$ and $\mathbf{Q}'$ according to the procedure described in sec. 3.4. We can then solve for $\mathbf{q}'$. After the $\mathbf{q}'$ have been determined, $(\mathbf{q}_1)_\mathrm{I}$ and $(\mathbf{q}_1)_\mathrm{II}$ can be evaluated from (3.60).

In practice, the reduced matrices are obtained directly from the submaster matrices without inverting $(\mathbf{K}_{11})_\mathrm{I}$ or $(\mathbf{K}_{11})_\mathrm{II}$. We shall illustrate this by using $\mathbf{K}_\mathrm{I}$ and $\mathbf{Q}_\mathrm{I}$ as an example. To express things simply, we drop the subscript I for $(\mathbf{K}_{11})_\mathrm{I}$, $(\mathbf{Q}_1)_\mathrm{I}$ in (3.56) and define

$$\mathbf{K}_a \equiv \mathbf{K}_b^T \equiv [\mathbf{K}_{12} \ \mathbf{K}_{13}]_\mathrm{I}, \quad \mathbf{K}_c \equiv \begin{bmatrix} \mathbf{K}_{22} & \mathbf{K}_{23} \\ \mathbf{K}_{32} & \mathbf{K}_{33} \end{bmatrix}_\mathrm{I}, \qquad \mathbf{Q}_c = \begin{Bmatrix} \mathbf{Q}_2 \\ \mathbf{Q}_3 \end{Bmatrix}_\mathrm{I}. \qquad (3.65)$$

Then (3.56) and (3.64) become

$$\mathbf{K}_I = \begin{bmatrix} \mathbf{K}_{11} & \mathbf{K}_a \\ \mathbf{K}_b & \mathbf{K}_c \end{bmatrix}, \qquad \mathbf{Q}_I = \begin{Bmatrix} \mathbf{Q}_1 \\ \mathbf{Q}_c \end{Bmatrix}, \tag{3.66}$$

$$\mathbf{K}'_I = \mathbf{K}_c - \mathbf{K}_b \mathbf{K}_{11}^{-1} \mathbf{K}_a, \qquad \mathbf{Q}'_I = \mathbf{Q}_c - \mathbf{K}_b \mathbf{K}_{11}^{-1} \mathbf{Q}_1.$$

Let the order of the matrix $\mathbf{K}_{11}$ be $N \times N$. Gaussian elimination is applied to $\mathbf{K}_I$ and $\mathbf{Q}_I$. After N steps—i.e., when $m = N + 1$ in (3.17)—we have

$$\mathbf{K}_I^{(N+1)} = \begin{bmatrix} \mathbf{K}_{11}^{(N+1)} & \mathbf{K}_a^{(N+1)} \\ 0 & \mathbf{K}_c^{(N+1)} \end{bmatrix}, \qquad \mathbf{Q}_I^{(N)} = \begin{Bmatrix} \mathbf{Q}_1^{(N+1)} \\ \mathbf{Q}_c^{(N+1)} \end{Bmatrix}, \tag{3.67}$$

where $\mathbf{K}_{11}^{(N+1)}$ is an upper triangular matrix.

We shall now show that (3.59a) is equivalent to

$$\mathbf{K}_{11}^{(N+1)}(\mathbf{q}_1)_I = \mathbf{Q}_1^{(N+1)} - \mathbf{K}_a^{(N+1)} \begin{Bmatrix} (\mathbf{q}_2)_I \\ (\mathbf{q}_3)_I \end{Bmatrix}, \tag{3.68}$$

and that the reduced matrices defined in (3.64) are simply

$$\begin{aligned} \mathbf{K}_c^{(N+1)} &= \mathbf{K}_c - \mathbf{K}_b \mathbf{K}_{11}^{-1} \mathbf{K}_a, \\ \mathbf{Q}_c^{(N+1)} &= \mathbf{Q}_c - \mathbf{K}_b \mathbf{K}_{11}^{-1} \mathbf{Q}_1, \end{aligned} \tag{3.69}$$

that is,

$$\mathbf{K}'_I = \mathbf{K}_c^{(N+1)}, \qquad \mathbf{Q}'_I = \mathbf{Q}_c^{(N+1)}.$$

Proof of 3.68 and 3.69: The proof of (3.68) is trivial, because it is simply reduced from (3.59) by Gaussian elimination. We shall prove (3.69) by induction. Equation (3.69) is certainly true if $N = 1$. We assume that it is also true for $N = M - 1$. In the case $N = M$ we partition

$$\mathbf{K}_{11} = \begin{bmatrix} \mathbf{K}_1 & \mathbf{K}_2 \\ \mathbf{K}_3 & \mathbf{K}_{NN} \end{bmatrix}, \quad \mathbf{Q}_1 = \begin{Bmatrix} \mathbf{Q}'_1 \\ \mathbf{Q}_N \end{Bmatrix}, \quad \mathbf{K}_a = \begin{Bmatrix} \mathbf{K}_4 \\ \mathbf{K}_5 \end{Bmatrix}, \quad \mathbf{K}_b = [\mathbf{K}_6 \; \mathbf{K}_7], \tag{3.70}$$

so that

$$\mathbf{K}_\mathrm{I} = \begin{bmatrix} \mathbf{K}_1 & \mathbf{K}_2 & \mathbf{K}_4 \\ \mathbf{K}_3 & \mathbf{K}_{NN} & \mathbf{K}_5 \\ \mathbf{K}_6 & \mathbf{K}_7 & \mathbf{K}_c \end{bmatrix},$$

where $\mathbf{K}_1$ is of the order $(M-1) \times (M-1)$. After $M-1$ steps, we have

$$\mathbf{K}_\mathrm{I}^{(M)} = \begin{bmatrix} \mathbf{K}_1^{(M)} & \mathbf{K}_2^{(M)} & \mathbf{K}_4^{(M)} \\ 0 & \mathbf{K}_{NN}^{(M)} & \mathbf{K}_5^{(M)} \\ 0 & \mathbf{K}_7^{(M)} & \mathbf{K}_c^{(M)} \end{bmatrix}, \quad \begin{Bmatrix} \mathbf{Q}_1^{(M)} \\ \mathbf{Q}_c^{(M)} \end{Bmatrix} = \begin{Bmatrix} \mathbf{Q}_1'^{(M)} \\ \mathbf{Q}_N^{(M)} \\ \mathbf{Q}_c^{(M)} \end{Bmatrix}. \tag{3.71}$$

By (3.69), we have

$$\begin{bmatrix} \mathbf{K}_{NN}^{(M)} & \mathbf{K}_5^{(M)} \\ \mathbf{K}_7^{(M)} & \mathbf{K}_c^{(M)} \end{bmatrix} = \begin{bmatrix} \mathbf{K}_{NN} & \mathbf{K}_5 \\ \mathbf{K}_7 & \mathbf{K}_c \end{bmatrix} - \begin{bmatrix} \mathbf{K}_3 \\ \mathbf{K}_6 \end{bmatrix} \mathbf{K}_1^{-1} [\mathbf{K}_2 \ \mathbf{K}_4]$$

$$= \begin{bmatrix} k & \mathbf{K}_5 - \mathbf{K}_3 \mathbf{K}_1^{-1} \mathbf{K}_4 \\ \mathbf{K}_7 - \mathbf{K}_6 \mathbf{K}_1^{-1} \mathbf{K}_2 & \mathbf{K}_c - \mathbf{K}_6 \mathbf{K}_1^{-1} \mathbf{K}_4 \end{bmatrix}, \tag{3.72}$$

$$\begin{Bmatrix} \mathbf{Q}_N^{(M)} \\ \mathbf{Q}_c^{(M)} \end{Bmatrix} = \begin{Bmatrix} \mathbf{Q}_N \\ \mathbf{Q}_c \end{Bmatrix} - \begin{Bmatrix} \mathbf{K}_3 \\ \mathbf{K}_6 \end{Bmatrix} \mathbf{K}_1^{-1} \mathbf{Q}_1',$$

where

$\mathbf{K}_{NN}^{(M)} = k = \mathbf{K}_{NN} - \mathbf{K}_3 \mathbf{K}_1^{-1} \mathbf{K}_2$.

From (3.71) and (3.72),

$$\mathbf{K}_c^{(M+1)} = \mathbf{K}_c - \mathbf{K}_6 \mathbf{K}_1^{-1} \mathbf{K}_4 - \frac{1}{k}(\mathbf{K}_7 - \mathbf{K}_6 \mathbf{K}_1^{-1} \mathbf{K}_2)(\mathbf{K}_5 - \mathbf{K}_3 \mathbf{K}_1^{-1} \mathbf{K}_4),$$

$$\mathbf{Q}_c^{(M+1)} = \mathbf{Q}_c - \mathbf{K}_6 \mathbf{K}_1^{-1} \mathbf{Q}_1' - [\mathbf{K}_7 - \mathbf{K}_6 \mathbf{K}_1^{-1} \mathbf{K}_2] \frac{\mathbf{Q}_N - \mathbf{K}_3 \mathbf{K}_1^{-1} \mathbf{Q}_1'}{k}. \tag{3.73}$$

Using (3.70) and the fact that

$$\mathbf{K}_{11}^{-1} = \begin{bmatrix} \mathbf{K}_1^{-1} + \mathbf{K}_1^{-1} \mathbf{K}_2 \mathbf{K}_3 \mathbf{K}_1^{-1}/k & -\mathbf{K}_1^{-1} \mathbf{K}_2/k \\ -\mathbf{K}_3 \mathbf{K}_1^{-1}/k & 1/k \end{bmatrix}, \tag{3.74}$$

we have accomplished our goal. Substituting (3.74) into (3.73), we can straightforwardly show that (3.73) is the same as (3.69) with M replaced by N.

It is obvious that using the Gaussian elimination technique, (i.e., using (3.69) to obtain the reduced submaster matrices) is much more efficient than using the formulas (3.63) and (3.64). After $\mathbf{q}'$ has been determined, $\mathbf{q}_1$ can be evaluated from (3.68) by simple back substitution ($\mathbf{K}_{11}^{(N+1)}$ is an upper triangular matrix).

A further improvement in efficiency can be achieved if we impose the constraint conditions of the substructures, using the procedure described in sec. 3.4, before performing Gaussian elimination to obtain the reduced master matrices. For example, for substructure I, since the $(\mathbf{q}_2)_\mathrm{I}$ are prescribed and are not in common with the rest of the structure, imposing constraint conditions is equivalent to altering $\mathbf{K}_\mathrm{I}$ and $\mathbf{Q}_\mathrm{I}$, respectively, to the forms

$$\begin{bmatrix} \mathbf{K}_{11} & 0 & \mathbf{K}_{13} \\ 0 & \mathbf{I} & 0 \\ \mathbf{K}_{13}^\mathrm{T} & 0 & \mathbf{K}_{33} \end{bmatrix}, \quad \begin{bmatrix} \mathbf{Q}_1 - \mathbf{K}_{12}\bar{\mathbf{q}}_2 \\ \bar{\mathbf{q}}_2 \\ \mathbf{Q}_3 - \mathbf{K}_{23}^\mathrm{T}\bar{\mathbf{q}}_2 \end{bmatrix}, \tag{3.75}$$

where $\bar{\mathbf{q}}_2$ is the prescribed value. Using these matrices instead of the original $\mathbf{K}_\mathrm{I}$ and $\mathbf{Q}_\mathrm{I}$ to perform Gaussian elimination is equivalent to using zero for $\mathbf{K}_{12}$ and $\mathbf{K}_{23}$ in (3.65)–(3.69).

As noted in sec. 3.4, in order to impose the boundary constraints $(\mathbf{q}_2)_\mathrm{I} = (\bar{\mathbf{q}}_2)_\mathrm{I}$, the ordering of components of $\mathbf{q}_\mathrm{I}$ in the sequence of $(\mathbf{q}_1)_\mathrm{I}$ and $(\mathbf{q}_2)_\mathrm{I}$ as given in (3.55) is not necessary. Also, any generalized coordinates that we do not wish to eliminate may be considered part of $(\mathbf{q}_3)_\mathrm{I}$. The generalized coordinates of substructure I may thus be partitioned into two parts:

$$\mathbf{q}_\mathrm{I} = \begin{Bmatrix} \mathbf{q}_1' \\ \mathbf{q}_3 \end{Bmatrix}_\mathrm{I}, \tag{3.76}$$

where the $(\mathbf{q}_1')$ are those not in common with the rest of the structure, which will be eliminated in the process of substructure condensation, and the $(\mathbf{q}_3)_\mathrm{I}$ are those in common with the rest of the structure or those we do not want to eliminate in the substructure condensation. After the submaster matrices have been assembled, the constraints associated with components of $(\mathbf{q}_1')_\mathrm{I}$ are imposed to obtain the constrained submaster matrices, which we might express as

$$\mathbf{K}_\mathrm{I}^* = \begin{bmatrix} \mathbf{K}_{11}^* & \mathbf{K}_{13}^* \\ \mathbf{K}_{13}^{*\mathrm{T}} & \mathbf{K}_{33}^* \end{bmatrix}_\mathrm{I}, \quad \mathbf{Q}_\mathrm{I}^* = \begin{Bmatrix} \mathbf{Q}_1^* \\ \mathbf{Q}_3^* \end{Bmatrix}. \tag{3.77}$$

Gaussian elimination is then performed on K_1^* and Q_1^* to reduce K_{11}^* to triangular form; in the meantime one obtains a result similar to (3.63), i.e.,

$$(K'_{33})_I = (K_{33} - K_{13}^{*T} K_{11}^{*-1} K_{13}^*)_I,$$
$$(Q'_3)_I = (Q_3^* - K_{13}^{*T} K_{11}^{*-1} Q_1^*)_I \tag{3.78}$$

directly. In carrying out the elimination process, we would use the fact that some rows and columns of K_{11} are zero, except on the diagonal, to save steps of multiplication.

If a similar procedure is applied to substructure II, Π of (3.58) can be written in the form

$$\Pi = \tfrac{1}{2} q'^T K' q' - q' Q' + C', \tag{3.79}$$

where

$$q' = (q_3)_I = (q_3)_{II},$$
$$K' = (K'_{33})_I + (K'_{33})_{II}, \tag{3.80}$$
$$Q' = (Q'_3)_I + (Q'_3)_{II},$$

and C' is some constant. If any components of q' are prescribed, one can impose the constraint condition by altering K' and Q' as described in sec. 3.4. We can then solve for q'.

The use of the substructure method is advantageous under many circumstances:

1 If the computer does not have enough core storage for the complete K simultaneously, we break K up into substructures. We can assemble the submaster matrices of a substructure in the core to obtain (3.56) and the reduced matrices, which are then placed in some form of external storage. The operation is then repeated for another substructure. After we have all the reduced matrices, we can read them back into core to assemble them into the master matrices K' and Q'. Since the order of K' will be much less than that of K, K' may be wholly stored in the core to solve for q'.

2 If we have to redesign only part of a structure, we can treat the part of the structure that is already designed as a substructure and obtain reduced matrices in terms of the generalized coordinates of the nodes that are common to the part to be designed. The many calculations required in the design will only need to include these generalized coordinates and those in the part being designed.

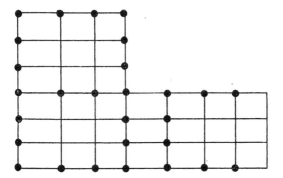

Fig. 3.3 Three subregions in an L-shaped domain.

3 For a structure made of many identical substructures, we need obtain the reduced submaster matrix only once. The total number of degrees of freedom involved in the calculation can be greatly reduced. An example is shown in fig. 3.3. We may divide the domain into three similar parts, and only one reduced submaster matrix will have to be generated. The final equation need involve only the degrees of freedom of the nodes with heavy dots.

Problems

1 If Gaussian elimination is applied to a symmetric matrix $\mathbf{K}$, the reduced matrix $\mathbf{K}^{(m)}$ is no longer symmetric for $m \geq 2$. But show that

$$K_{ij}^{(m)} = K_{ji}^{(m)} \qquad \text{for} \qquad i, j \geq m.$$

2 Derive the element matrices and assemble the master matrices $\mathbf{K}$ and $\mathbf{Q}$

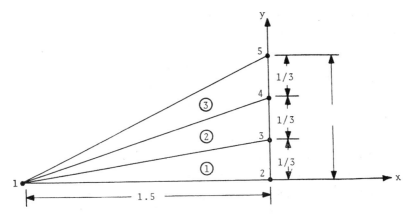

Fig. 3.4

according to the method described in sec. 3.2 for the problem shown in fig. 3.4. The differential equation is

$$\frac{\partial^2 \phi}{\partial x^2} + \frac{\partial^2 \phi}{\partial y^2} = 2$$

and the boundary conditions are

$\phi = 5 + xy + y^2$ between nodes 1 and 5,

$\dfrac{\partial \phi}{\partial x} = y$ on $x = 0$,

$\dfrac{\partial \phi}{\partial y} = x$ on $y = 0$.

3 Obtain the constrained master matrix and the constrained load vector by means of the techniques outlined in sec. 3.3 for prob. 2, assuming the constraints

$\phi_1 = 5.0, \quad \phi_5 = 6.0.$

4 Solve prob. 3 by Gaussian elimination.
5 Solve prob. 3 by triple factoring. Count the number of necessary multiplications and divisions.
6 Redo prob. 4 with the boundary condition on the x-axis changed to $\phi = 5$.
7 In using the substructure technique, one can eliminate all constrained degrees of freedom in the reduced matrices before assembling the final master matrices (i.e. $\mathbf{q}_3$ in (3.76) will include only the unconstrained degrees of freedom of a given substructure). Describe the process.

4 Implementation of Assembly and Solution Schemes for Large Systems on High-Speed Computers

4.1 Introduction

The major engineering advantage of the finite-element method is that it can really be used to solve complex practical problems. This success, of course, follows from the availability of high-speed digital computers that can perform millions of operations in a matter of seconds or less. However, the finite-element method also owes a great deal of its success to the ingenious engineers who make the computer work efficiently at the organization and handling of large amounts of data, the execution of the proper programs, and the systematic presentation of output.

The *efficient* use of the present generation of computers is still the key to the successful application of the finite-element method. For example, if one has to solve a system of 2,000 equations, there will be about two million entries for a symmetric $\mathbf{K}$; the number of operations required to obtain the solution will be of order 10^{10} according to (3.22), and the computer time required will thus be of order 10^4 seconds on an IBM 370/155. Present computers do not have enough core space for two million pieces of information; besides this, 10^4 seconds is a lot of computer time.

Various research groups have developed methods for the treatment of large systems of equations, but many of these are not available for general use. In this chapter we shall discuss the essential organizational steps needed for the solution of a large system of equations, and we shall outline a direct method of solution based on a computer system we ourselves have used. These steps involve assembling the element matrices and storing them in the computer, constraining the master matrices in a simple and straightforward manner so as to satisfy rigid boundary conditions, and determining the unknowns efficiently from the set of algebraic equations. Since in most finite-element analyses the matrix $\mathbf{K}$ is symmetric, we shall restrict our discussion to such cases.

4.2 Semiband Width of a Symmetric K

We start by noting that, in (3.2), many entries of $\mathbf{K}$ are zero; that is, the matrix $\mathbf{K}$ is only sparsely populated (it is even more so for a large system of equations). These zero entries are not used in the determination of the solution vector. As can be seen in (3.36), (3.37), and (3.38), any steps involving multiplication by a zero can be omitted. If the zero entries of **K** can be recognized ahead of time, they will not have to be stored in the computer. This will save core storage in the computer. If multiplications by these zero entries can be skipped, this will, of course also save tremendous amounts of computing time. It is therefore essential that our computer system be set up to take advantage of the sparseness of the matrix **K**.

An essential piece of information that enables us to recognize most of these zero entries in a symmetric **K** is the column number of the first nonzero entry, which we shall call $I(i)$ for the ith row. The quantity $I(i)$ is so defined that

$$K_{ij} = 0 \quad \text{for} \quad j < I(i). \tag{4.1}$$

We shall also denote $I(i)$ by I_i.

In the finite-element method the determination of I_i is relatively straightforward. We generally assume that every element-stiffness matrix is fully populated. Let the ith degree of freedom (of the global system) be associated with node n_i. The number n_i is, in general, different from i because usually each node has more than one degree of freedom. The entry K_{ij} will be nonzero only for those js that are degrees of freedom associated with the elements having node n_i as a common node. The number $I(i)$ is therefore the smallest value of all those js. To take an example, let $i = 5$ in the finite-element division of fig. 2.5, which is reproduced here (with the nodal coordinates omitted) as fig. 4.1. In this case there is one degree of freedom per node, so that n_i and i are the same; that is, node 5 is associated with the fifth degree of freedom. Node 5 is common to elements 3 and 4. The degrees of freedom associated with elements 3 and 4 are 2, 4, 5, and 6. The minimum value of 2, 4,

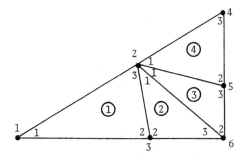

Fig. 4.1 Finite-element construction used to illustrate the determination of I_i.

5, and 6 is, of course, 2; thus $I_5 = 2$. In other words, $K_{5j} = 0$ for all $j < I(5)$; that is, $K_{51} = 0$, as given in (2.44).

Consider another example, using the same finite-element division of fig. 4.1. Suppose that each node has two degrees of freedom such that the $(2k - 1)$th and $2k$th degrees of freedom are associated with node k. Let us find the first nonzero entry of the tenth row. In this case $i = 10$, and the node 5 is associated with the tenth degree of freedom, (i.e., $n_i = 5$). Just as in the previous example, node 5 is common only to elements 3 and 4. However, the degrees of freedom associated with elements 3 and 4 are 3, 4, 7, 8, 9, 10, 11, and 12, because these two elements have nodes 2, 4, 5, and 6. The minimum of 3, 4, and 7 through 12 is, of course, 3, so that $I_{10} = 3$. In other words, $K_{10,j} = 0$ for $j < 3$.

In practice, I_i is determined by the following search process. We set $I_i^{(0)} = i$, where $i = 1, 2, \cdots, n$ before the search, n being the total number of degrees of freedom. Because we are assuming that all the element-stiffness matrices are fully populated, the diagonal term K_{ii} is nonzero. In other words, we always have $I_i \leq i$. Let $I_i^{(m)}$ denote the value of I_i after the mth element has been searched. Let $i_1, i_2, \cdots, i_p$ be the global degree numbers of the mth element and p be the number of degrees of freedom of the mth element. Define

$$I \equiv \min(i_1, i_2, \cdots, i_p).$$

Then

$$I_{i_r}^{(m)} = \min(I_{i_r}^{(m-1)}, I), \qquad r = 1, 2, \cdots p, \tag{4.2}$$

and the remaining $I_i^{(m)}$ (i. e., those for which $i \neq i_1, \cdots, i_p$) are the same as the $I_i^{(m-1)}$). In other words, if the ith degree of freedom is not associated with the mth element, the corresponding I_i will not be altered when the mth element is being searched. After we search through all the elements, we have the array I_i for $i = 1, 2, \cdots, n$ established.

To illustrate the search procedure for the determination of I_i, we again use the finite-element division shown in fig. 4.1. Recall that each node has only one degree of freedom, so that the global degree number is the same as the global nodal number and that n_i and i are the same. There are a total of six degrees of freedom. We set

$$I_i^{(0)} = i, \qquad i = 1, 2, \cdots, 6.$$

From the first element we have $i_1 = 1$, $i_2 = 3$, $i_3 = 2$, and

$I = \min(i_1, i_2, i_3) = \min(1, 3, 2) = 1$.

Therefore,

$I_{i_1}^{(1)} (= I_1^{(1)}) = \min(I_1^{(0)}, I) = \min(1, 1) = 1,$
$I_{i_2}^{(1)} (= I_3^{(1)}) = \min(I_2^{(0)}, I) = \min(3, 1) = 1,$
$I_{i_3}^{(1)} (= I_2^{(1)}) = \min(I_3^{(0)}, I) = \min(2, 1) = 1,$

and the rest of the $I_i^{(1)}$ are the same as $I_i^{(0)}$; that is,

$I_i^{(1)} = I_i^{(0)} = i$ for $i = 4, 5, 6$.

From the second element, we have $i_1 = 2$, $i_2 = 3$, $i_3 = 6$, and

$I = \min(2, 3, 6) = 2$.

Then

$I_2^{(2)} = \min(I_2^{(1)}, I) = 1,$ $I_1^{(2)} = I_1^{(1)} = 1,$
$I_3^{(2)} = \min(I_3^{(1)}, I) = 1,$ and $I_4^{(2)} = I_4^{(1)} = 4,$
$I_6^{(2)} = \min(I_6^{(1)}, I) = 2,$ $I_5^{(2)} = I_5^{(1)} = 5.$

From the third element, we have $i_1 = 2$, $i_2 = 6$, $i_3 = 5$, and $I = 2$.
Then

$I_2^{(3)} = \min(I_2^{(2)}, I) = 1,$ $I_1^{(3)} = I_1^{(2)} = 1,$
$I_6^{(3)} = \min(I_6^{(2)}, I) = 2,$ and $I_3^{(3)} = I_3^{(2)} = 1,$
$I_5^{(3)} = \min(I_5^{(2)}, I) = 2,$ $I_4^{(3)} = I_4^{(2)} = 4.$

From the fourth element we have $i_1 = 2$, $i_2 = 5$, $i_3 = 4$, and $I = 2$.

Then

$I_2^{(4)} = \min(I_2^{(3)}, I) = \min(1, 2) = 1,$ $I_1^{(4)} = I_1^{(3)} = 1,$
$I_5^{(4)} = \min(I_5^{(3)}, I) = \min(2, 2) = 2,$ and $I_3^{(4)} = I_3^{(3)} = 1,$
$I_4^{(4)} = \min(I_4^{(3)}, I) = \min(4, 2) = 2,$ $I_6^{(4)} = I_6^{(3)} = 2.$

The array $I_i^{(4)}$ is now the array I_i, with

$I_1 = I_2 = I_3 = 1,$
$I_4 = I_5 = I_6 = 2.$

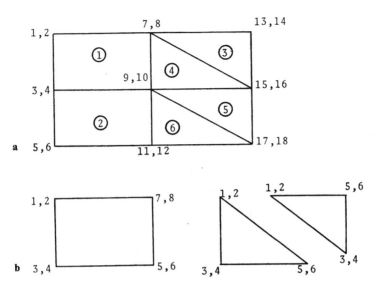

Fig. 4.2 A domain with elements having two degrees of freedom per node. (a) Entire domain and its global degree-of-freedom numbers. (b) Elements and local numbers.

We shall next consider the example shown in fig. 4.2. The nodes are taken at the vertices of the elements, and each node has two degrees of freedom. The uncircled numbers in fig. 4.2a are the global degree numbers and those in fig. 4.2b are the local degree numbers.* We see that there are a total of 18 degrees of freedom, while there are 8 degrees of freedom for each rectangular element and 6 for each triangular one. To determine $I_i (i = 1, 2, \cdots, 18)$ we first set

$$I_i^{(0)} = i, \quad i = 1, 2, \cdots, 18.$$

From the first element,

$$i_1 = 1, \quad i_2 = 2, \quad i_3 = 3, \quad i_4 = 4, \quad i_5 = 9, \quad i_6 = 10, \quad i_7 = 7, \quad i_8 = 8.$$

We have

$$I = \min(i_1, i_2, \cdots, i_8) = 1,$$

and, therefore, according to (4.2) with $m = 1$,

---

*The local degree number can be arbitrarily assigned. Of course, once it is assigned, the transformation between the local number and the global number, and the entries of the element matrices k_{ij}, Q_i must be evaluated accordingly. The counterclockwise assignment of local degree numbers, as in fig. 4.2b, is conventional.

$$I_1^{(1)} = I_2^{(1)} = I_3^{(1)} = I_4^{(1)} = I_9^{(1)} = I_{10}^{(1)} = I_7^{(1)} = I_8^{(1)} = 1,$$

and the rest of $I_i^{(1)}$ is the same as $I_i^{(0)}$. We then proceed to the second element to determine $I_i^{(2)}$, the third element to determine $I_i^{(3)}$, etc. For example, when we are determining $I_i^{(5)}$, we have, from element 5,

$$i_1 = 9, \quad i_2 = 10, \quad i_3 = 17, \quad i_4 = 18, \quad i_5 = 15, \quad i_6 = 16.$$

Therefore,

$$I = \min(i_1, i_2, \cdots, i_6) = 9$$

and

$$I_{i_r}^{(5)} = \min(I_{i_r}^{(4)}, I) = \min(I_{i_r}^{(4)}, 9), \qquad r = 1, 2, \cdots, 6,$$

and the rest of $I_i^{(5)}$ is the same as $I_i^{(4)}$. Finally, after element 6 has been searched,

$$I_1 = I_2 = I_3 = I_4 = I_7 = I_8 = I_9 = I_{10} = 1,$$
$$I_5 = I_6 = I_{11} = I_{12} = 3,$$
$$I_{13} = I_{14} = I_{15} = I_{16} = 7,$$
$$I_{17} = I_{18} = 9.$$

In practice, we do not form directly a different array $I_i^{(m)}$ for each different m; instead, we define $I_i = i$ at the beginning and keep altering I_i according to (4.2) as the computer searches through the elements. In fact, the array I_i can be obtained by observation once all the global degree numbers are assigned, as can be seen in fig. 4.2a. For a problem involving many elements, it is much more convenient to let the computer do the work.

The left semiband width, say B_i, of the ith row is simply

$$B_i = i - I_i + 1, \tag{4.3}$$

and the maximum semiband width B is

$$B = \max B_i, \qquad i = 1, 2, \cdots, n. \tag{4.4}$$

The left semiband width of each row depends on how we number the generalized coordinates. As we shall see later, the storage space for **K** and the number of operations required to obtain the solution depend heavily on the

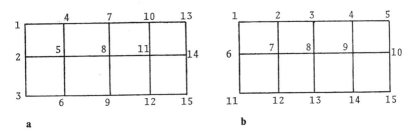

Fig. 4.3 Two numbering schemes for a single degree of freedom per node.

left semiband width. To minimize the band width of each row is a topological problem. In practice, the rule of thumb is to number the generalized coordinates so that the degree numbers within all elements have only small-digit differences. In the example shown in fig. 4.3 the numbering scheme of part a is much better than that of part b, because it gives a much smaller left semiband width for each row.

4.3 Banded Matrices and Gaussian Elimination

Variations in any computer program involving Gaussian elimination depend on the different ways of recognizing the zero entries of **K** and skipping the multiplication operation by these entries. The most commonly used practice is to utilize the fact that **K** is banded. Let B denote maximum semi-band width; that is, let

$$K_{ij} = 0 \quad \text{for} \quad \begin{cases} i \leq j - B \\ j \leq i - B \end{cases} \text{or} \quad \begin{matrix} i \geq j + B, \\ j \geq i + B. \end{matrix} \tag{4.5}$$

It can be easily shown that (4.5) is also true for all the reduced matrices. In what follows m takes on the values from 2 to n. Equation (3.17) thus becomes

$$K_{ij}^{(m)} = \begin{cases} K_{ij}^{(m-1)} & \text{for } i \leq m-1, \text{ or } i \geq m + B - 1, \text{ or } \\ & \quad j \geq m + B - 1, \\ 0 & \text{for } i \geq m, \text{ and } j \leq m - 1, \\ K_{ij}^{(m-1)} - \dfrac{K_{i,m-1}^{(m-1)} K_{m-1,j}^{(m-1)}}{K_{m-1,m-1}^{(m-1)}} & \text{for } \begin{cases} m \leq i \leq m + B - 2, \text{ and} \\ m \leq j \leq m + B - 2; \end{cases} \end{cases} \tag{4.6}$$

$$Q_i^{(m)} = \begin{cases} Q_i^{(m-1)} & \text{for } i \leq m-1, \text{ or } i \geq m + B - 1, \\ Q_i^{(m-1)} - \dfrac{K_{i,m-1}^{(m-1)}}{K_{m-1,m-1}^{(m-1)}} Q_{m-1}^{(m-1)} & \text{for } m \leq i \leq m + B - 2. \end{cases}$$

Because $K_{ij}^{(n)} = 0$ for $j \geq B + i$, (3.19) becomes

$$q_n = \frac{Q_n^{(n)}}{K_{nn}^{(n)}},$$
$$q_i = \frac{1}{K_{ii}^{(n)}}\left[Q_i^{(n)} - \sum_{j=i+1}^{J} K_{ij}^{(n)} q_j\right], \qquad i = n-1, n-2, \cdots, 1, \tag{4.7}$$

where

$$J = \min(n, i + B - 1).$$

The total number of multiplications and divisions is of order nB^2. In practice, $B \ll n$, and a tremendous savings in the computing time can be made by recognizing the banded property of **K**.

4.3.1 Storing a Symmetric Banded Matrix in the Computer and its Solution Algorithm

In (4.6), the off-diagonal terms of the lower triangle of **K** are eliminated one by one; and in the case of the banded matrix, the semiband width to the right of the diagonal term is unchanged as the elimination procedure progresses. In the final steps it reduces to an upper triangular matrix which is needed for the back substitution.

We can show that for a symmetric **K** at any stage of the elimination, only the information in the upper triangular portion of $\mathbf{K}^{(m)}$ is needed: For $m = 1$, $\mathbf{K}^{(1)}(=\mathbf{K})$ is symmetric. The proof is trivial. For $m \geq 2$, $\mathbf{K}^{(m)}$ is no longer symmetric. However, $K_{ij}^{(m)}$, with i or j less than or equal to $m - 1$, will no longer participate in any further elimination—see (4.6)—because $K_{ij}^{(m)} = 0$ for all $j \leq m - 1$ and $i > j$ and only those $K_{ij}^{(m)}$ with $j \geq i$, and $i \leq m - 1$ (in the region of the upper triangle) are needed for back substitution. Furthermore, for $i, j \geq m$ we still have

$$K_{ij}^{(m)} = K_{ji}^{(m)}; \tag{4.8}$$

thus only the information in the upper triangular portion of the matrix is needed.* Thus, for a banded **K**, we need only store the upper semiband of **K** (see fig 4.4)

We can define a rectangular $n \times B$ matrix **A** to represent the nonzero band of **K** such that

$$A_{i,j-i+1} = K_{ij}, \qquad i = 1, 2, \cdots, n, \qquad j = i, i+1, \cdots, i+B-1,$$

---

*Equation (4.8) can be shown to be true by the method of induction used in prob. 1 of ch. 3.

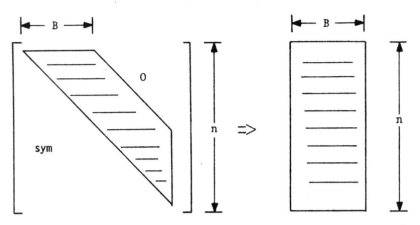

Fig. 4.4 The upper semiband of a symmetric banded matrix stored as a two-dimensional array.

or, in the elimination process,

$$A^{(m)}_{i,j-i+1} = K^{(m)}_{ij}.$$

Using the fact that $K^{(m-1)}_{i,m-1} = K^{(m-1)}_{m-1,i}$ for $i \geq m$, we find, upon substitution into (4.6), that

$$A^{(m)}_{i,j-i+1} = \begin{cases} A^{(m-1)}_{i,j-i+1} & \text{for } i \leq m-1 \text{ or } i \geq m+B-1 \\ & \text{or } j \geq m+B-1 \\ A^{(m-1)}_{i,j-i+1} - \dfrac{A^{(m-1)}_{m-1,i-m+2}}{A^{(m-1)}_{m-1,1}} A^{(m-1)}_{m-1,j-m+2} \\ & \text{for } m \leq i \leq m+B-2, \text{ and } i \leq j \leq m+B-2 \end{cases} \quad (4.9a)$$

$$Q^{(m)}_i = \begin{cases} Q^{(m-1)}_i & \text{for } i \leq m-1 \text{ or } i \geq m+B-1 \\ Q^{(m-1)}_i - \dfrac{A^{(m-1)}_{m-1,i-m+2}}{A^{(m-1)}_{m-1,1}} Q^{(m-1)}_{m-1} & \text{for } m \leq i \leq m+B-2. \end{cases}$$

By replacing $j - i + 1$ by l and using the fact that the range of l must be nonnegative, we have

$$A^{(m)}_{i,l} = \begin{cases} A^{(m-1)}_{i,l} & \text{for } i \leq m-1 \text{ or } i \geq m+B-1, \text{ or } l \geq m+B-i \\ A^{(m-1)}_{i,l} - \dfrac{A^{(m-1)}_{m-1,i-m+2}}{A^{(m-1)}_{m-1,1}} A^{(m-1)}_{m-1,l+i-m+1} \\ & \text{for } m \leq i \leq m+B-2, \text{ and } 1 \leq l \leq m+B-i-1. \end{cases} \quad (4.9b)$$

For practical convenience, we introduce $B - 1$ additional variables, $q_{n+1}, \ldots, q_{n+B-1}$, and set

$q_{n+i} = 0, \quad i = 1, 2, \cdots, B - 1.$

Then (4.7) with $K_{ij}^{(n)}$ expressed in terms of $A_{ij}^{(n)}$ gives

$$q_n = \frac{Q^{(n)}}{A_{n,1}^{(n)}},$$
$$q_i = \frac{1}{A_{i,1}^{(n)}}\left[Q_i^{(n)} - \sum_{l=2}^{B} A_{il}^{(n)} q_{i+l-1} \right], \quad i = n - 1, n - 2, \cdots, 1. \tag{4.10}$$

In performing the elimination on the computer, the location of $A_{i,l}^{(m-1)}$ and $Q_i^{(m-1)}$ is used to store $A_{i,l}^{(m)}$, and $Q_i^{(m)}$; that is, we only have to update A_{il} for $m \le i \le m + B - 2$, and $1 \le l \le m + B - i - 1$, according to (4.9) and leave the rest of A_{il} and Q_i untouched in the mth step of elimination.

The number of operations are $\frac{1}{2}B(B - 1)$ for (4.9b) and $B - 1$ for (4.9a) except for $m \ge n - B + 2$, in which case the number of the operations is less. For $m = 2, \cdots, n$, the total number of operations is given roughly by

$$\left[\frac{B(B - 1)}{2} + (B - 1) \right](n - 1).$$

With a triangularized **K**, and the updating of **Q**, the total number of operations for (4.10) is $B(n - 1) + 1$. Therefore, the total number of operations required for obtaining the solution to a system of equations with a banded symmetric matrix is roughly

$$\left(\frac{B^2}{2} + \frac{3B}{2} \right)(n - 1). \tag{4.11}$$

We note without proof that if $K_{m-1,m-1}^{(m-1)}$ is zero or near zero, the elimination procedure will fail or lead to numerical error. The remedy is, instead of using the $(m - 1)$th equation to eliminate the variable q_{m-1} from the rest of the equations, to select the maximum of $\left| K_{m-1,j}^{(m-1)} \right|, j > m - 1$, and eliminate the corresponding variable. For the mathematical operations involved, the reader is referred to Isaacson and Keller (1966).

4.4 External Storage. Varieties of Gaussian Elimination

Even though we only need to store the nonzero band of the upper triangle of a symmetric **K**, the computer may still not have enough core to store all the information simultaneously. For instance, if $n = 2,000$ and $B = 100$, the number of entries of **K** which must be stored is $2{,}000 \times 100 = 200{,}000$ words. This is a large core requirement, though it is already very small compared to the *total* population of **K**, which is 4×10^6 words.

To resolve this problem we can utilize the external storage of the computer. It is recognized that in (4.9b) only those $A_{i,l}^{(m)}$ in rows with $m \leq i \leq m + B - 2$ are different from $A_{i,l}^{(m-1)}$. The information needed to evaluate those $A_{i,l}^{(m)}$ are the $A_{i,l}^{(m-1)}$ in the rows $m - 1 \leq i \leq m + B - 2$. Therefore, in the mth elimination step, all the entries involved are in the $(m - 1)$th through $(m + B - 2)$th rows of **A**, and it will therefore be desirable to have these entries simultaneously in core for fast access. The rest of the information can be stored in the external storage and read into core only when it is needed. It is common practice to assign a fixed amount of core to store as much information as possible, and use the external storage only when the assigned space is filled, in order to reduce the frequency of shuffling the data in and out of core.

Various techniques have been developed for utilizing the external storage, and we shall discuss two here: substructures and front solution.

4.4.1 Substructures

The substructure technique has been described in sec. 3.6. We assemble the submaster matrices of a substructure, perform Gaussian elimination to obtain the reduced submaster matrices, and save the latter in the external storage. Equation (3.68) is used to evaluate the eliminated degrees of freedom later on. The process is repeated for each substructure. Since these reduced matrices involve many fewer degrees of freedom, all the information required for the solutions can usually be stored simultaneously in core. It should be noted that the band width of these final equations can be larger than that of the straight forward approach, that is, larger than the band width of the system of equations for all degrees of freedom of the entire domain.

4.4.2 Front Solution

Front solution is a variation of Gaussian elimination that makes the utilization of the external storage easy. It is customary to think of the processes of assembling **K** and **Q**, of imposing boundary constraints, and of solving the equation

Kq = Q

as distinct phases occurring one after the other. However, these processes

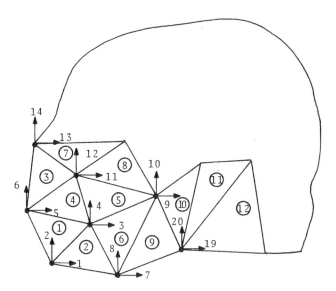

Fig. 4.5 Finite-element divisions for front solution.

can be performed in parallel in the Gaussian elimination method. Consider the example shown in fig. 4.5, with two degrees of freedom per node, as indicated by the arrows. The element numbers are circled. All the information contained in **K** and **Q** for the degrees of freedom 1 to 6 has been assembled after the data for elements 1 to 6 have been generated. Thus, with the Gaussian technique, it is possible to impose any constraint conditions which may occur at 1 to 6 and to eliminate these degrees of freedom; that is, the unconstrained degrees of freedom among 1 to 6 can be expressed in terms of the degrees of freedom 7 to 14 before the data for elements 7, 8, etc., are generated. If this is done, if each condensed row required for the back-substitution phase is saved in external storage, and if their core locations in the computer are used to store the new information being generated from elements 7, 8, etc., the core storage requirement for a large problem may be reduced considerably.

Reference to fig. 4.5 shows that some care must be taken in programming to realize fully the potential savings of Gaussian elimination. For example, information does not begin to appear for degrees of freedom 19 and 20 until the data for element 9 are generated, and these degrees may be eliminated after the data for element 12 have been assembled. Thus a requirement exists, not only for a table of the degrees of freedom to which each element connects, but also for some flags to mark the first and last appearance of each degree. The flags serve two purposes: they permit a calculation to be made of the maximum storage requirement, in terms of the maximum number of degrees of freedom for which information must be held in core simultaneously, and they are used to reserve subareas of storage as the information associated with various degrees of freedom is shuffled into and out of core. Each of the two phases (assembly–constraint–forward elimination and back substitution) propagates through the structure from node to node like a wave; hence, the front solution is also referred to as the "wave front" technique (Irons 1970).

A useful comparison of core storage required for the wave front and the regular methods is obtained by considering a domain of $M \times N$ rectangular elements ($M \leq N$) with corner nodes only and d degrees of freedom per node (fig. 4.6). The band width of **K**, which primarily affects storage in the regular methods, is controlled by the coupling between the degrees of freedom of the node and that of the nodes of the elements surrounding that node. If the degrees of freedom are numbered sequentially in one column of nodes after another, the smallest nodal number of the nodes coupled with node $I + M + 2$ is I. The degree numbers for these two nodes are, respectively, $d(I + M + 1) + i$ and $d(I - 1) + i$, where $i = 1, 2, \cdots, d$. The semiband width for the $[d(I + M + 1) + d]$th degree of freedom will be

$$B = [d(I + M + 1) + d] - [d(I - 1) + 1] + 1 = d(M + 3). \tag{4.12}$$

Since fig. 4.6 is a regular mesh, B can be used as the average semiband width

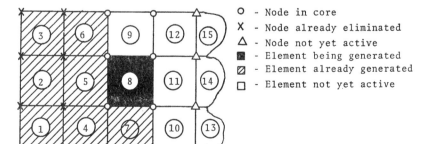

Fig. 4.6 Node status in the core.

to estimate the storage requirement. Also, the total number of degrees of freedom in the domain may be approximated by

$$D = d(M + 1)(N + 1) \approx dMN.$$

Thus the minimum storage requirement for **K** in the regular methods approaches

$$S_R = DB \approx d^2 M^2 N. \tag{4.13}$$

On the other hand, it is easily shown that the minimum number of nodes which must be kept in core simultaneously is $M + 3$ when the wave front method is used (see fig. 4.6). Therefore, the minimum-storage requirement for this method is a full lower triangle for $d(M + 3)$ degrees of freedom:

$$S_W = \frac{d}{2}(M + 3)[d(M + 3) + 1] \approx \frac{d^2}{2} M^2. \tag{4.14}$$

Thus

$$\frac{S_R}{S_W} \approx 2N. \tag{4.15}$$

Equation (4.15) indicates that the use of the wave front method, compared to the regular method, can achieve orders-of-magnitude savings in storage when N is large. Of course, there is a price to pay for this advantage because the wave front method requires many shuffles into and out of core, which means longer execution time, and a table for tracing the degrees of freedom currently in core, etc., which means more complicated programming.

4.5 Storage of a Symmetric K as a One-Dimensional Array

There are still other methods which can help to resolve the core storage problem: the method of sec. 4.3 made use of the maximum semiband width. In practice, the band width can vary widely from one row (or column) to another. The upper semiband width B_i of the ith column (which is equal to the left semiband width of the ith row for a symmetric matrix) can be much less than B, the matrix semiband width. It can be shown that in the Gaussian elimination process B_i is unchanged. In the next section we shall show that B_i of the factorized matrix **L** is the same as that of **K**. Therefore, only B_i storage locations are actually needed for the ith row (or column). Significant

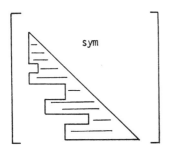

Fig. 4.7 Schematic diagram of a symmetric matrix with variable band width. The shaded area shows nonzero entries.

savings in the storage space for **K** and in the number of operations required for solving the algebraic equations can be achieved if, instead of using the matrix **A** of sec. 4.3, which is a two-dimensional array, we use a one-dimensional arrays to store entries within the semiband widths B_i for each of the n rows. The bookkeeping, of course, becomes slightly more complex.

A schematic diagram is shown in fig. 4.7, the shaded area being the location of nonzero entries.* We shall only store the shaded portion of the matrix, one row after another, in a one-dimensional array **R**. There is one entry, K_{11}, in the first row, and this will be R_1. The semiband width of the second row is $B_2 = 2 - I_2 + 1$, where I_2 is the first nonzero entry of the second row. The nonzero entries are K_{2,I_2} through $K_{2,2}$ in the lower triangle of **K**. These we shall store as R_2 through $R_{S(2)}$, where

$$S(2) = 1 + B_2 = 1 + 2 - I_2 + 1.$$

In general, there are $B_i = i - I_i + 1$ nonzero entries—K_{i,I_i} through $K_{i,i}$—in the ith row of the lower triangle; we shall store them as $R_{S(i)-i+I_i}$ through $R_{S(i)}$; that is,

$$R_{S(i)-i+j} = K_{i,j}, \quad j = I_i, I_i + 1, \cdots, i, \tag{4.16}$$

and

$$S(i) = S(i-1) + B_i = S(i-1) + i - I_i + 1, \quad i = 2, 3, \cdots, n, \tag{4.17}$$

---

*What this actually means is that the leading entry of each row of the shaded area is nonzero. There are still some or even many zero entries within this area. However, many of these zero entries will become nonzero after factorization or Gaussian elimination.

with $S(1) = 1$. Obviously, $S(i)$ is the location of the diagonal terms K_{ii} in the array R; that is, $S(i)$ is the total number of nonzero entries in the first i rows of the lower triangle of K.

In practice, it is more convenient to use the quantity $J(i)$ defined by

$$J(i) \equiv S(i) - i. \tag{4.18}$$

Thus $J(1) = 0$, and, in general, from (4.17), $J(i)$ can be evaluated from $I(i)$ using the following recurrence formula

$$\begin{aligned} J(i) &= S(i-1) - I_i + 1 \\ &= S(i-1) - (i-1) + i - I_i = J_{i-1} + i - I_i \end{aligned} \tag{4.19}$$

for $i = 2, 3, \cdots, n$. Equation (4.16) can now be written

$$R_{J(i)+j} = K_{ij} \quad \text{for} \quad I_i \le j \le i; \tag{4.20}$$

for the other values of j in the lower triangle of K, $K_{ij} = 0$ and is not stored. Therefore, with $J(i)$ defined as in (4.19), it is very easy to find K_{ij} in the array R. The total storage requirement for K is

$$S(n) = J(n) + n.$$

4.6 Skipping Multiplications by Zero Entries in Triple Factoring

In sec. 4.5 we showed how to store K as a one-dimensional array R to reduce the storage requirements. Such a reduction is only fruitful if, in the solution process, no additional core space is required. In this section we shall show that in the triple-factoring method the nonzero entries of L lie in the shaded area; that is,

$$L_{ij} = 0 \quad \text{for all} \quad j < I_i, \tag{4.21}$$

where I_i is the first nonzero entry of the ith row of K. Therefore, in the factoring process, we can use the same storage location* of R to store $K^{(n)}$ of L and D.

We shall prove (4.21) by the method of induction. It is trivial if $I_i = 1$ because all the subscripts of L are positive. If $I_i \ge 2$, then $L_{i1} = 0$ because

---
*This is also true for Gaussian elimination.

$$L_{i1} = \frac{K_{i1}}{D_1}. \tag{4.22}$$

Let us assume that (4.21) is true for $j = k - 1$, where $k < I_i$:

$$L_{i,k-1} = 0.$$

Then L_{ik} is also zero because

$$L_{ik} = \frac{1}{D_k}\left[K_{i,k} - \sum_{l=1}^{k-1} L_{il}D_l L_{kl}\right] = \frac{1}{D_k} K_{ik} = 0.$$

Using (4.21), we can now skip the multiplication by zero entries and write (3.35) as

$$D_j = K_{jj} - \sum_{l=I'}^{j-1} L_{jl}D_l L_{jl} \tag{4.23a}$$

and

$$L_{ij} = \frac{1}{D_j}\left[K_{ij} - \sum_{l=I}^{j-1} L_{il}D_l L_{jl}\right], \quad i = j+1, j+2, \cdots, n, \tag{4.23b}$$

where $I = \max(I_i, I_j)$ and $j = 1, 2, \cdots, n$. The summation terms are zero when $j = 1$ or when the lower range is larger than the upper range of the summation; that is, if

$$j - 1 < I_j,$$

there are no summation terms in (4.23a), and if

$$j - 1 < I,$$

there are none for (4.23b).

In practice, it is more convenient to perform the factorization steps row by row in the computer for a sparsely populated matrix. We express (4.23a,b) as a sequence

$$\begin{aligned}D_1 &= K_1, \\ L_{ij} &= \frac{1}{D_i}\left[K_{ij} - \sum_{l=I}^{j-1} L_{il}D_l L_{jl}\right],\end{aligned} \tag{4.24a}$$

where $j = I_i, I_i + 1, \cdots, i - 1;\ i = 1, 2, \cdots, n$; and

$I = \max [I_i, I_j]$,

with the summation terms being zero if $I > j - 1$, and

$$D_i = K_{ii} - \sum_{l=I_i}^{i-1} L_{il}D_l L_{il}, \tag{4.24b}$$

where the summation terms are zero if $i = 1$ or $I_i = i$, and $i = 1, 2, \cdots, n$. The forward substitution (3.27) becomes

$$g_i = Q_i - \sum_{j=I_i}^{i-1} L_{ij}g_j, \quad i = 1, 2, 3, \cdots, n, \tag{4.25}$$

in which the summed terms are 0 if $i = 1$ or $I_i = i$. The back substitution becomes

$$\begin{aligned} q_n &= \frac{g_n}{D_n}, \\ q_i &= \frac{g_i}{D_i} - \sum_{j=i+1}^{J} L_{ji}g_j, \quad i = n-1, n-2, \cdots, 1, \end{aligned} \tag{4.26}$$

where

$J = \min(n, i + B - 1)$,

B being the semiband width. Since $L_{ji} = 0$ for $i < I_j$, these entries are not stored in the computer; therefore, a logical instruction is incorporated to skip summation terms when $i < I_j$ in (4.26).

4.7 Assembling, Constraining, and Solving for a Symmetric K Using the One-Dimensional Array

In Sec. 4.5 the location of K_{ij} of the lower triangle of **K** in the one-dimensional array **R** is given. That means we can directly assemble all the element-stiffness matrices in **R**, impose constraint conditions, and obtain the solution. The entire procedure involves five steps.

1 Set up a finite-element subdivision and assign the degree numbers of the generalized coordinates. That is, we must determine the relation between the local degree number $1, 2, \cdots, p$ and the global degree number $i_1, i_2, \cdots, i_p$ for each element.

2 Search through the list $i_1, i_2, \cdots, i_p$ of each element to determine I_1,

$I_2, \cdots, I_n$—the column of the first nonzero entry of each row of **K**—and evaluate $J(i)$ according to (4.19). Then $J(i) + j$ is the location of K_{ij} for $j \geq I_i$ in the one-dimensional array **R**. The total length of **R** is $J(n) + n$, where n is the order of **K**.

3 Assemble element-stiffness matrices and load vectors into one-dimensional arrays **R** and **Q**, respectively. Let $\mathbf{R}^{(0)}$ and $\mathbf{R}^{(m)}$ denote the array **R** before any and after the mth element has been assembled. Similarly, let $\mathbf{Q}^{(0)}$ and $\mathbf{Q}^{(m)}$ denote before and after values of the load vector **Q**. Set $R_i^{(0)} = 0$ ($i = 1, 2, \cdots, J(n) + n$) and $Q_i^{(0)} = 0$ ($i = 1, 2, \cdots, n$). In assembling the mth element use (3.1a) and (4.20):

$$R^{(m)}_{J(M)+N} = R^{(m-1)}_{J(M)+N} + k_{rs},$$
$$Q^{(m)}_{i_r} = Q^{(m-1)}_{i_r} + (Q_r)_m, \qquad (4.27)$$
$$r = 1, 2, \cdots, p, \quad s = 1, 2, \cdots, r,$$

where

$$M = \max(i_r, i_s), \quad N = \min(i_r, i_s).$$

The rest of $R_i^{(m)}$ and $Q_i^{(m)}$ are, respectively, the same as $R_i^{(m-1)}$ and $Q_i^{(m-1)}$. (Recall that k_{rs} and $(Q_r)_m$ are, respectively, the components of the element-stiffness matrix and loading vector, p denotes the number of degrees of freedom, i_r is the global degree number of the rth generalized coordinate of the mth element, and $J(i) + j$ is the location of $K_{ij}(I_i \leq j < i)$ in the array **R**.)

Physically, we do not create new arrays $\mathbf{R}^{(m)}$ and $\mathbf{Q}^{(m)}$; instead, we just alter **R** and **Q** according to (4.27) until we have all elements assembled.

4 Impose constraint conditions. Suppose we have to constrain the jth component of **q** so that it is equal to $\bar{q}_j$. (We use **q** as the generalized coordinate vector here in place of Φ as in secs. 3.1 and 3.2). Using the fact that $K_{ij} = K_{ji}$, we can rewrite (3.8) as

$$Q_i^* = Q_i - K_{ji}\bar{q}_j,$$
$$K_{ji}^* = 0, \qquad i = 1, 2, \cdots, j-1,$$

for $i < j$,

$$Q_i^* = Q_i - K_{ij}\bar{q}_j,$$
$$K_{ij}^* = 0, \qquad i = j+1, j+2, \cdots, n,$$

for $i > j$, and

$$Q_i^* = \bar{q}_j,$$

$K^*_{ij} = 1$

for $i = j$. Using $K_{ji} = 0$ for $i < I_j$ and $K_{ij} = 0$ for $j < I_i$ (which are not stored in the computer) and using (4.20), we have

$$Q^*_i = Q_i - R_{J(j)+i}\bar{q}_j,$$
$$R^*_{J(j)+i} = 0, \qquad i = I_j, I_j + 1, \cdots, j - 1, \qquad (4.28a)$$

for $i < j$ (skip this operation if $I_j = j$),

$$Q^*_i = Q_i - R_{J(i)+j}\bar{q}_j,$$
$$R^*_{J(i)+j} = 0, \qquad i = j + 1, \; j + 2, \cdots, n, \qquad (4.28b)$$

for $i > j$ (skip this operation if $j = n$ or $j < J(i)$),

$$Q^*_i = \bar{q}_j,$$
$$R^*_{J(i)+j} = 1, \qquad (4.28c)$$

for $i = j$, and the rest of R^*_i and Q^*_i are the same as R_i and Q_i, respectively.

In practice, we do not physically create new **R*** and **Q***; instead, we alter **R** and **Q** according to (4.28). Therefore, we simply repeat the operation of (4.28) until we have imposed all the constraint conditions. It is an element of good planning in the computer system to create an array to indicate the degree of freedom that is going to be constrained and its corresponding prescribed value before the constraining phase of operation starts.

Another possibililty may arise in practice. Instead of having a constraint condition of the form $q_j = \bar{q}_j$, we may require a combination of several generalized coordinates to be equal to a specific value. For example, we may require

$$a_1 q_{i_1} + a_2 q_{i_2} + \cdots + a_r q_{i_r} = \bar{q}, \qquad (4.29)$$

with $a_1 \neq 0$. In this case a coordinate transformation can be performed first so that $a_1 q_{i_1} + \cdots + a_r q_{i_r}$ itself becomes an independent degree of freedom. This transformation, usually called *rotation*, will, of course, require alterations in **R** and **Q**. Then the constraint conditions can be imposed according to (4.28a,b,c). We shall examine such a rotation in the next chapter when we discuss the plane stress-strain problem.

5a Solve the problem by Gaussian elimination. Let $\mathbf{R}^{(m)}$ be the one-dimensional array used to store the upper triangle of the reduced $\mathbf{K}^{(m)}$ of

(4.6) in the form*

$$R_{J(j)+i}^{(m)} = K_{ij}^{(m)}, \tag{4.30}$$

where m ranges from 2 to n. We can write (4.6) as

$$R_{J(j)+i}^{(m)} = R_{J(j)+i}^{(m-1)} - \frac{R_{J(j)+m-1}^{(m-1)} R_{J(i)+m-1}^{(m-1)}}{R_{J(m-1)+m-1}^{(m-1)}} \tag{4.31a}$$

for $m \leq i \leq \min(n, m + B - 2)$, $m \leq j \leq i$, and $m - 1 \geq \max(I_i, I_j)$; and

$$Q_i^{(m)} = Q_i^{(m-1)} - Q_{m-1}^{(m-1)} \frac{R_{J(i)+m-1}^{(m-1)}}{R_{J(m-1)+m-1}^{(m-1)}} \tag{4.31b}$$

for $m \leq i \leq \min(n, m + B - 2)$ and $m - 1 \geq I_i$. The rest of $R_i^{(m)}$ and $Q_i^{(m)}$ are the same as $R_i^{(m-1)}$ and $Q_i^{(m-1)}$, respectively. [Recall that B is the semiband width and I_i is the first nonzero entry of the ith column (same as the ith row) of **K**]. From (4.7), we have the solution

$$\begin{aligned} q_n &= \frac{Q_n^{(n)}}{R_{J(n)+n}^{(n)}}, \\ q_i &= \frac{1}{R_{J(i)+i}^{(n)}} \left[Q_i^{(n)} - \sum_{j=i+1}^{J} {}' R_{J(j)+i}^{(n)} q_j \right], \quad i = n-1, n-2, \cdots, 1, \end{aligned} \tag{4.32}$$

where

$$J = \min(n, i + B - 1)$$

and the prime indicates that there are no summation terms if $J = i$ and that for those values of j such that $i < I_j$, the summation operation is skipped.

It should be noted that, in Sec 4.7.3, we use $R_i^{(m)}$ and $Q_i^{(m)}$ to denote, respectively, the arrays for **K** and **Q** after the mth element has been assembled. Here $R_i^{(m)}$ and $Q_i^{(m)}$ are used to denote, respectively, the arrays for reduced **K** and **Q** after $m - 1$ Gaussian elimination steps have been performed.

5b Solve the problem by the method of triple factoring. Let **R** and **R*** be the one-dimensional arrays used to store the lower triangle of the matrix **K** and the factorized matrices **L** and **D** with

---

*This is done to store the upper triangle of K columnwise.

$$R^*_{J(i)+i} = D_i,$$
$$R^*_{J(i)+j} = L_{ij}, \qquad I_i \le j \le i, \tag{4.33}$$

where $i = 1, 2, \cdots, n$. A substitution into (4.24a,b) yields

$$R^*_1 = R_1;$$

Let*

$$U_{J(i)+j} = R_{J(i)+j} - \sum_{l=I}^{j-1} U_{J(i)+l} R^*_{J(j)+l} \qquad \text{for} \qquad I_i \le j \le i \text{ and} \tag{4.34a}$$
$$i = 2, 3, \cdots, n$$

where

$I = \max(I_i, I_j)$.

There are no summation terms if $I > j - 1$. Then

$$R^*_{J(i)+j} = U_{J(i)+j}/R^*_{J(j)+j} \tag{4.34b}$$
$$R^*_{J(i)+i} = U_{J(i)+i}$$

where $i = 2, 3, \cdots, n$, and $j = I_i, I_i + 1, \cdots, i - 1$. From (4.25) we have

$$g_1 = Q_1,$$
$$g_i = Q_i - \sum_{j=I_i}^{i-1} R^*_{J(i)+j} g_j, \qquad i = 2, 3, \cdots, n. \tag{4.35}$$

Now, from (4.26), let†

$$q_i^{(0)} = g_i/R^*_{J(i)+i}, \qquad i = 1, 2, \cdots, n; \tag{4.36}$$

then

$$\begin{aligned} q_i^{(m)} &= q_i^{(m-1)} \qquad \text{for} \qquad i \le I_m - 1 \quad \text{or} \quad i \ge n - m + 1 \\ &= q_i^{(m-1)} - R^*_{J(n-m+1)+i} q_{n-m+1}^{(m)} \qquad \text{for} \qquad I_m \le i \le n - m, \end{aligned} \tag{4.37}$$

where $m = 1, 2, \cdots, n$. Then we have the solution

$$\mathbf{q} = \mathbf{q}^{(n)}. \tag{4.38}$$

*See prob. 4.1 for the reasoning behind the intermediate step (4.34a).
†See prob. 4.2.

In practice, the same storage locations are used for $\mathbf{R}^*$, $\mathbf{U}$ and $\mathbf{R}$ in (4.34a, b). In creating $\mathbf{R}^*$ and $\mathbf{U}$ in the computer, we simply replace the value R_k by the value U_k in (4.34a), then the value U_k by the value R_k^* in (4.34b) for $k = J(i) + j$, $j = I_i$, $I_i + 1$, $\cdots$, i, and $i = 1, 2, \cdots, n$. Common storage locations are also used to store the load vector $\mathbf{Q}$, the intermediate solution vector $\mathbf{g}$, the vector $\mathbf{q}^{(m)}$, and the final solution vector $\mathbf{q}$.

In the case when the computer cannot use dynamic memory allocation, instead of defining the dimensions of all the arrays for $\mathbf{R}$, $\mathbf{Q}$, I_i, $J(i)$, etc., it is more convenient to pack all these arrays in a single array. The only additional information needed is pointers to indicate the starting location of each vector in this array. For example, let $\mathbf{S}$ be the single array for storing the vector in the order $\mathbf{R}$, $\mathbf{Q}$, I_i, etc. Then $\mathbf{R}$ is stored in S_k where $k = 1, 2, \cdots, J(n) + n$, $\mathbf{Q}$ is stored in S_k where $k = J(n) + n + 1, \cdots, J(n) + 2n$, the I_i's are stored in S_k where $k = J(n) + 2n + 1, \cdots, J(n) + 3n$, etc. The pointers would then have the value zero for $\mathbf{R}$, the value $J(n) + n$ for $\mathbf{Q}$ and the value $J(n) + 2n$ for the I_i's. In fact, the pointers plus one are the starting address locations of the vectors in the array.

Problems

1 Show that in comparing (4.24a, b) and (4.34a, b) the use of the intermediate step (4.34a) will save approximately half of the multiplications in factoring a large matrix during actual computation.

2 Show that (4.26) is the same as (4.36) through (4.38), but with a different computing sequence. In computer calculation the latter sequence will eliminate many logical tests for skipping the summation terms in (4.26) when $i < I_j$.

5 Applications to Solid Mechanics

5.1 Basic Equations for Solid Mechanics

A major portion of the literature on the finite-element method treats problems in solid mechanics in detail. The detailed formulation of the general solid-mechanics problem is beyond the scope of this book; it may be found in a number of excellent texts.* In this chapter, we shall summarize the basic relations used to derive the finite-element equations for a solid continuum. Discussion will be limited to the case of linear, elastic material.

5.1.1 Stress and Equilibrium Equations

Imagine a surface within a body. The interaction between the material on one side of the surface and that on the other side can be described in terms of the forces exerted across the surface. A stress is defined as the force per unit area exerted by that part of the body on one side of the surface on that of the other side. If couples are disregarded, the state of stress at a point is uniquely defined by six stress components.

$$\sigma^T = \{\sigma_x \quad \sigma_y \quad \sigma_z \quad \sigma_{xy} \quad \sigma_{xz} \quad \sigma_{yz}\} \tag{5.1}$$

in Cartesian coordinates (fig. 5.1), where σ_x, σ_y, and σ_z are the *normal stresses* and σ_{xy}, σ_{xz}, and σ_{yz} are the *shear stresses*. All of these stresses act on the faces of an elemental volume as shown in fig. 5.1. Positive values of stresses imply that stresses point to a positive coordinate direction for a positive face and point to a negative direction for a negative face. A positive face is one with its outer normal pointing in a positive coordinate direction.

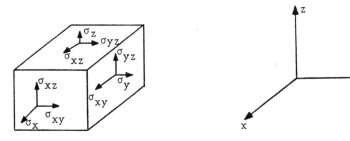

Fig. 5.1 Notation for stress components.

*For example: Timoshenko and Goodier (1970); Fung (1965); Love (1963); Landau and Lifshitz (1959); Sokolnikoff (1956); and Sechler (1952).

The *traction*

$$\mathbf{T}_\nu = \begin{Bmatrix} T_x \\ T_y \\ T_z \end{Bmatrix}_\nu \tag{5.2}$$

is the force per unit area acting on the surface with unit outward normal

$$\boldsymbol{\nu} = \begin{Bmatrix} \nu_x \\ \nu_y \\ \nu_z \end{Bmatrix}, \tag{5.3}$$

where ν_x, ν_y, and ν_z are the direction cosines of the normal. The tractions are related to the stresses by

$$\mathbf{T}_\nu = \begin{Bmatrix} T_x \\ T_y \\ T_z \end{Bmatrix}_\nu = \begin{Bmatrix} \nu_x \sigma_x + \nu_y \sigma_{xy} + \nu_z \sigma_{xz} \\ \nu_x \sigma_{xy} + \nu_y \sigma_y + \nu_z \sigma_{yz} \\ \nu_x \sigma_{xz} + \nu_y \sigma_{yz} + \nu_z \sigma_z \end{Bmatrix}. \tag{5.4}$$

Rotated Coordinate System The stress components in (5.1) are in a Cartesian system (x, y, z). Consider a second set of Cartesian coordinates (x', y', z') with the same origin but with different orientation. The coordinates are then related by

$$\mathbf{x} = \boldsymbol{\beta}\mathbf{x}', \tag{5.5}$$

in which

$$\mathbf{x} = \begin{Bmatrix} x \\ y \\ z \end{Bmatrix}, \quad \mathbf{x}' = \begin{Bmatrix} x' \\ y' \\ z' \end{Bmatrix}, \quad \boldsymbol{\beta} = \begin{bmatrix} \beta_{11} & \beta_{12} & \beta_{13} \\ \beta_{21} & \beta_{22} & \beta_{23} \\ \beta_{31} & \beta_{32} & \beta_{33} \end{bmatrix}.$$

The rows of $\boldsymbol{\beta}$ are vectors along the x-, y-, and z-axes; for example, $\{\beta_{11}\ \beta_{12}\ \beta_{13}\}$ is a unit vector along the x-axis with the βs the direction cosines with respect to the x'-, y'-, and z'-axes. (It should be noted that $\boldsymbol{\beta}^{-1} = \boldsymbol{\beta}^T$). The stress components in the new reference system can be expressed as

$$\begin{bmatrix} \sigma'_x & \sigma'_{xy} & \sigma'_{xz} \\ & \sigma'_y & \sigma'_{yz} \\ \text{sym} & & \sigma'_z \end{bmatrix} = \boldsymbol{\beta} \begin{bmatrix} \sigma_x & \sigma_{xy} & \sigma_{xz} \\ & \sigma_y & \sigma_{yz} \\ \text{sym} & & \sigma_z \end{bmatrix} \boldsymbol{\beta}^T. \tag{5.6}$$

Principal Stresses If the coordinate axes are the principal axes, the stress vector becomes

$$\sigma^T = \{\sigma_1 \quad \sigma_2 \quad \sigma_3 \quad 0 \quad 0 \quad 0\}. \tag{5.7}$$

The components σ_1, σ_2, and σ_3 are the roots of

$$-\sigma^3 + I_1\sigma^2 - I_2\sigma + I_3 = 0, \tag{5.8}$$

in which the Is are the stress invariants defined by

$$\begin{aligned} I_1 &= \sigma_x + \sigma_y + \sigma_z = \sigma_1 + \sigma_2 + \sigma_3, \\ I_2 &= \sigma_x\sigma_z + \sigma_y\sigma_z + \sigma_z\sigma_x - \sigma_{xy}^2 - \sigma_{xz}^2 - \sigma_{yz}^2 = \sigma_1\sigma_2 + \sigma_2\sigma_3 + \sigma_3\sigma_1. \\ I_3 &= |\sigma_{ij}| = \sigma_1\sigma_2\sigma_3 \end{aligned} \tag{5.9}$$

The direction cosines of the principal axes can be obtained by solving

$$\begin{bmatrix} \sigma_x & \sigma_{xy} & \sigma_{xz} \\ & \sigma_y & \sigma_{yz} \\ \text{sym} & & \sigma_z \end{bmatrix} \begin{Bmatrix} \nu_1 \\ \nu_2 \\ \nu_3 \end{Bmatrix} = \sigma_i \begin{Bmatrix} \nu_1 \\ \nu_2 \\ \nu_3 \end{Bmatrix}, \quad i = 1, 2, 3, \tag{5.10}$$

where the σ_i are the *principal stresses*.

Equilibrium Equations The equilibrium of the body requires that the stresses satisfy the equations

$$\begin{aligned} \frac{\partial \sigma_x}{\partial x} + \frac{\partial \sigma_{xy}}{\partial y} + \frac{\partial \sigma_{xz}}{\partial z} + \bar{F}_x &= 0, \\ \frac{\partial \sigma_{xy}}{\partial x} + \frac{\partial \sigma_y}{\partial y} + \frac{\partial \sigma_{yz}}{\partial z} + \bar{F}_y &= 0, \\ \frac{\partial \sigma_{xz}}{\partial x} + \frac{\partial \sigma_{yz}}{\partial y} + \frac{\partial \sigma_z}{\partial z} + \bar{F}_z &= 0, \end{aligned} \tag{5.11}$$

where the $\bar{F}$s are the components of the body-force vector,

$$\bar{F}^T = \{\bar{F}_x \quad \bar{F}_y \quad \bar{F}_z\}.$$

Two-Dimensional Stress In many practical situations, some components of stress are either zero or can be expressed in terms of other stress components.

A typical case often encountered is that of plane stress, a two-dimensional problem in which we only have to consider σ_x, σ_y, and σ_{xy}. In this case, we write the stress vector as

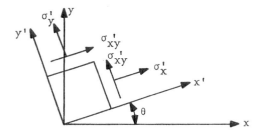

Fig. 5.2 Rotation of coordinates for two-dimensional problems.

$$\boldsymbol{\sigma}^T = \{\sigma_x \quad \sigma_y \quad \sigma_{xy}\}; \tag{5.12}$$

the rotation of one coordinate system about another can be expressed in terms of a single angle θ (fig. 5.2); then (5.6) can be written as

$$\begin{Bmatrix} \sigma'_x \\ \sigma'_y \\ \sigma'_{xy} \end{Bmatrix} = \begin{bmatrix} \cos^2\theta & \sin^2\theta & 2\sin\theta\cos\theta \\ \sin^2\theta & \cos^2\theta & -2\sin\theta\cos\theta \\ -\sin\theta\cos\theta & \sin\theta\cos\theta & \cos^2\theta - \sin^2\theta \end{bmatrix} \begin{Bmatrix} \sigma_x \\ \sigma_y \\ \sigma_{xy} \end{Bmatrix}. \tag{5.13}$$

The principle stresses are

$$\begin{Bmatrix} \sigma_1 \\ \sigma_2 \end{Bmatrix} = \frac{\sigma_x + \sigma_y}{2} \pm \sqrt{\left(\frac{\sigma_x - \sigma_y}{2}\right)^2 + \sigma_{xy}^2}, \tag{5.14}$$

with the principal direction given by

$$\tan 2\theta = \frac{2\sigma_{xy}}{\sigma_x - \sigma_y}. \tag{5.15}$$

The maximum shear stress on the (x, y)-plane is

$$\tau = \sqrt{\left(\frac{\sigma_x - \sigma_y}{2}\right)^2 + \sigma_{xy}^2}, \tag{5.16}$$

directed at $\pm 45°$ from the principal direction given by (5.15).

5.1.2 Strain. The Strain-Displacement Relation

The deformation at a point is described uniquely by the six components of *strain*,

$$\mathbf{e}^T = \{e_x \quad e_y \quad e_z \quad e_{xy} \quad e_{xz} \quad e_{yz}\}. \tag{5.17}$$

The principal strains occur for principal strain directions (as in the case of stress), so for the principal coordinate axes we have

$$\mathbf{e}^T = \{e_1 \quad e_2 \quad e_3 \quad 0 \quad 0 \quad 0\}.$$

We shall restrict ourselves to the case of small strain and linear elasticity. In this case the relations between the components of strain and those of the displacements

$$\mathbf{u}^T = \{u \quad v \quad w\}$$

at a point are given by

$$\begin{aligned} e_x &= \frac{\partial u}{\partial x}, & e_y &= \frac{\partial v}{\partial y}, & e_z &= \frac{\partial w}{\partial z}, \\ e_{xy} &= \frac{\partial u}{\partial y} + \frac{\partial v}{\partial x}, & e_{xz} &= \frac{\partial u}{\partial z} + \frac{\partial w}{\partial x}, & e_{yz} &= \frac{\partial v}{\partial z} + \frac{\partial w}{\partial y}. \end{aligned} \tag{5.18}$$

In two-dimensional problems, we only have to deal with e_x, e_y, e_{xy}. For these problems, write the strain vector as

$$\mathbf{e}^T = \{e_x \quad e_y \quad e_{xy}\}.$$

The coordinate transformation for the strain system is similar to that for the stress:

$$\begin{bmatrix} e'_x & \frac{1}{2}e'_{xy} & \frac{1}{2}e'_{xz} \\ & e'_y & \frac{1}{2}e'_{yz} \\ \text{sym} & & e'_z \end{bmatrix} = \boldsymbol{\beta}^T \begin{bmatrix} e_x & \frac{1}{2}e_{xy} & \frac{1}{2}e_{xz} \\ & e_y & \frac{1}{2}e_{yz} \\ \text{sym} & & e_z \end{bmatrix} \boldsymbol{\beta},$$

in which the x'-coordinates are related to the x-coordinates by (5.5).

The foregoing equation may be written in a different form, namely,

$$\mathbf{e} = \mathbf{B}\mathbf{e}', \tag{5.19a}$$

in which

$$\mathbf{B} = \begin{bmatrix} \beta_{11}^2 & \beta_{12}^2 & \beta_{13}^2 & \beta_{11}\beta_{12} & \beta_{11}\beta_{13} & \beta_{12}\beta_{13} \\ \beta_{21}^2 & \beta_{22}^2 & \beta_{23}^2 & \beta_{21}\beta_{22} & \beta_{21}\beta_{23} & \beta_{22}\beta_{23} \\ \beta_{31}^2 & \beta_{32}^2 & \beta_{33}^2 & \beta_{31}\beta_{32} & \beta_{31}\beta_{33} & \beta_{32}\beta_{33} \\ 2\beta_{11}\beta_{21} & 2\beta_{12}\beta_{22} & 2\beta_{13}\beta_{23} & \beta_{11}\beta_{22}+\beta_{12}\beta_{21} & \beta_{11}\beta_{23}+\beta_{13}\beta_{21} & \beta_{12}\beta_{23}+\beta_{13}\beta_{22} \\ 2\beta_{11}\beta_{31} & 2\beta_{12}\beta_{32} & 2\beta_{13}\beta_{33} & \beta_{11}\beta_{32}+\beta_{12}\beta_{31} & \beta_{11}\beta_{33}+\beta_{13}\beta_{31} & \beta_{12}\beta_{33}+\beta_{13}\beta_{32} \\ 2\beta_{21}\beta_{31} & 2\beta_{22}\beta_{32} & 2\beta_{23}\beta_{33} & \beta_{21}\beta_{32}+\beta_{22}\beta_{31} & \beta_{21}\beta_{33}+\beta_{23}\beta_{31} & \beta_{22}\beta_{33}+\beta_{23}\beta_{32} \end{bmatrix}.$$

$$\tag{5.19b}$$

In the two-dimensional case, **B** reduces to

$$\mathbf{B} = \begin{bmatrix} \cos^2\theta & \sin^2\theta & -\sin\theta\cos\theta \\ \sin^2\theta & \cos^2\theta & \sin\theta\cos\theta \\ 2\sin\theta\cos\theta & -2\sin\theta\cos\theta & \cos^2\theta - \sin^2\theta \end{bmatrix}, \qquad (5.19c)$$

with θ as shown in fig. 5.2.

5.1.3 Linear Stress-Strain Relations

Hooke's law states that the power of any spring body is in the same proportion as the extension. Cauchy generalized Hooke's law into the statement that the components of stress are linearly related to the components of strain, that is,

$$\begin{Bmatrix} \sigma_x \\ \sigma_y \\ \sigma_z \\ \sigma_{xy} \\ \sigma_{xz} \\ \sigma_{yz} \end{Bmatrix} = \begin{bmatrix} c_{11} & c_{12} & c_{13} & c_{14} & c_{15} & c_{16} \\ & c_{22} & c_{23} & c_{24} & c_{25} & c_{26} \\ & & c_{33} & c_{34} & c_{35} & c_{36} \\ & & & c_{44} & c_{45} & c_{46} \\ & \text{sym} & & & c_{55} & c_{56} \\ & & & & & c_{66} \end{bmatrix} \begin{Bmatrix} e_x \\ e_y \\ e_z \\ e_{xy} \\ e_{xz} \\ e_{yz} \end{Bmatrix}, \qquad (5.20a)$$

or

$$\boldsymbol{\sigma} = \mathbf{Ce}, \qquad (5.20b)$$

where **C** is the *elastic-coefficient matrix*.* The coordinate transformation of the elastic-coefficient matrix is

$$\mathbf{C}' = \mathbf{B}^\mathrm{T}\mathbf{CB}. \qquad (5.20c)$$

Equations (5.20a, b) represent the constitutive law (stress-strain relation) for anisotropic material. There are 21 independent elastic constants in **C**. If the elastic coefficients of a material exhibit symmetries, i.e., if the coefficients for certain different directions are the same, the number of independent constants will be less than 21 (Love 1963; Landau and Lifshitz 1959; Sokolnikoff 1956). For an orthotropic material, that is, where there is symmetry with respect to three mutually perpendicular planes, (5.20) becomes

---

*The reason that engineering strain components e_{xy}, e_{xz}, and e_{yz} are used in (5.18) is so that **C** will have the symmetric form of (5.20a).

$$\begin{Bmatrix} \sigma_x \\ \sigma_y \\ \sigma_z \\ \sigma_{xy} \\ \sigma_{xz} \\ \sigma_{yz} \end{Bmatrix} = \begin{bmatrix} c_{11} & c_{12} & c_{13} & 0 & 0 & 0 \\ & c_{22} & c_{23} & 0 & 0 & 0 \\ & & c_{33} & 0 & 0 & 0 \\ & \text{sym} & & c_{44} & 0 & 0 \\ & & & & c_{55} & 0 \\ & & & & & c_{66} \end{bmatrix} \begin{Bmatrix} e_x \\ e_y \\ e_z \\ e_{xy} \\ e_{xz} \\ e_{yz} \end{Bmatrix}, \quad (5.21)$$

in which there are, at most, nine independent elastic constants.

For an isotropic material, that is, one for which the elastic properties are identical in all directions, the number of independent elastic constants reduces to two and (5.20) becomes

$$\begin{Bmatrix} \sigma_x \\ \sigma_y \\ \sigma_z \\ \sigma_{xy} \\ \sigma_{xz} \\ \sigma_{yz} \end{Bmatrix} = \frac{E}{(1+\nu)(1-2\nu)} \begin{bmatrix} 1-\nu & \nu & \nu & 0 & 0 & 0 \\ & 1-\nu & \nu & 0 & 0 & 0 \\ & & 1-\nu & 0 & 0 & 0 \\ & & & \frac{1-2\nu}{2} & 0 & 0 \\ & \text{sym} & & & \frac{1-2\nu}{2} & 0 \\ & & & & & \frac{1-2\nu}{2} \end{bmatrix} \begin{Bmatrix} e_x \\ e_y \\ e_z \\ e_{xy} \\ e_{xz} \\ e_{yz} \end{Bmatrix},$$

(5.22)

in which E is Young's modulus of elasticity and ν is Poisson's ratio.

5.1.4 Plane Stress and Strain

Often, geometry and loading conditions are such as to reduce a three-dimensional problem to a two-dimensional one. The most common cases will now be discussed.

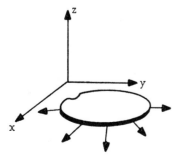

Fig. 5.3 A thin plate in plane stress.

Plane Stress If a thin plate is loaded by forces applied at its boundary, parallel to the plane of the plate and distributed uniformly over the thickness (fig. 5.3), there is no loading of either face of the plate, and it may be assumed that σ_z, σ_{xz}, and σ_{yz} are zero within the plate; the state of stress is then called plane stress. In this case, the stress-strain relation is

$$\begin{Bmatrix} \sigma_x \\ \sigma_y \\ \sigma_{xy} \end{Bmatrix} = \begin{bmatrix} c_{11} & c_{12} & c_{13} \\ & c_{22} & c_{23} \\ \text{sym} & & c_{33} \end{bmatrix} \begin{Bmatrix} e_x \\ e_y \\ e_{xy} \end{Bmatrix}. \tag{5.23}$$

For material symmetric about the x- and y-axes, we have

$$\begin{Bmatrix} \sigma_x \\ \sigma_y \\ \sigma_{xy} \end{Bmatrix} = \begin{bmatrix} c_{11} & c_{12} & 0 \\ & c_{22} & 0 \\ \text{sym} & & c_{33} \end{bmatrix} \begin{Bmatrix} e_x \\ e_y \\ e_{xy} \end{Bmatrix}; \tag{5.24a}$$

or, more commonly, we write

$$\begin{Bmatrix} \sigma_x \\ \sigma_y \\ \sigma_{xy} \end{Bmatrix} = \frac{1}{1 - \nu_{xy}\nu_{yx}} \begin{bmatrix} E_x & E_x\nu_{xy} & 0 \\ & E_y & 0 \\ \text{sym} & & G_{xy} \end{bmatrix} \begin{Bmatrix} e_x \\ e_y \\ e_{xy} \end{Bmatrix}, \tag{5.24b}$$

in which G_{xy} may or may not be related to E_x, E_y, and ν_{xy}, but where $E_x\nu_{yx} = E_y\nu_{yx}$.

For an isotropic material, the stress-strain relation becomes

$$\begin{Bmatrix} \sigma_x \\ \sigma_y \\ \sigma_{xy} \end{Bmatrix} = \frac{E}{1 - \nu^2} \begin{bmatrix} 1 & \nu & 0 \\ \nu & 1 & 0 \\ 0 & 0 & \frac{1-\nu}{2} \end{bmatrix} \begin{Bmatrix} e_x \\ e_y \\ e_{xy} \end{Bmatrix}. \tag{5.25}$$

Plane Strain If a long cylindrical or prismatic body with its axis is the z-direction is loaded by forces that are perpendicular to its axis and that do not vary along the length, and if the material is also homogeneous along the axis, then any cross section will be in the same state of strain as any other. If the end sections are confined between fixed rigid planes, so that axial displacements are prevented, we have

$$e_z = e_{xz} = e_{yz} = 0. \tag{5.26}$$

In this case, the stress-strain relations of an anisotropic material can be expressed in a form similar to (5.23). For an isotropic material, we have

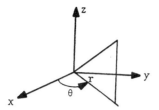

Fig. 5.4 Cylindrical polar coordinates.

$$\begin{Bmatrix} \sigma_x \\ \sigma_y \\ \sigma_{xy} \end{Bmatrix} = \frac{E}{(1+\nu)(1-2\nu)} \begin{bmatrix} 1-\nu & \nu & 0 \\ \nu & 1-\nu & 0 \\ 0 & 0 & \frac{1-2\nu}{2} \end{bmatrix} \begin{Bmatrix} e_x \\ e_y \\ e_{xy} \end{Bmatrix}. \quad (5.27)$$

Solid of Revolution For an axisymmetric body, it will be much more convenient to describe the problem in terms of cylindrical coordinates r, θ, z (fig. 5.4). Let the displacement components in the r-, θ-, and z-directions be u, v, and w, respectively. The strain-displacement relations are

$$e_r = \frac{\partial u}{\partial r}, \qquad e_\theta = \frac{1}{r}\frac{\partial v}{\partial \theta} + \frac{u}{r}, \qquad e_z = \frac{\partial w}{\partial z}, \qquad (5.28)$$

$$e_{r\theta} = \frac{1}{r}\frac{\partial u}{\partial \theta} + \frac{\partial v}{\partial r} - \frac{v}{r}, \qquad e_{\theta z} = \frac{\partial v}{\partial z} + \frac{\partial w}{r\partial \theta}, \qquad e_{rz} = \frac{\partial u}{\partial z} + \frac{\partial w}{\partial r}. \quad (5.29)$$

The constitutive law relating the stresses σ_r, σ_θ, σ_z, $\sigma_{r\theta}$, $\sigma_{\theta z}$, and σ_{rz} to the strains is the same as that in (5.20), or (5.22) for an isotropic material.

If the material is homogeneous and the loading can be represented in terms of harmonics in θ, the solution can also be expressed in terms of harmonics, namely, as

$$u = \sum_j u_j(r,z)\cos j\theta,$$

$$v = \sum_j v_j(r,z)\sin j\theta + v_0(r,z), \qquad (5.30)$$

$$w = \sum_j w_j(r,z)\cos j\theta.$$

All the harmonics are independent of each other so that (5.29) can be used to write

$$(e_r)_j = \frac{\partial u_j}{\partial r}, \qquad (e_\theta)_j = \frac{jv_j + u_j}{r}, \qquad (e_z)_j = \frac{\partial w_j}{\partial z},$$

$$(e_{r\theta})_j = \frac{\partial v_j}{\partial r} - \frac{v_j + ju_j}{r}, \qquad (e_{\theta z})_j = \frac{\partial v_j}{\partial z} - \frac{jw_j}{r}, \qquad (5.31)$$

$$(e_{rz})_j = \frac{\partial u_j}{\partial z} + \frac{\partial w_j}{\partial r}.$$

In solving for u, v, and w, we only have to treat one harmonic at a time, that is, u_j, v_j, and w_j, which are functions of r and z only.

Axisymmetric Problems An important class of problems is that where the axisymmetric body is loaded by axially symmetric forces and the material of the body is homogeneous in θ. The stress and the strain components will be independent of the angular coordinate; hence all the derivatives with respect to θ vanish and in addition, $\sigma_{r\theta}, \sigma_{\theta z}, e_{r\theta}, e_{\theta z}$, and v are zero.* The strain-displacement relations of (5.29) or (5.31) for the nonzero components reduce to

$$e_r = \frac{\partial u}{\partial r}, \qquad e_\theta = \frac{u}{r}, \qquad e_z = \frac{\partial w}{\partial z}, \qquad e_{rz} = \frac{\partial u}{\partial z} + \frac{\partial w}{\partial r}. \qquad (5.32)$$

The stress-strain relations are given by

$$\begin{Bmatrix} \sigma_r \\ \sigma_\theta \\ \sigma_z \\ \sigma_{rz} \end{Bmatrix} = \begin{bmatrix} c_{11} & c_{12} & c_{13} & c_{14} \\ & c_{22} & c_{23} & c_{24} \\ & & c_{33} & c_{34} \\ \text{sym} & & & c_{44} \end{bmatrix} \begin{Bmatrix} e_r \\ e_\theta \\ e_z \\ e_{rz} \end{Bmatrix}. \qquad (5.33)$$

For an isotropic material, (5.33) reduces to

$$\begin{Bmatrix} \sigma_r \\ \sigma_\theta \\ \sigma_z \\ \sigma_{rz} \end{Bmatrix} = \frac{E}{(1+\nu)(1-2\nu)} \begin{bmatrix} 1-\nu & \nu & \nu & 0 \\ & 1-\nu & \nu & 0 \\ & & 1-\nu & 0 \\ \text{sym} & & & \frac{1-2\nu}{2} \end{bmatrix} \begin{Bmatrix} e_r \\ e_\theta \\ e_z \\ e_{rz} \end{Bmatrix}. \qquad (5.34)$$

5.1.5 Principle of Minimum Potential Energy

Let a body V be in a state of static equilibrium under the action of specified body and surface forces (fig. 5.5). The boundary surface may be divided into two parts ∂V_u and ∂V_σ, where the displacement and the surface traction are prescribed, respectively:

---

*Another case of axisymmetry is pure torsion, where $u = w = 0$ and v is independent of, or at most linearly proportional to, θ.

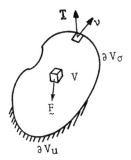

Fig. 5.5 A body in static equilibrium.

$$\mathbf{u} = \bar{\mathbf{u}} \quad \text{on} \quad \partial V_u, \tag{5.35}$$

$$\mathbf{T} = \bar{\mathbf{T}} \quad \text{on} \quad \partial V_\sigma, \tag{5.36}$$

in which the overbar denotes a prescribed quantity. The principle of minimum potential energy states that, *of all displacements satisfying the given boundary conditions (5.35), those which satisfy the equations of equilibrium and the traction boundary conditions (5.36) alone give a stationary (extreme) value of the potential energy Π*, where

$$\Pi = \int_V [\tfrac{1}{2}\mathbf{e}^T\mathbf{C}\mathbf{e} - \mathbf{u}^T\bar{\mathbf{F}}]\,dV - \int_{\partial V_\sigma} \mathbf{u}^T\bar{\mathbf{T}}\,dS, \tag{5.37}$$

in which **e** and **u** are the strain vector and the displacement vector, respectively (sec. 5.1.2), **C** is the elastic-coefficient matrix (sec. 5.1.3), **F** is the body-force vector, and V is the domain occupied by the solid continuum. For the proof of this statement, the reader is referred to Fung (1965).

For an isotropic material, (5.37) reduces to

$$\Pi = \int_A \left\{ \frac{E}{2(1-\nu^2)} \left[e_x^2 + e_y^2 + 2\nu e_x e_y + \frac{1-\nu}{2} e_{xy}^2 \right] + u\bar{F}_x + v\bar{F}_y \right\} t\,dx\,dy$$

$$- \int_{\partial A_\sigma} t(u\bar{T}_x + v\bar{T}_y)\,ds \tag{5.38}$$

for the plane-stress problem, and to

$$\Pi = \int_A \left\{ \frac{E}{2(1+\nu)(1-2\nu)} \left[(1-\nu)(e_x^2 + e_y^2) + 2\nu e_x e_y + \frac{1-2\nu}{2} e_{xy}^2 \right] \right.$$

$$\left. + u\bar{F}_x + v\bar{F}_y \right\} t\,dx\,dy - \int_{\partial A_\sigma} t(u\bar{T}_x + v\bar{T}_y)\,ds \tag{5.39}$$

for the plane-strain problem, where t is the thickness and ∂A_σ is the portion of the boundary over which the traction is prescribed.

5.2 Finite-Element Formulation

In the finite-element analysis, the domain V is divided in a finite number of elements, and (5.37) is written in the form

$$\Pi = \sum_{\substack{\text{all} \\ \text{elements}}} \pi_n, \tag{5.40}$$

in which

$$\pi_n = \int_{V_n} [\tfrac{1}{2}\mathbf{e}^T\mathbf{C}\mathbf{e} - \mathbf{u}^T\bar{\mathbf{F}}]\, dV - \int_{(\partial V_\sigma)_n} \mathbf{u}^T\bar{\mathbf{T}}\, dS, \tag{5.41}$$

where V_n denotes the region of the nth element and $(\partial V_\sigma)_n$ is the part of ∂V_σ that is on V_n. It should be noted that if the displacement field is continuous over the entire body, that is, if $\mathbf{u}$ of one element is the same as that of all the adjacent elements over their common boundaries, then Π of (5.40) is equal to that of (5.37). Since π_n is nothing but the potential energy of the nth element, the sum of the potential energy of all the elements is the same as the potential energy of the entire domain.

To construct the finite-element solution, we represent $\mathbf{u}$ in the form

$$\mathbf{u} = \mathbf{D}(\mathbf{x})\mathbf{q}_n \tag{5.42}$$

within the nth element. The components of $\mathbf{D}$ are the interpolation functions and the components of $\mathbf{q}_n$ are the generalized coordinates, which are the values of the displacements and sometimes also derivatives of the displacements at a finite number of nodal points of the element. It is required that $\mathbf{u}$ be continuous over the entire domain; therefore the interpolation function must be such that when the displacements at the nodes along the interelement boundary of two neighboring elements are compatible, the displacements all along the corresponding interelement boundary are also compatible. Using the strain-displacement relations given in sec. 5.1.2 or 5.1.4 we can express $\mathbf{e}$ in terms of $\mathbf{q}_n$ for the corresponding cases, that is,

$$\mathbf{e} = \mathbf{E}\mathbf{q}_n, \tag{5.43}$$

in which $\mathbf{E}$ is a matrix which is, in general, a function of the spatial coordinates. Using the stress-strain relation, we have

$$\boldsymbol{\sigma} = \mathbf{C}\mathbf{e} = \mathbf{C}\mathbf{E}\mathbf{q}_n. \tag{5.44}$$

A substitution of (5.42) and (5.43) into (5.41) yields

$$\pi_n = \tfrac{1}{2}\mathbf{q}_n^T\mathbf{k}_n\mathbf{q}_n - \mathbf{q}_n^T\mathbf{Q}_n, \tag{5.45}$$

in which

$$\mathbf{k}_n = \int_{V_n} \mathbf{E}^T\mathbf{C}\mathbf{E}\, dV \tag{5.46}$$

and

$$\mathbf{Q}_n = \int_{V_n} \mathbf{D}^T\bar{\mathbf{F}}\, dV + \int_{(\partial V_\sigma)_n} \mathbf{D}^T\bar{\mathbf{T}}\, dS. \tag{5.47}$$

For the nth element, $\tfrac{1}{2}\mathbf{q}_n^T\mathbf{k}_n\mathbf{q}_n$ is the strain energy in terms of the generalized nodal displacements, with $\mathbf{k}_n$ by definition the element-stiffness matrix. The vector (column matrix) $\mathbf{Q}_n$ gives the generalized forces due to prescribed loads. Such generalized forces are equivalent nodal forces defined from the work done by the external loads and are consistent with the assumed displacements; for this reason $\mathbf{Q}_n$ is called the consistent element-loading vector.

One may visualize that (5.42) is an approximate solution within the element based on interpolating the displacements in terms of the generalized nodal displacements. The individual element-stiffness matrix is derived directly by such an approximate solution. We may say that it characterizes the behavior of the element, because, when there is only one isolated element, the equation

$$\mathbf{k}_n\mathbf{q}_n = \mathbf{Q}_n$$

is the approximate relation between the generalized nodal displacements $\mathbf{q}_n$ and the generalized nodal forces $\mathbf{Q}_n$.

A substitution of (5.45) into (5.40) yields

$$\Pi = \sum_{\substack{\text{all} \\ \text{elements}}} [\tfrac{1}{2}\mathbf{q}_n^T\mathbf{k}_n\mathbf{q}_n - \mathbf{q}_n^T\mathbf{Q}_n]. \tag{5.48}$$

The generalized coordinates can be determined by requiring that (5.35) be

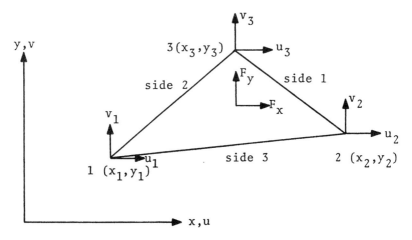

Fig. 5.6 A constant-strain triangle.

satisfied at the boundary nodes over ∂V_n and that Π of (5.48) be a minimum with respect to all the unknown generalized coordinates. Equation (5.48) is in the same form as (2.36). We can apply the same procedures of assembly, constraint, and solution techniques to evaluate the unknowns. After the qs are determined, the stresses and the strains of each element can be evaluated according to (5.43) and (5.44).

5.3 Evaluation of the Element Matrices.

The form of **D**, **E**, $\mathbf{k}_n$, and $\mathbf{Q}_n$ depends on the shape and the number of nodal points of the element and what the generalized coordinates are. We shall discuss in detail a few simple cases in this section. More complicated cases will be discussed in the next chapter.

5.3.1 Triangular Element for Plane Stress or Strain

Consider the simplest triangular element with the vertices being the nodal points and only the values of the displacements at the nodes being the generalized coordinates. The typical nth element is shown in fig. 5.6. Since there are three generalized coordinates for u and v separately, both u and v can be put in the form

$$
\begin{aligned}
u &= \alpha_1 + \alpha_2 x + \alpha_3 y, \\
v &= \alpha_4 + \alpha_5 x + \alpha_6 y,
\end{aligned}
\tag{5.49}
$$

which is the same form as that for ϕ in (2.15). The procedure in sec. 2.5.1 applies here, and on expressing the αs in terms of the nodal values of u and v we obtain

$$u = u_1 f_1(x, y) + u_2 f_2(x, y) + u_3 f_3(x, y),$$
$$v = v_1 f_1(x, y) + v_2 f_2(x, y) + v_3 f_3(x, y), \tag{5.50}$$

where the f_is are the same as those in (2.19), that is,

$$f_i(x, y) = \frac{a_i + b_i x + c_i y}{2\Delta}, \qquad i = 1, 2, 3, \tag{5.51}$$

in which

$$\Delta = \text{Area of the triangle} = \frac{1}{2} \begin{vmatrix} 1 & 1 & 1 \\ x_1 & x_2 & x_3 \\ y_1 & y_2 & y_3 \end{vmatrix},$$

$a_1 = x_2 y_3 - x_3 y_2, \qquad a_2 = x_3 y_1 - x_1 y_3, \qquad a_3 = x_1 y_2 - x_2 y_1,$
$b_1 = y_2 - y_3, \qquad b_2 = y_3 - y_1, \qquad b_3 = y_1 - y_2,$
$c_1 = x_3 - x_2, \qquad c_2 = x_1 - x_3, \qquad c_3 = x_2 - x_1.$

Write (5.50) in matrix form,

$$\mathbf{u} = \begin{Bmatrix} u \\ v \end{Bmatrix} = \mathbf{D}\mathbf{q}_n, \tag{5.52}$$

where

$$\underset{2\times 4}{\mathbf{D}} = \begin{bmatrix} f_1 & 0 & f_2 & 0 & f_3 & 0 \\ 0 & f_1 & 0 & f_2 & 0 & f_3 \end{bmatrix},$$

$$\mathbf{q}_n^T = \{u_1 \quad v_1 \quad u_2 \quad v_2 \quad u_3 \quad v_3\}.$$

Using the strain-displacement relations (5.18), we have

$$\underset{3\times 1}{\mathbf{e}} = \begin{Bmatrix} e_x \\ e_y \\ e_{xy} \end{Bmatrix} = \begin{bmatrix} \dfrac{\partial}{\partial x} & 0 \\ 0 & \dfrac{\partial}{\partial y} \\ \dfrac{\partial}{\partial y} & \dfrac{\partial}{\partial x} \end{bmatrix} \begin{Bmatrix} u \\ v \end{Bmatrix} = \underset{3\times 6}{\mathbf{E}} \underset{6\times 1}{\mathbf{q}_n}, \tag{5.53}$$

where

$$\mathbf{E} = \frac{1}{2\Delta}\begin{bmatrix} b_1 & 0 & b_2 & 0 & b_3 & 0 \\ 0 & c_1 & 0 & c_2 & 0 & c_3 \\ c_1 & b_1 & c_2 & b_2 & c_3 & b_3 \end{bmatrix}. \tag{5.54}$$

On substituting into (5.46), $\mathbf{k}_n$ becomes

$$\mathbf{k}_n = \int_\Delta \mathbf{E}^T \mathbf{C} \mathbf{E} t \, dxdy, \tag{5.55}$$

where Δ is the area of the triangle, t is the thickness of the elastic body, and $\mathbf{C}$ is the elastic-coefficient matrix. Since $\mathbf{E}$ is constant for a material that is homogeneous within the element, we have

$$\mathbf{k}_n = t \Delta \, \mathbf{E}^T \mathbf{C} \mathbf{E}. \tag{5.56}$$

From (5.53), it is clear that the strains are constant, so this element is usually called the *constant-strain triangle* (CST). Equation (5.56) can even be used for nonhomogeneous material in which $\mathbf{C}$ will be the mean value over the element.

If the material is isotropic, then

$$\mathbf{k}_n = \gamma \times$$
$$\begin{bmatrix} \alpha b_1^2 + \beta c_1^2 & (\nu+\beta)b_1c_1 & \alpha b_1 b_2 + \beta c_1 c_2 & \nu b_1 c_2 + \beta b_2 c_1 & \alpha b_1 b_3 + \beta c_1 c_3 & \nu b_1 c_3 + \beta b_3 c_1 \\ & \alpha c_1^2 + \beta b_1^2 & \nu b_2 c_1 + \beta b_1 c_2 & \alpha c_1 c_2 + \beta b_1 b_2 & \nu b_3 c_1 + \beta b_1 c_3 & \alpha c_1 c_3 + \beta b_1 b_3 \\ & & \alpha b_2^2 + \beta c_2^2 & (\nu+\beta)b_2 c_2 & \alpha b_2 b_3 + \beta c_2 c_3 & \nu b_2 c_3 + \beta b_3 c_2 \\ & & & \alpha c_2^2 + \beta b_2^2 & \nu b_3 c_2 + \beta b_2 c_3 & \alpha c_2 c_3 + \beta b_2 b_3 \\ & \text{sym} & & & \alpha b_3^2 + \beta c_3^2 & (\nu+\beta)b_3 c_3 \\ & & & & & \alpha c_3^2 + \beta b_3^2 \end{bmatrix},$$
$$\tag{5.57}$$

where

$$\gamma = \frac{Et}{4(1-\nu^2)\Delta}, \qquad \alpha = 1, \qquad \beta = \frac{1-\nu}{2} \tag{5.58}$$

for plane stress problems ($\mathbf{C}$ is given in 5.25), and

$$\gamma = \frac{Et}{4(1+\nu)(1-2\nu)\Delta}, \qquad \alpha = 1-\nu, \qquad \beta = \frac{1-2\nu}{2} \tag{5.59}$$

for plane strain problems ($\mathbf{C}$ is given in 5.27).

5.3 Evaluation of the Element Matrices

If the generalized coordinates are arranged in the order

$$\mathbf{q}_n^* = \{u_1 \quad u_2 \quad u_3 \quad v_1 \quad v_2 \quad v_3\}, \tag{5.60}$$

the element-stiffness matrix can be expressed in more compact form as

$$\mathbf{k}_n^* = \gamma \begin{bmatrix} \alpha \mathbf{b}\mathbf{b}^T + \beta \mathbf{c}\mathbf{c}^T & \nu \mathbf{b}\mathbf{c}^T + \beta \mathbf{c}\mathbf{b}^T \\ \text{sym} & \alpha \mathbf{c}\mathbf{c}^T + \beta \mathbf{b}\mathbf{b}^T \end{bmatrix}, \tag{5.61}$$

where

$$\mathbf{b}^T = \{b_1 \quad b_2 \quad b_3\}, \qquad \mathbf{c}^T = \{c_1 \quad c_2 \quad c_3\}.$$

However, for convenience of assembly, the form (5.57) is more commonly used.

To evaluate the consistent element-loading vector according to (5.47), we in principle have to know the distribution of $\bar{\mathbf{F}}$ and $\bar{\mathbf{T}}$. For the case where $\bar{\mathbf{F}}$ and $\bar{\mathbf{T}}$ are constant within the element the integration of (5.47) can be carried out explicitly. Noting that

$$\int_\Delta f_i(x, y)\,dx\,dy = \frac{\Delta}{3},$$

$$\int_{\text{side } j} f_i(x, y)\,ds = \frac{l_j}{2}(1 - \delta_{ij}), \tag{5.62}$$

where δ_{ij} is the Kronecker delta ($\delta_{ij} = 1$ if $i = j$, and $\delta_{ij} = 0$ if $i \neq j$) and l_j is the length of side j (e.g., for $j = 1$, $l_1 = [b_1^2 + c_1^2]^{1/2} = [(y_2 - y_3)^2 + (x_2 - x_3)^2]^{1/2}$) we have

$$\mathbf{Q}_n^T = t\mathbf{F}^T \int_\Delta \mathbf{f}(x, y)\,dx\,dy = t\{\bar{F}_x \quad \bar{F}_y\}\begin{bmatrix} 1 & 0 & 1 & 0 & 1 & 0 \\ 0 & 1 & 0 & 1 & 0 & 1 \end{bmatrix}\frac{\Delta}{3}$$

$$= \frac{t\Delta}{3}\{\bar{F}_x \quad \bar{F}_y \quad \bar{F}_x \quad \bar{F}_y \quad \bar{F}_x \quad \bar{F}_y\} \tag{5.63}$$

for an element without any prescribed boundary traction, and

$$\mathbf{Q}_n^T = t\bar{\mathbf{F}}^T \int_\Delta \mathbf{f}(x, y)\,dx\,dy + t\bar{\mathbf{T}}^T \int_{\text{side } j} \mathbf{f}(x, y)\,ds$$

$$= \frac{t\Delta}{3}\{\bar{F}_x \quad \bar{F}_y \quad \bar{F}_x \quad \bar{F}_y \quad \bar{F}_x \quad \bar{F}_y\}$$

$$+ \frac{tl_j}{2}\{\bar{T}_x(1 - \delta_{1j}) \quad \bar{T}_y(1 - \delta_{1j}) \quad \bar{T}_x(1 - \delta_{2j}) \quad \bar{T}_y(1 - \delta_{2j}) \tag{5.64}$$

$$\bar{T}_x(1 - \delta_{3j}) \quad \bar{T}_y(1 - \delta_{3j})\}$$

for an element with boundary tractions prescribed on side j. If side 1 is subjected to $\bar{\mathbf{T}}$, we have

$$\begin{aligned}\mathbf{Q}_n^T &= \frac{t\triangle}{3}\{\bar{F}_x \quad \bar{F}_y \quad \bar{F}_x \quad \bar{F}_y \quad \bar{F}_x \quad \bar{F}_y\} \\ &\quad + \frac{tl_1}{2}\{0 \quad 0 \quad \bar{T}_x \quad \bar{T}_y \quad \bar{T}_x \quad \bar{T}_y\}.\end{aligned} \quad (5.65)$$

In practice, even if $\bar{\mathbf{F}}$ or $\bar{\mathbf{T}}$ are not constant, they can be approximated by using their mean value over the element in (5.63) and (5.64). Such an approximation is consistent with the approximate representation of the constant strains within the element.

5.3.2 Rectangular Element for Plane Stress or Strain

Consider the typical rectangular element shown in fig. 5.7. There are four generalized coordinates each for u and v, so we can assume that

$$\begin{aligned}u &= \alpha_1 + \alpha_2 x + \alpha_3 y + \alpha_4 xy, \\ v &= \alpha_5 + \alpha_6 x + \alpha_7 y + \alpha_8 xy,\end{aligned} \quad (5.66)$$

which is the same form as that of (2.20). Equation (5.66) can be expressed in terms of nodal values by

$$\begin{aligned}u &= \sum_{i=1}^{4} u_i f_i(x, y), \\ v &= \sum_{i=1}^{4} v_i f_i(x, y),\end{aligned} \quad (5.67)$$

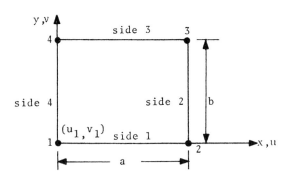

Fig. 5.7 A rectangular element.

in which the f_is are the interpolation functions defined in (2.24). For the convenient evaluation of the element matrices, we will first denote the generalized-coordinate vector by $\mathbf{q}_n^*$,

$$\mathbf{q}_n^{*T} = \{u_1 \quad u_2 \quad u_3 \quad u_4 \quad v_1 \quad v_2 \quad v_3 \quad v_4\}. \tag{5.68}$$

Then

$$\mathbf{u} = \begin{Bmatrix} u \\ v \end{Bmatrix} = \begin{bmatrix} \mathbf{f}^T & 0 \\ 0 & \mathbf{f}^T \end{bmatrix} \mathbf{q}_n^* = \mathbf{D}^*\mathbf{q}^*,$$

where $\tag{5.69}$

$$\mathbf{f}^T = \{f_1 \quad f_2 \quad f_3 \quad f_4\}.$$

From the strain displacement relations we see that

$$\underset{3\times 1}{\mathbf{e}} = \begin{bmatrix} \dfrac{\partial}{\partial x} & 0 \\ 0 & \dfrac{\partial}{\partial y} \\ \dfrac{\partial}{\partial y} & \dfrac{\partial}{\partial x} \end{bmatrix} \begin{Bmatrix} u \\ v \end{Bmatrix} = \mathbf{E}^*\mathbf{q}_n^*, \tag{5.70}$$

where

$$\underset{3\times 8}{\mathbf{E}^*} = \begin{bmatrix} \dfrac{\partial \mathbf{f}^T}{\partial x} & 0 \\ 0 & \dfrac{\partial \mathbf{f}^T}{\partial y} \\ \dfrac{\partial \mathbf{f}^T}{\partial y} & \dfrac{\partial \mathbf{f}^T}{\partial x} \end{bmatrix}. \tag{5.71}$$

Substituting into (5.46) we have

$$\mathbf{k}_n^* = \int_0^b \int_0^a t\mathbf{E}^{*T}\mathbf{C}\mathbf{E}^* \, dx dy. \tag{5.72}$$

Note that

$$\frac{\partial \mathbf{f}^T}{\partial x} = \frac{\{y - b \quad b - y \quad y \quad -y\}}{ab},$$

$$\frac{\partial \mathbf{f}^T}{\partial y} = \frac{\{x - a \quad -x \quad x \quad a - x\}}{ab},$$

and define

$$\mathbf{A}_{xx} = \frac{a}{b} \int_0^b \int_0^a \frac{\partial \mathbf{f}}{\partial x} \frac{\partial \mathbf{f}^T}{\partial x} \, dxdy = \begin{bmatrix} \frac{1}{3} & -\frac{1}{3} & -\frac{1}{6} & \frac{1}{6} \\ & \frac{1}{3} & \frac{1}{6} & -\frac{1}{6} \\ & & \frac{1}{3} & -\frac{1}{3} \\ \text{sym} & & & \frac{1}{3} \end{bmatrix},$$

$$\mathbf{A}_{yy} = \frac{b}{a} \int_0^b \int_0^a \frac{\partial \mathbf{f}}{\partial y} \frac{\partial \mathbf{f}^T}{\partial y} \, dxdy = \begin{bmatrix} \frac{1}{3} & \frac{1}{6} & -\frac{1}{6} & -\frac{1}{3} \\ & \frac{1}{3} & -\frac{1}{3} & -\frac{1}{6} \\ & & \frac{1}{3} & \frac{1}{6} \\ \text{sym} & & & \frac{1}{3} \end{bmatrix}, \quad (5.73)$$

$$\mathbf{A}_{xy} = \int_0^b \int_0^a \frac{\partial \mathbf{f}}{\partial x} \frac{\partial \mathbf{f}^T}{\partial y} \, dxdy = \frac{1}{4} \begin{bmatrix} 1 & 1 & -1 & -1 \\ -1 & -1 & 1 & 1 \\ -1 & -1 & 1 & 1 \\ 1 & 1 & -1 & -1 \end{bmatrix}.$$

For a material which is homogeneous within the element, (5.72) can then be written as

$$\mathbf{k}_n^* = t \begin{bmatrix} \frac{b}{a} c_{11} \mathbf{A}_{xx} + c_{13}(\mathbf{A}_{xy} + \mathbf{A}_{xy}^T) + \frac{a}{b} c_{33} \mathbf{A}_{yy} \\ \frac{b}{a} c_{13} \mathbf{A}_{xx} + c_{12} \mathbf{A}_{xy} + c_{33} \mathbf{A}_{xy}^T + \frac{a}{b} c_{23} \mathbf{A}_{yy} \\ \text{sym} \\ \frac{b}{a} c_{33} \mathbf{A}_{xx} + c_{23}(\mathbf{A}_{xy} + \mathbf{A}_{xy}^T) + \frac{a}{b} c_{22} \mathbf{A}_{yy} \end{bmatrix}. \quad (5.74)$$

From (5.47), the element-loading vector becomes

$$\mathbf{Q}_n^* = \int_0^b \int_0^a \begin{Bmatrix} t\bar{F}_x \\ t\bar{F}_y \end{Bmatrix} t\,dxdy + \int_{\text{side}\,j} \begin{Bmatrix} t\bar{T}_x \\ t\bar{T}_y \end{Bmatrix} ds, \quad (5.75)$$

where side j has prescribed surface tractions. If the $\bar{F}$s and $\bar{T}$s are constant, (5.75) becomes

$$\mathbf{Q}_n^{*T} = \{Q_1^* \quad Q_2^* \cdots Q_8^*\}, \tag{5.76}$$

where

$$Q_i^* = \frac{tab}{4}\bar{F}_x + \frac{tl_j}{2}\bar{T}_x\delta_i,$$

$$Q_{i+4}^* = \frac{tab}{4}\bar{F}_y + \frac{tl_j}{2}\bar{T}_y\delta_i \qquad i = 1, 2, 3, 4,$$

where l_j is the length of side j (which is either a or b) and where

$$\delta_i = \begin{cases} 1 & \text{if the } i\text{th node is on side } j, \\ 0 & \text{if the } i\text{th node is not on side } j. \end{cases}$$

For example, if side 4 has a prescribed traction, since nodes 1 and 4 are on side 4, then

$$\mathbf{Q}_n^{*T} = \frac{tab}{4}\{\bar{F}_x \bar{F}_x \bar{F}_x \bar{F}_x \bar{F}_y \bar{F}_y \bar{F}_y \bar{F}_y\} + \frac{tb}{2}\{\bar{T}_x 0 0 \bar{T}_x \bar{T}_y 0 0 \bar{T}_y\}. \tag{5.77}$$

Even so, the ordering of the generalized coordinates in (5.68) is not convenient for assembly,* and we therefore define

$$\mathbf{q}_n^T = \{u_1 \quad v_1 \quad u_2 \quad v_2 \quad u_3 \quad v_3 \quad u_4 \quad v_4\}. \tag{5.78}$$

Comparing (5.78) and (5.68), we have, between the components of the two generalized-coordinate vectors, the relation

$$q_i = q_{I(i)}^*, \tag{5.79}$$

where

$$I(2i - 1) = i, \qquad I(2i) = i + 4, \qquad i = 1, 2, 3, 4. \tag{5.80}$$

In other words, $\mathbf{q}_n$ is related to $\mathbf{q}_n^*$ by

$$\mathbf{q}_n^* = \mathbf{T}\mathbf{q}_n,$$

---

*In practice one uses the same degrees of freedom per node. On using the ordering of (5.78) one can easily establish the relations between local and global numbers for the degrees of freedom, once the relations between local and global numbers for nodes are given.

where

$$T = \begin{bmatrix} 1 & 0 & 0 & 0 & 0 & 0 & 0 & 0 \\ 0 & 0 & 1 & 0 & 0 & 0 & 0 & 0 \\ 0 & 0 & 0 & 0 & 1 & 0 & 0 & 0 \\ 0 & 0 & 0 & 0 & 0 & 0 & 1 & 0 \\ 0 & 1 & 0 & 0 & 0 & 0 & 0 & 0 \\ 0 & 0 & 0 & 1 & 0 & 0 & 0 & 0 \\ 0 & 0 & 0 & 0 & 0 & 1 & 0 & 0 \\ 0 & 0 & 0 & 0 & 0 & 0 & 0 & 1 \end{bmatrix}.$$

Since

$$\mathbf{q}_n^T \mathbf{k}_n \mathbf{q}_n = \mathbf{q}_n^{*T} \mathbf{k}_n^* \mathbf{q}_n^* = \mathbf{q}_n^T \mathbf{T}^T \mathbf{k}_n^* \mathbf{T} \mathbf{q}_n,$$

$$\mathbf{q}_n^T \mathbf{Q}_n = \mathbf{q}_n^{*T} \mathbf{Q}_n^* = \mathbf{q}_n^T \mathbf{T}^T \mathbf{Q}_n^*,$$

we have

$$\begin{aligned} \mathbf{k}_n &= \mathbf{T}^T \mathbf{k}_n^* \mathbf{T} = [k_{ij}], \\ \mathbf{Q}_n &= \mathbf{T}^T \mathbf{Q}_n^* = \{Q_i\}. \end{aligned} \tag{5.81}$$

It can be shown that the components are related by

$$k_{ij} = k^*_{I(i)I(j)}, \qquad Q_i = Q^*_{I(i)}, \qquad i, j = 1, 2, \cdots 8, \tag{5.82}$$

where $I(i)$ is given by (5.80).

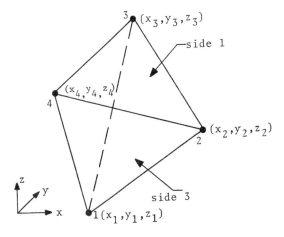

Fig. 5.8 A tetrahedral element.

5.3.3. Tetrahedral Element

Consider the nth element shown in fig. 5.8, with the values of the displacements at the four vertices as the generalized coordinates. We can express u, v, w as linear polynomials within the element, that is, as

$$u = \alpha_1 + \alpha_2 x + \alpha_3 y + \alpha_4 z,$$
$$v = \alpha_5 + \alpha_6 x + \alpha_7 y + \alpha_8 z, \tag{5.83}$$
$$w = \alpha_9 + \alpha_{10} x + \alpha_{11} y + \alpha_{12} z.$$

By expressing the αs in terms of the nodal values of the displacements, (5.83) becomes

$$u = \sum_{i=1}^{4} u_i f_i(x, y, z),$$
$$v = \sum_{i=1}^{4} v_i f_i(x, y, z), \tag{5.84}$$
$$w = \sum_{i=1}^{4} w_i f_i(x, y, z),$$

where

$$f_i(x, y, z) = \frac{a_i + b_i x + c_i y + d_i z}{6V}, \tag{5.85}$$

in which

$$V = \text{volume of the element} = \frac{1}{6} \begin{vmatrix} 1 & 1 & 1 & 1 \\ x_1 & x_2 & x_3 & x_4 \\ y_1 & y_2 & y_3 & y_4 \\ z_1 & z_2 & z_3 & z_4 \end{vmatrix},$$

$$a_1 = \begin{vmatrix} x_2 & x_3 & x_4 \\ y_2 & y_3 & y_4 \\ z_2 & z_3 & z_4 \end{vmatrix}, \quad b_1 = -\begin{vmatrix} 1 & 1 & 1 \\ y_2 & y_3 & y_4 \\ z_2 & z_3 & z_4 \end{vmatrix}, \tag{5.86}$$

$$c_1 = \begin{vmatrix} 1 & 1 & 1 \\ x_2 & x_3 & x_4 \\ z_2 & z_3 & z_4 \end{vmatrix}, \quad d_1 = -\begin{vmatrix} 1 & 1 & 1 \\ x_2 & x_3 & x_4 \\ y_2 & y_3 & y_4 \end{vmatrix},$$

and the other a, b, c, and ds are obtained by the cyclic permutation of subscripts. Write (5.84) in matrix form,

$$\underset{3\times1}{\mathbf{u}} = \begin{Bmatrix} u \\ v \\ w \end{Bmatrix} = \underset{3\times12}{\mathbf{D}} \underset{12\times1}{\mathbf{q}_n},$$

in which

$$\mathbf{D} = \begin{bmatrix} f_1 & 0 & 0 & f_2 & 0 & 0 & f_3 & 0 & 0 & f_4 & 0 & 0 \\ 0 & f_1 & 0 & 0 & f_2 & 0 & 0 & f_3 & 0 & 0 & f_4 & 0 \\ 0 & 0 & f_1 & 0 & 0 & f_2 & 0 & 0 & f_3 & 0 & 0 & f_4 \end{bmatrix}, \tag{5.87}$$

$$\mathbf{q}_n^\mathrm{T} = \{u_1 \quad v_1 \quad w_1 \quad \cdots \quad u_4 \quad v_4 \quad w_4\}.$$

From the strain-displacement relations (5.18), we have

$$\underset{6\times 1}{\mathbf{e}} = \{e_x \quad e_y \quad e_z \quad e_{xy} \quad e_{xz} \quad e_{yz}\}^\mathrm{T} = \underset{6\times 12}{\mathbf{E}} \underset{12\times 1}{\mathbf{q}_n},$$

where

$$\mathbf{E} = \frac{1}{6V} \begin{bmatrix} b_1 & 0 & 0 & b_2 & 0 & 0 & b_3 & 0 & 0 & b_4 & 0 & 0 \\ 0 & c_1 & 0 & 0 & c_2 & 0 & 0 & c_3 & 0 & 0 & c_4 & 0 \\ 0 & 0 & d_1 & 0 & 0 & d_2 & 0 & 0 & d_3 & 0 & 0 & d_4 \\ c_1 & b_1 & 0 & c_2 & b_2 & 0 & c_3 & b_3 & 0 & c_4 & b_4 & 0 \\ d_1 & 0 & b_1 & d_2 & 0 & b_2 & d_3 & 0 & b_3 & d_4 & 0 & b_4 \\ 0 & d_1 & c_1 & 0 & d_2 & c_2 & 0 & d_3 & c_3 & 0 & d_4 & c_4 \end{bmatrix}. \tag{5.88}$$

Substituting in (5.46) and (5.47), we obtain

$$\mathbf{k}_n = V \mathbf{E}^\mathrm{T} \mathbf{C} \mathbf{E} \tag{5.89}$$

for the material which is homogeneous within the element; and

$$\mathbf{Q}_n = \{Q_1 \quad Q_2 \quad \cdots \quad Q_{12}\}^\mathrm{T}, \tag{5.90}$$

where

$$Q_{3i-2} = \frac{V}{4}\bar{F}_x + \frac{S_j}{3}\bar{T}_x(1 - \delta_{ij}),$$

$$Q_{3i-1} = \frac{V}{4}\bar{F}_y + \frac{S_j}{3}\bar{T}_y(1 - \delta_{ij}),$$

$$Q_{3i} = \frac{V}{4}\bar{F}_z + \frac{S_j}{3}\bar{T}_z(1 - \delta_{ij}), \qquad i = 1, 2, 3, 4,$$

for the case of constant body force $\bar{\mathbf{F}}$ and constant prescribed traction $\bar{\mathbf{T}}$ over side j with S_j the area of side j. Equations (5.89) and (5.90) can be used even for inhomogeneous materials and a nonuniform distributed load within the element, in which case $\mathbf{C}$, $\bar{\mathbf{F}}$, and $\bar{\mathbf{T}}$ are some mean values over the element.

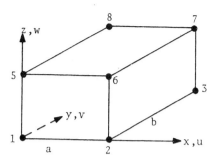

Fig. 5.9 A hexahedral element.

5.3.4. Hexahedral Element

The procedures for constructing the element matrices for a hexahedral element (fig. 5.9) are very similar to those for a rectangular element. The displacements within the element are expressed in terms of the nodal displacements:

$$u = \sum_{i=1}^{8} u_i f_i(x, y, z),$$
$$v = \sum_{i=1}^{8} v_i f_i(x, y, z), \quad (5.91)$$
$$w = \sum_{i=1}^{8} w_i f_i(x, y, z).$$

where

$$f_1(x, y, z) = \frac{(a - x)(b - y)(c - z)}{abc}, \quad f_5(x, y, z) = \frac{(a - x)(b - y)z}{abc},$$
$$f_2(x, y, z) = \frac{x(b - y)(c - z)}{abc}, \quad f_6(x, y, z) = \frac{x(b - y)z}{abc},$$
$$f_3(x, y, z) = \frac{xy(c - z)}{abc}, \quad f_7(x, y, z) = \frac{xyz}{abc},$$
$$f_4(x, y, z) = \frac{(a - x)y(c - z)}{abc}, \quad f_8(x, y, z) = \frac{(a - x)yz}{abc}.$$

These interpolation functions are often called *trilinear functions*. In matrix form,

$$\mathbf{u}_{3\times 1} = \left\{\begin{array}{c} u \\ v \\ w \end{array}\right\} = \mathbf{D}^*_{3\times 24} \mathbf{q}^*_{n}_{24\times 1}, \tag{5.92}$$

in which

$$\mathbf{q}^*_n = \left\{\begin{array}{c} \mathbf{q}_u \\ \mathbf{q}_v \\ \mathbf{q}_w \end{array}\right\}, \qquad \mathbf{D}^*_{3\times 24} = \begin{bmatrix} \mathbf{f}^\mathrm{T} & 0 & 0 \\ 0 & \mathbf{f}^\mathrm{T} & 0 \\ 0 & 0 & \mathbf{f}^\mathrm{T} \end{bmatrix}, \tag{5.93}$$

$$\mathbf{q}_u^\mathrm{T} = \{u_1\, u_2 \cdots u_8\}, \qquad \mathbf{q}_v^\mathrm{T} = \{v_1\, v_2 \cdots v_8\}, \qquad \mathbf{q}_w^\mathrm{T} = \{w_1\, w_2 \cdots w_8\}.$$

Substituting (5.92) into (5.18), we obtain

$$\mathbf{e}_{6\times 1} = \mathbf{E}^*_{6\times 24} \mathbf{q}^*_{n}_{24\times 1},$$

where

$$\mathbf{E}^*_{6\times 24} = \begin{bmatrix} \dfrac{\partial \mathbf{f}^\mathrm{T}}{\partial x} & 0 & 0 \\ 0 & \dfrac{\partial \mathbf{f}^\mathrm{T}}{\partial y} & 0 \\ 0 & 0 & \dfrac{\partial \mathbf{f}^\mathrm{T}}{\partial z} \\ \dfrac{\partial \mathbf{f}^\mathrm{T}}{\partial y} & \dfrac{\partial \mathbf{f}^\mathrm{T}}{\partial x} & 0 \\ \dfrac{\partial \mathbf{f}^\mathrm{T}}{\partial z} & 0 & \dfrac{\partial \mathbf{f}^\mathrm{T}}{\partial x} \\ 0 & \dfrac{\partial \mathbf{f}^\mathrm{T}}{\partial z} & \dfrac{\partial \mathbf{f}^\mathrm{T}}{\partial y} \end{bmatrix}. \tag{5.94}$$

Substituting (5.94) into (5.46) and (5.47), we have

$$\begin{aligned} \mathbf{k}^*_n{}_{24\times 24} &= \int_{V_*} \mathbf{E}^{*\mathrm{T}} \mathbf{C} \mathbf{E}^* dx dy dz, \\ \mathbf{Q}^*_n{}_{24\times 1} &= \int_{V_*} \mathbf{D}^{*\mathrm{T}} \bar{\mathbf{F}} dx dy dz + \int_{(\partial V_\sigma)_*} \mathbf{D}^{*\mathrm{T}} \bar{\mathbf{T}} ds. \end{aligned} \tag{5.95}$$

For a homogeneous material, $\mathbf{k}^*_n$ can be integrated out explicitly. In particular, for an orthotropic material with the axes of symmetry the x-, y-, and z-axes, we have

$$\mathbf{k}_n^* = \begin{bmatrix} \dfrac{bc}{a}c_{11}\mathbf{a}_{xx} + \dfrac{ac}{b}c_{44}\mathbf{a}_{yy} & c(c_{12}\mathbf{a}_{xy}+c_{44}\mathbf{a}_{xy}^T) & b(c_{13}\mathbf{a}_{xz}+c_{55}\mathbf{a}_{xz}^T) \\ + \dfrac{ab}{c}c_{55}\mathbf{a}_{zz} & & \\ & \dfrac{bc}{a}c_{44}\mathbf{a}_{xx} + \dfrac{ac}{b}c_{22}\mathbf{a}_{yy} & a(c_{23}\mathbf{a}_{yz}+c_{66}\mathbf{a}_{yz}^T) \\ & + \dfrac{ab}{c}c_{66}\mathbf{a}_{zz} & \\ \text{sym} & & \dfrac{bc}{a}c_{55}\mathbf{a}_{xx} + \dfrac{ac}{b}c_{66}\mathbf{a}_{yy} \\ & & + \dfrac{ab}{c}c_{33}\mathbf{a}_{zz} \end{bmatrix},$$

(5.96)

in which

$$\mathbf{a}_{xx} = \frac{a}{bc}\int_0^c\int_0^b\int_0^a \frac{\partial \mathbf{f}}{\partial x}\frac{\partial \mathbf{f}^T}{\partial x}\,dxdydz = \begin{bmatrix} \tfrac{1}{3}\mathbf{A}_{xx} & \tfrac{1}{6}\mathbf{A}_{xx} \\ \tfrac{1}{6}\mathbf{A}_{xx} & \tfrac{1}{3}\mathbf{A}_{xx} \end{bmatrix},$$

$$\mathbf{a}_{yy} = \frac{b}{ac}\int_0^c\int_0^b\int_0^a \frac{\partial \mathbf{f}}{\partial y}\frac{\partial \mathbf{f}^T}{\partial y}\,dxdydz = \begin{bmatrix} \tfrac{1}{3}\mathbf{A}_{yy} & \tfrac{1}{6}\mathbf{A}_{yy} \\ \tfrac{1}{6}\mathbf{A}_{yy} & \tfrac{1}{3}\mathbf{A}_{yy} \end{bmatrix},$$

$$\mathbf{a}_{zz} = \frac{c}{ab}\int_0^c\int_0^b\int_0^a \frac{\partial \mathbf{f}}{\partial z}\frac{\partial \mathbf{f}^T}{\partial z}\,dxdydz = \begin{bmatrix} \mathbf{A}_{zz} & -\mathbf{A}_{zz} \\ -\mathbf{A}_{zz} & \mathbf{A}_{zz} \end{bmatrix},$$

(5.97)

$$\mathbf{a}_{xy} = \frac{1}{c}\int_0^c\int_0^b\int_0^a \frac{\partial \mathbf{f}}{\partial x}\frac{\partial \mathbf{f}^T}{\partial y}\,dxdydz = \begin{bmatrix} \tfrac{1}{3}\mathbf{A}_{xy} & \tfrac{1}{6}\mathbf{A}_{xy} \\ \tfrac{1}{6}\mathbf{A}_{xy} & \tfrac{1}{3}\mathbf{A}_{xy} \end{bmatrix},$$

$$\mathbf{a}_{xz} = \frac{1}{b}\int_0^c\int_0^b\int_0^a \frac{\partial \mathbf{f}}{\partial x}\frac{\partial \mathbf{f}^T}{\partial z}\,dxdydz = \begin{bmatrix} \mathbf{A}_{xz} & -\mathbf{A}_{xz} \\ -\mathbf{A}_{xz} & \mathbf{A}_{xz} \end{bmatrix},$$

$$\mathbf{a}_{yz} = \frac{1}{a}\int_0^c\int_0^b\int_0^a \frac{\partial \mathbf{f}}{\partial y}\frac{\partial \mathbf{f}^T}{\partial z}\,dxdydz = \begin{bmatrix} \mathbf{A}_{yz} & -\mathbf{A}_{yz} \\ -\mathbf{A}_{yz} & \mathbf{A}_{yz} \end{bmatrix},$$

where

$$\mathbf{A}_{xx} = \begin{bmatrix} \tfrac{1}{3} & -\tfrac{1}{3} & -\tfrac{1}{6} & \tfrac{1}{6} \\ & \tfrac{1}{3} & \tfrac{1}{6} & -\tfrac{1}{6} \\ & & \tfrac{1}{3} & -\tfrac{1}{3} \\ \text{sym} & & & \tfrac{1}{3} \end{bmatrix}, \quad \mathbf{A}_{yy} = \begin{bmatrix} \tfrac{1}{3} & \tfrac{1}{6} & -\tfrac{1}{6} & -\tfrac{1}{3} \\ & \tfrac{1}{3} & -\tfrac{1}{3} & -\tfrac{1}{6} \\ & & \tfrac{1}{3} & \tfrac{1}{6} \\ \text{sym} & & & \tfrac{1}{3} \end{bmatrix},$$

$$\mathbf{A}_{zz} = \begin{bmatrix} \frac{1}{9} & \frac{1}{18} & \frac{1}{36} & \frac{1}{18} \\ & \frac{1}{9} & \frac{1}{18} & \frac{1}{36} \\ & & \frac{1}{9} & \frac{1}{18} \\ \text{sym} & & & \frac{1}{9} \end{bmatrix}, \tag{5.98}$$

$$\mathbf{A}_{xy} = \frac{1}{4} \begin{bmatrix} 1 & 1 & -1 & -1 \\ -1 & -1 & 1 & 1 \\ -1 & -1 & 1 & 1 \\ 1 & 1 & -1 & -1 \end{bmatrix}, \quad \mathbf{A}_{xz} = \frac{1}{12} \begin{bmatrix} 1 & 1 & \frac{1}{2} & \frac{1}{2} \\ -1 & -1 & -\frac{1}{2} & -\frac{1}{2} \\ -\frac{1}{2} & -\frac{1}{2} & -1 & -1 \\ \frac{1}{2} & \frac{1}{2} & 1 & 1 \end{bmatrix},$$

$$\mathbf{A}_{yz} = \frac{1}{12} \begin{bmatrix} 1 & \frac{1}{2} & \frac{1}{2} & 1 \\ \frac{1}{2} & 1 & 1 & \frac{1}{2} \\ -\frac{1}{2} & -1 & -1 & -\frac{1}{2} \\ -1 & -\frac{1}{2} & -\frac{1}{2} & -1 \end{bmatrix}.$$

Note that $\mathbf{A}_{xx}$, $\mathbf{A}_{yy}$, and $\mathbf{A}_{xy}$ have the same numerical value in (5.98) as in (5.73).

In the case of uniformly distributed loads within the element, we have

$$\mathbf{Q}_n^{*\mathrm{T}} = \{Q_1^* \; Q_2^* \; \cdots \; Q_{24}^*\},$$

in which

$$\begin{aligned} Q_i^* &= \frac{V}{8} \bar{F}_x + \frac{S_j \bar{T}_x \delta_i}{4}, \\ Q_{i+8}^* &= \frac{V}{8} \bar{F}_y + \frac{S_j \bar{T}_y \delta_i}{4}, \\ Q_{i+16}^* &= \frac{V}{8} \bar{F}_z + \frac{S_j \bar{T}_z \delta_i}{4}, \qquad i = 1, 2, \cdots, 8, \end{aligned} \tag{5.99}$$

where

$$\delta_i = \begin{cases} 1 & \text{if node } i \text{ is on side } j \\ 0 & \text{if node } i \text{ is not on side } j. \end{cases}$$

It should be remembered that (5.96) and (5.99) can be used for a nonhomogeneous material in which the loads are nonuniformly distributed within the elements. In such cases, the mean values of $\mathbf{C}$, $\bar{\mathbf{F}}$, and $\bar{\mathbf{T}}$ within the elements should be used. Such an approximation is consistent with the approximation of the strain within the element.

The ordering of the matrices $\mathbf{k}_n^*$ and $\mathbf{Q}_n^*$ associated with the generalized-coordinate vector $\mathbf{q}_n^*$ defined in (5.93) is not convenient for assembly. We shall define

$$\mathbf{q}_n^T = \{u_1\ v_1\ w_1\ u_2\ \cdots\ u_4\ v_4\ w_4\}. \tag{5.100}$$

The relation between the components of q_n and q_n^* is

$$q_i = q_{I(i)}^*, \qquad i = 1, 2, \cdots, 24,$$

where

$$\begin{aligned}
I(3i - 2) &= i, \\
I(3i - 1) &= i + 8, \\
I(3i) &= i + 16, \qquad i = 1, 2, \cdots, 8.
\end{aligned} \tag{5.101}$$

Let the element-stiffness matrix and loading vector associated with $\mathbf{q}_n$ be denoted by $\mathbf{k}_n$ and $\mathbf{Q}_n$, respectively. As for the rectangular element (sec. 5.3.2), it can be shown that the components of $\mathbf{k}_n$ and $\mathbf{Q}_n$ are, respectively,

$$\begin{aligned}
k_{ij} &= k_{I(i)I(j)}^*, \\
Q_i &= Q_{I(i)}^*, \qquad i, j = 1, 2, \cdots, 24.
\end{aligned} \tag{5.102}$$

5.4 Rotation of Local Coordinates about Global Coordinates

In order to use the assembly procedures described in chs. 3 and 4, all the generalized coordinates of different elements must have the same physical meaning at a common node. For example, let the generalized coordinates be the nodal displacements. The components of the displacement vectors of different elements at the same node must be referred to a common coordinate

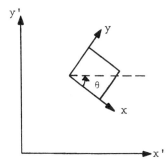

Fig. 5.10 Coordinate rotation.

system. In practice, only one such coordinate system, called the *global* system, is used for all the nodes of the entire body. However, in constructing the element matrices, the global system may not be convenient to use. For instance, in fig 5.10 let the (x', y')-coordinates be the global system. It will be much more convenient to use the (x, y)-coordinates,—the *local* system—to construct the element matrices for the rectangular element. The local system is related to the global system by

$$\mathbf{x} = \beta \mathbf{x}', \tag{5.103}$$

as given in (5.5). In particular,

$$\beta = \begin{bmatrix} \cos\theta & -\sin\theta \\ \sin\theta & \cos\theta \end{bmatrix}$$

for the two-dimensional case.

The transformation for the displacements at the *l*th node will be

$$\mathbf{u}_l = \beta \mathbf{u}'_l \quad \text{or} \quad \mathbf{u}'_l = \beta^T \mathbf{u}_l. \tag{5.104}$$

For the *n*th element, the generalized coordinates can be partitioned such that

$$\mathbf{q}_n = \begin{Bmatrix} \mathbf{u}_1 \\ \mathbf{u}_2 \\ \vdots \\ \mathbf{u}_m \end{Bmatrix} = \mathbf{R}\mathbf{q}'_n, \tag{5.105}$$

in which m is the number of nodes, $\mathbf{u}_1, \mathbf{u}_2, \cdots, \mathbf{u}_m$ are the generalized coordinates of the first, second, ... , and mth node of the element, respectively, and

$$\mathbf{R} = \begin{bmatrix} \beta & & 0 \\ & \beta & \\ & & \ddots \\ 0 & & & \beta \end{bmatrix}; \tag{5.106}$$

this last is called the rotation matrix.* The number of submatrices on the

---

*If the nodal values of the partial derivatives of the displacements are also used as the generalized coordinates, **R** must be modified to take this into account.

diagonal of **R** is equal to the number of nodes in the corresponding element. The element matrices associated with $\mathbf{q}'_n$ are

$$\mathbf{k}'_n = \mathbf{R}^T \mathbf{k}_n \mathbf{R}, \qquad \mathbf{Q}'_n = \mathbf{R}^T \mathbf{Q}_n, \tag{5.107}$$

while $\mathbf{k}_n$ and $\mathbf{Q}_n$ are associated with the generalized coordinates $\mathbf{q}_n$. The matrices $\mathbf{k}'_n$ and $\mathbf{Q}'_n$ are then used for assembly.

5.5 Mixed Boundary Conditions

There are other circumstances when one must also perform a coordinate transformation. In mixed-boundary problems, for instance, the components of the displacement can be prescribed in certain directions while the components of traction are prescribed in other directions, and these directions are not the same as those of the global system. As shown in fig. 5.11, we may have the displacement prescribed in the z'-direction, and the traction prescribed in the x'- and y'-directions, while the generalized coordinates are the nodal displacements in the x-, y-, and z-directions. Under such circumstances, one must use the nodal displacements in the x'- y'-, and z'-directions as the generalized coordinates in order to make it easy to impose the boundary constraints. This will require a transformation of coordinates for all these boundary nodes by altering the stiffness matrix and the loading vector. The procedure is the same as outlined in (5.103) through (5.107), except that the submatrices on the diagonal of **R** [(5.105) and (5.106)] are the identity matrix of the same order as β for those nodes which are not these type of boundary nodes. The corresponding β for different nodes can be different depending on

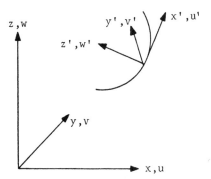

Fig. 5.11 Mixed boundary conditions.

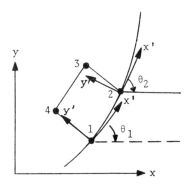

Fig. 5.12 Example.

the relation between the x-system and the x'-system at the node. Referring to fig. 5.12, suppose only the displacements in the y'-direction at nodes 1 and 2 of the element are prescribed; then

$$\mathbf{R} = \begin{bmatrix} \beta_1 & & & 0 \\ & \beta_2 & & \\ & & \mathbf{I} & \\ 0 & & & \mathbf{I} \end{bmatrix}, \tag{5.108}$$

where

$$\beta_i = \begin{bmatrix} \cos\theta_i & -\sin\theta_i \\ \sin\theta_i & \cos\theta_i \end{bmatrix}, \quad i = 1, 2.$$

In practice, it is more convenient to perform such a transformation on the master (assembled) stiffness matrix and loading vector, if all the components of these are in the computer core simultaneously. Suppose such a transformation is required for the lth nodes. Then

$$\mathbf{K}' = \mathbf{R}^T\mathbf{K}\mathbf{R}, \qquad \mathbf{Q}' = \mathbf{R}^T\mathbf{Q}, \tag{5.109}$$

where

$$\mathbf{R} = \begin{bmatrix} \mathbf{I} & & & & & \\ & \mathbf{I} & & & & \\ & & \ddots & & & \\ & & & \beta_l & & \\ & & & & \mathbf{I} & \\ & & & & & \ddots \\ & & & & & & \mathbf{I} \end{bmatrix},$$

where β_l is the transformation matrix for the direction cosines which relate the two coordinate systems at the lth node. The number of submatrices on the diagonal of **R** is equal to the number of nodes in the assemblage. In (5.109), let the degrees of freedom associated with the lth node be $L + 1, \cdots, L + m$, where $m = 2$ for two-dimensional problems and $m = 3$ for three-dimensional problems. We have

$$K'_{i,L+j} = K'_{L+j,i} = \sum_{r=1}^{m} \beta_{rj} K_{L+r,i},$$
$$i = 1, 2, \cdots, L + m,$$

$$K'_{i,L+j} = K'_{L+j,i} = \sum_{r=1}^{m} K_{i,L+r} \beta_{rj}, \quad (5.110)$$
$$i = L + 1, L + 2, \cdots, N,$$

$$Q'_{L+j} = \sum_{r=1}^{m} \beta_{rj} Q_{L+r},$$

in which $j = 1, \cdots, m$, and N is the total number of degrees of freedom. In terms of the one-dimensional array (4.20) we have

$$R'_{J(L+j)+i} = \sum_{r=1}^{m} \beta_{rj} R_{J(L+r)+i},$$
$$i = I_{L+j}, \cdots, L \text{ (no summation if } I_{L+j} > L),$$

$$R'_{J(i)+L+j} = \sum_{r=1}^{m} R_{J(i)+L+r} \beta_{rj},$$
$$i = L + m + 1, \cdots, N \text{ (no summation if } L + m = N \text{ or } I_i > L + j),$$

$$R'_{J(L+j)+L+i} = \sum_{r=1}^{m} \sum_{s=1}^{m} \beta_{rj} M_{rs} \beta_{si}, \quad i = 1, 2, \cdots, j, \quad (5.111)$$

where $j = 1, 2, \cdots, m$, and where I_i and $J(i)$ are defined in (4.1), (4.17), and (4.18) and

$$M_{rs} = M_{sr} = R_{J(M_a)+M_i},$$

in which

$$M_a = \max(L + r, L + s),$$
$$M_i = \min(L + r, L + s).$$

After the transformation outlined in (5.110) or (5.111) is performed for every boundary node which involves mixed boundary conditions, one can then impose the constraint conditions in the usual manner as outlined in sec. 4.6.4.

For more general types of constraint conditions, the transformation of the stiffness matrix may be too time-consuming for even a fast computer. In this case, one may treat the constraint condition as an additional equation to be solved simultaneously.

6 Interpolation Functions, Numerical Integration, and Higher-Order Elements

6.1 Introduction

In using the finite-element method, the most essential step is the choosing of the element. This includes two basic tasks.

First, one must approximate the field variables by a set of shape functions in terms of the selected generalized coordinates. That is, one must approximate a function $u(\mathbf{x})$ within the element A_n, in the form

$$\tilde{u}(\mathbf{x}) = \sum_{i=1}^{m} f_i(\mathbf{x}) q_i = \mathbf{f}^T(\mathbf{x})\mathbf{q}, \qquad (6.1)$$

where q_i is a generalized coordinate which can be the value of u or a derivative of u at some nodal points on A_n, and f_i is a function so constructed that when the generalized coordinates at any node along an interelement boundary are the same regardless of from which element the node is approached—in which case we say that the generalized coordinates are *compatible* along that boundary—the function u (and in some cases also the derivatives of u) is (are) the same regardless of the direction (element) from which the node is approached. That is, if q_i is compatible along an interelement boundary, then f_i is such that u (and sometimes its derivatives) is (are) also compatible along that boundary. Such a function, in the terminology of numerical analysis, is called an interpolation function (see, e.g., Collatz 1966; Kopal 1961; Hidebrand 1956). In the finite-element formulation the functions $f_i(\mathbf{x})$ are almost exclusively polynomials. The accuracy of the interpolation depends upon the order of completeness of the polynomials, which in general is directly related to the number of nodal points for the interpolation functions.

The second task is to evaluate the element matrices, such as the element-stiffness matrix and loading vector, etc.; this will entail the evaluation of many integrals. It may be difficult or impossible to integrate these integrals analytically, in particular for more complex elements of higher order. With the aid of a digital computer, the integration can be done numerically. This numerical, rather than analytic, integration eliminates many tedious algebraic operations and often reduces computing time and storage requirements (Irons 1969, 1966).

In the present chapter we will first discuss the important interpolation functions in general and the numerical integration techniques used in the finite-element formulation; then we will construct various higher-order elements. Some applications will also be discussed.

For convenience, the interpolation function and the numerical integration will be expressed in terms of *normalized* local coordinates ξ. That is to say, the limits of the element will lie within the range ± 1 in ξ. The physical system will be denoted by x-coordinates and is related to ξ-coordinates by a transformation. The notations

$$D_\xi^{(n)} u \quad \text{and} \quad D_x^{(n)} u$$

denote the nth order derivative of u with respect to ξ-coordinates and x-coordinates, respectively.

6.2 Interpolation in One Dimension

One of the general interpolation schemes is Hermite interpolation. A function $u(\xi)$ is approximated by a polynomial $\tilde{u}(\xi)$ which is expressed in terms of the values of $u(\xi)$ and all the first N derivatives of $u(\xi)$ at m discrete points by

$$\tilde{u}(\xi) = \sum_{j=1}^{m} [H_{0j}^{(N)}(\xi) u(\xi_j) + H_{1j}^{(N)}(\xi) u'(\xi_j) + \cdots + H_{Nj}^{(N)}(\xi) u^{(N)}(\xi_j)], \tag{6.2}$$

where $H_{ij}^{(N)}(\xi)$ is called the m-point Nth order Hermite polynomial and $u^{(l)}(\xi_j)$ is the value of the lth derivative of u at $\xi = \xi_j$; there are $N + 1$ terms in the bracket, and the total number of terms for $\tilde{u}$ is $m(N + 1)$.

The Hermite polynomials have the property that

$$\frac{d^l H_{ij}^{(N)}}{d\xi^l}(\xi_k) = \delta_{jk}\delta_{il}, \quad j, k = 1, 2, \cdots, m, \quad i, l = 0, 1, \cdots, N, \tag{6.3}$$

where δ_{ik} is the Kronecker delta. There are $m(N + 1)$ equations in (6.3) for $H_{ij}^{(N)}(\xi)$. Therefore, it is in general a polynomial of degree $m(N + 1) - 1$:

$$H_{ij}^{(N)}(\xi) = \alpha_1 + \alpha_2 \xi + \cdots + \alpha_{m(N+1)} \xi^{m(N+1)-1}. \tag{6.4}$$

The αs can be evaluated by using (6.3) for each pair of i, j. The Hermite polynomial constructed from (6.4) is called $m(N + 1) - 1$ degree complete, since (6.2) can exactly represent any polynomial of that degree.

6.2.1 Zeroth-Order Interpolation

We have zeroth-order Hermite interpolation when $N = 0$. For zeroth-order interpolation, (6.2) is

$$\tilde{u}(\xi) = \sum_{i=1}^{m} f_i(\xi) u(\xi_i). \tag{6.5}$$

This is commonly known as Lagrange's interpolation formula. It can be easily shown that the Lagrange interpolation functions f_i are given by

$$f_i(\xi) = \frac{L(\xi)}{(\xi - \xi_i) L'(\xi_i)}, \tag{6.6}$$

where

$$L(\xi) = (\xi - \xi_1)(\xi - \xi_2) \cdots (\xi - \xi_m), \tag{6.7}$$

which statisfies (6.3); that is,

$$f_i(\xi_k) = \delta_{ik}, \quad i, k = 1, 2, \cdots, m. \tag{6.8}$$

Equation (6.5) is in general a polynomial of $(m - 1)$th degree, which can be viewed as a truncated Taylor series. For a smooth function which possesses the first m derivatives, the error of Lagrange's interpolation formula is

$$u(\xi) - \tilde{u}(\xi) = E(\xi) = \frac{u^m(\zeta)}{m!} L(\xi), \tag{6.9}$$

where ζ is some value between ξ_1 and ξ_m. Thus if $u(\xi)$ is a polynomial of order $m - 1$ or less, this interpolation is exact. For the case $m = 2$ ($\xi_1 = -1$ and $\xi_2 = 1$), the interpolation functions are linear:

$$f_1(\xi) = \tfrac{1}{2}(1 - \xi), \tag{6.10}$$
$$f_2(\xi) = \tfrac{1}{2}(1 + \xi). \tag{6.11}$$

When $m = 3$ (with the three points $\xi_1 = -1$, $\xi_2 = 0$, and $\xi_3 = 1$), the Hermite polynomials are quadratic:

$$\begin{aligned}
f_1(\xi) &= H_{01}^{(0)}(\xi) = -\tfrac{1}{2}(1 - \xi)\xi, \\
f_2(\xi) &= H_{02}^{(0)}(\xi) = (1 - \xi)(1 + \xi), \\
f_3(\xi) &= H_{03}^{(0)}(\xi) = \tfrac{1}{2}(1 + \xi)\xi.
\end{aligned} \tag{6.12}$$

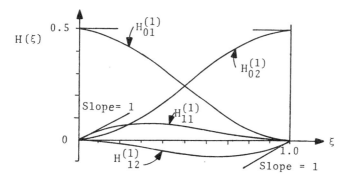

Fig. 6.1 Two-point first-order Hermite interpolation functions.

Similarly, cubic Hermite polynomials can be written out using (6.6) for $m = 4$, etc.

6.2.2. First-Order Interpolation

For first-order interpolation, $N = 1$. The interpolation can be expressed in terms of Lagrange's interpolation functions. It can be shown that

$$H_{0i}^{(1)}(\xi) = [1 - 2f_i'(\xi_i)(\xi - \xi_i)][f_i(\xi)]^2,$$
$$H_{1i}^{(1)}(\xi) = (\xi - \xi_i)[f_i(\xi)]^2. \tag{6.13}$$

The error of using (6.1) with a ζ within ξ_1 to ξ_m is given by

$$E(\xi) = \frac{u^{2m}(\zeta)}{(2m)!} [L(\xi)]^2. \tag{6.14}$$

Thus, if $u(\xi)$ is a polynomial of order $2m - 1$ or less, this interpolation is exact.

For the case $m = 2$ in first-order interpolation, with $\xi_1 = -1$ and $\xi_2 = 1$, the Hermite polynomials are cubic functions given by

$$\begin{aligned} H_{01}^{(1)}(\xi) &= (2 + \xi)(1 - \xi)^2/4, \\ H_{02}^{(1)}(\xi) &= (2 - \xi)(1 + \xi)^2/4, \\ H_{11}^{(1)}(\xi) &= (1 + \xi)(1 - \xi)^2/4, \\ H_{12}^{(1)}(\xi) &= -(1 - \xi)(1 + \xi)^2/4. \end{aligned} \tag{6.15}$$

These functions are shown in fig. 6.1.

6.2.3 Second-Order Interpolation

For second-order interpolation, $N = 2$. As before, we can express second-order interpolation in terms of the Lagrange interpolation functions,

$$H_{0i}^{(2)}(\xi) = \{1 - 3f'_i(\xi_i)(\xi - \xi_i) + [6(f'_i(\xi_i))^2 - \tfrac{3}{2}f''_i(\xi_i)](\xi - \xi_i)^2\}[f_i(\xi)]^3,$$
$$H_{1i}^{(2)}(\xi) = (\xi - \xi_i)[1 - 3f'_i(\xi_i)(\xi - \xi_i)][f_i(\xi)]^3, \qquad (6.16)$$
$$H_{2i}^{(2)}(\xi) = \frac{(\xi - \xi_i)^2}{2}[f_i(\xi)]^3.$$

The error of (6.1) in this case for a ζ within ξ_1 to ξ_m is

$$E(\xi) = \frac{u^{(3m)}(\zeta)}{(3m)!}[L(\xi)]^3. \qquad (6.17)$$

For $m = 2$,

$$H_{01}^{(2)}(\xi) = [1 + \tfrac{3}{2}(1 + \xi) + \tfrac{3}{2}(1 + \xi)^2](1 - \xi)^3/8,$$
$$H_{02}^{(2)}(\xi) = [1 + \tfrac{3}{2}(1 - \xi) + \tfrac{3}{2}(1 - \xi)^2](1 + \xi)^3/8,$$
$$H_{11}^{(2)}(\xi) = [1 + \tfrac{3}{2}(1 + \xi)](1 + \xi)(1 - \xi)^3/8,$$
$$H_{12}^{(2)}(\xi) = -[1 + \tfrac{3}{2}(1 - \xi)](1 - \xi)(1 + \xi)^3/8, \qquad (6.18)$$
$$H_{21}^{(2)}(\xi) = (1 + \xi)^2(1 - \xi)^3/16,$$
$$H_{22}^{(2)}(\xi) = (1 - \xi)^2(1 + \xi)^3/16.$$

A one-dimensional interpolation for $u(x)$ with the m discrete points x_1, $x_2, \cdots, x_m$ can be expressed in terms of ξ with limits ± 1 by a coordinate transformation

$$\xi = \frac{x - (x_m + x_1)/2}{h}, \qquad (6.19)$$

where

$$h = (x_m - x_1)/2.$$

The nth derivatives in the ξ- and x-systems are related by

$$u^{(n)}(\xi) = h^n \frac{d^n u}{dx^n},$$

which may be written as

$$D_\xi^{(n)} u = h^n D_x^{(n)} u. \qquad (6.20)$$

Integration in the two systems is related by

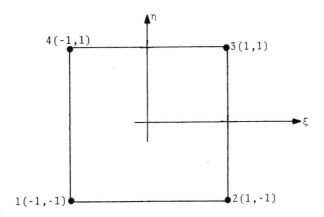

Fig. 6.2 A four-node square element.

$$\int_{x_1}^{x_2} u(x)\, dx = h \int_{-1}^{1} u(\xi)\, d\xi. \tag{6.21}$$

6.3 Interpolation in Two Dimensions

6.3.1 Square Element

Let a four-node square element have sides $\xi = \pm 1$ and $\eta = \pm 1$, and label the corners 1 to 4 as shown in fig. 6.2. The function u expressed in terms of its values at the four corner nodes is written as

$$\tilde{u}(\xi, \eta) = \sum_{i=1}^{4} f_i(\xi, \eta) u_i = \mathbf{f}^T \mathbf{u}, \tag{6.22}$$

where

$$f_i(\xi, \eta) = \tfrac{1}{4}(1 + \xi\xi_i)(1 + \eta\eta_i), \tag{6.23}$$

is a bilinear function similar to that given in (2.24) and (5.67). Since $\tilde{u}(\xi, \eta)$ varies linearly along all the sides of the square, it will be continuous at the interelement boundaries if $\tilde{u}(\xi, \eta)$ of the adjacent elements also varies linearly along its sides and is compatible at the two common corners. It should be noted that f_i is the product of the vertical line $1 + \xi\xi_i = 0$ and the horizontal line $1 + \eta\eta_i = 0$.

If a function over a square is to be expressed in terms of the values of the

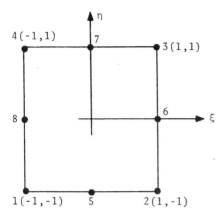

Fig. 6.3 An eight-node square element.

function at the four corner points *and* the four midpoints of the four sides, as shown in fig. 6.3, we can use the interpolation function

$$\tilde{u}(\xi, \eta) = \sum_{i=1}^{8} f_i(\xi, \eta) u_i = \mathbf{f}^T \mathbf{u}, \qquad (6.24)$$

in which

$$\begin{aligned} f_i(\xi, \eta) &= \tfrac{1}{4}(1 + \xi\xi_i)(1 + \eta\eta_i)(\xi\xi_i + \eta\eta_j - 1), & i &= 1, 2, 3, 4, \\ &= \tfrac{1}{2}(1 - \xi^2)(1 + \eta\eta_i), & i &= 5, 7, \\ &= \tfrac{1}{2}(1 - \eta^2)(1 + \xi\xi_i) & i &= 6, 8. \end{aligned} \qquad (6.25)$$

The function $f_i(\xi, \eta)$ is the product of three linear functions, which are straight lines on the (ξ, η)-plane. For example, f_1 is the product of the functions representing straight lines passing nodes 2 and 3, nodes 3 and 4, and nodes 5 and 8. The highest-order term in f_i is either $\xi^2\eta$ or $\eta^2\xi$, and all the f_i, $i = 1, 2, \cdots, 8$ are linearly independent (why?); thus $\tilde{u}(\xi, \eta)$ of (6.24) can represent a complete quadratic polynomial

$$u(\xi, \eta) = \alpha_1 + \alpha_2 \xi + \alpha_3 \eta + \alpha_4 \xi^2 + \alpha_5 \xi\eta + \alpha_6 \eta^2 + \alpha_7 \xi^2\eta + \alpha_8 \xi\eta^2. \qquad (6.26)$$

The error in using (6.24) to represent a general function $u(\xi, \eta)$ will be of order $D_\xi^{(3)} u$. This is because (6.26) can be viewed as a truncated Taylor series, in which the neglected terms are ξ^3, η^3, etc.; their coefficients are related to the third derivatives of u. The order of the magnitude of $D_\xi^{(3)} u$ will

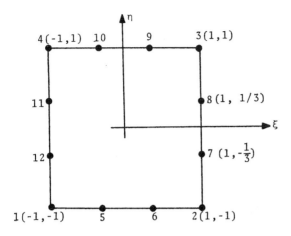

Fig. 6.4 A twelve-node square element.

depend on the relation between ξ and the physical coordinates x of the element. One can easily construct a nine-node interpolation which can represent a complete biquadratic function. In this case, $\tilde{u}$ is also complete up to quadratic terms; thus the additional node cannot improve the order of accuracy of $\tilde{u}$.

It is seen that $\tilde{u}$ is a quadratic function along each side of the rectangle; hence the interpolation function will maintain continuity at the interelement boundary if the nodal values are compatible.

A twelve-node interpolation can be constructed over a square area in the (ξ, η)-plane, as shown in fig. 6.4. Here

$$\tilde{u}(\xi, \eta) = \sum_{i=1}^{12} f_i(\xi, \eta) u_i = \mathbf{f}^T \mathbf{u}, \tag{6.27}$$

where

$$f_i(\xi, \eta) = \tfrac{1}{32}(1 + \xi\xi_i)(1 + \eta\eta_i)[-10 + 9(\xi^2 + \eta^2)] \tag{6.28a}$$

for the corner nodes, that is, for $i = 1, 2, 3, 4$, and

$$f_i(\xi, \eta) = \tfrac{9}{32}(1 + \xi\xi_i)(1 - \eta^2)(1 + 9\eta\eta_i) \tag{6.28b}$$

for the nodes along the sides $\xi_i = \pm 1$ with $\eta_i = \pm 1/3$, or for $i = 7, 8, 11, 12$, and

$$f_i(\xi, \eta) = \tfrac{9}{32}(1 + \eta\eta_i)(1 - \xi^2)(1 + 9\xi\xi_i) \tag{6.28c}$$

for the nodes $\eta_i = \pm 1$, with $\xi_i = \pm 1/3$, or for $i = 5, 6, 9, 10$. The interpolation $f_i (i = 1, 2, 3, 4)$ is the product of two straight lines and the circle passing through nodes 5 through 12. The interpolation $f_i (i = 5, \cdots, 12)$ is the product of straight lines parallel to the ξ- or η-axis.

The highest-order terms in f_i are $\xi^3\eta$ or $\xi\eta^3$, and all twelve f_i are linearly independent; thus $\tilde{u}(\xi, \eta)$ of (6.24) can be represented by a complete cubic polynomial

$$u(\xi, \eta) = \alpha_1 + \alpha_2 \xi + \alpha_3 \eta + \cdots + \alpha_{10} \eta^3 + \alpha_{11} \xi^3 \eta + \alpha_{12} \xi^3.$$

The error in using (6.27) to represent a general function u will be of order $D_\xi^4 u$.

Similarly, one can construct a sixteen-node interpolation with four additional interior nodes within the element. It can be shown that such an interpolation can only represent a complete cubic polynomial exactly. Therefore, the order of accuracy is not improved over that of using (6.27). In practice, it may not be worthwhile to introduce such additional nodes in the interior.

An interpolation can also be expressed in terms of the value of u and the derivatives of u at the nodes. One such interpolation is the bicubic interpolation

$$\begin{aligned}\tilde{u}(\xi, \eta) = \sum_{i=1}^{2} \sum_{j=1}^{2} [&H_{0i}^{(1)}(\xi) H_{0j}^{(1)}(\eta) u(\xi_i, \eta_j) \\ &+ H_{1i}^{(1)}(\xi) H_{0j}^{(1)}(\eta) u_{,\xi}(\xi_i, \eta_j) \\ &+ H_{0i}^{(1)}(\xi) H_{1j}^{(1)}(\eta) u_{,\eta}(\xi_i, \eta_j) \\ &+ H_{1i}^{(1)}(\xi) H_{1j}^{(1)}(\eta) u_{,\xi\eta}(\xi_i, \eta_j)]\end{aligned} \tag{6.29}$$

for a four-node square. It can be easily verified that such an interpolation can only represent exactly a complete bicubic polynomial as the sixteen-node interpolation; therefore, the order of approximation is in general the same as that of the twelve-node interpolation. However, the use of the interpolation (6.29) will not only insure compatibility in u but also in all the first partial derivatives of u along the interelement boundaries.

It should be remarked that although a smooth function can still be formed if the last term of (6.29) is omitted, it is not a desirable one for finite-element applications. In this case $\tilde{u}(\xi, \eta)$ cannot represent a term $\xi\eta$ exactly, since the cross derivative of $\tilde{u}_{\xi\eta}$ will always vanish at the four corners. That is, $\tilde{u}$ can only represent a complete linear polynomial and hence it only has the same order of accuracy as that of the four-node interpolation (6.22). However, the present case will have twelve degrees of freedom per element while in (6.22) there are only four. Such incomplete interpolations for plate bending problems have been found to converge to incorrect answers. The nonconvergence of the solutions is due to the fact that $\tilde{u}(\xi, \eta)$ cannot represent a complete quadratic polynomial exactly.

6.3.2 Right-Triangular Element

Interpolation functions can be constructed easily over right triangles with nodes labeled as shown in fig. 6.5. In order to simplify the expression for the interpolation functions, it is convenient to denote coordinates ξ, η by ζ_1 and ζ_2, respectively, and introduce a new coordinate

$$\zeta_3 = 1 - \zeta_1 - \zeta_2. \tag{6.30}$$

The coordinates $\zeta_1, \zeta_2, \zeta_3$ are called the triangular coordinates. It can be seen that ζ_1, ζ_2, and ζ_3 are related to the areas of the subtriangles around the point p within the element triangle (fig. 6.5a). Therefore, they are also called area coordinates.

Linear Interpolation A three-node interpolation with the corners of the triangle taken as the nodes (fig. 6.5a) is simply

$$\tilde{u}(\zeta_1, \zeta_2, \zeta_3) = \sum_{i=1}^{3} \zeta_i u_i = \mathbf{f}^T \mathbf{u}, \tag{6.31}$$

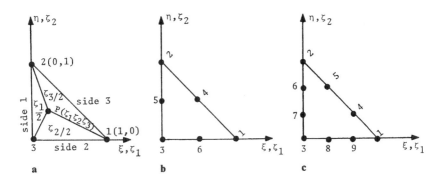

Fig. 6.5 Right-triangular elements.

which is the same as (2.19). It can be seen that (6.31) can form a complete linear polynomial in ξ and η.

Quadratic Interpolation With the three corners as well as the midpoints of the three sides as nodes (fig. 6.5b), a six-node interpolation function can be easily formed:

$$\tilde{u}(\zeta_1, \zeta_2, \zeta_3) = \sum_{i=1}^{6} f_i(\zeta_1, \zeta_2, \zeta_3) u_i = \mathbf{f}^T \mathbf{u}, \tag{6.32}$$

where

$$\mathbf{f} = \begin{Bmatrix} \zeta_1(2\zeta_1 - 1) \\ \zeta_2(2\zeta_2 - 1) \\ \zeta_3(2\zeta_3 - 1) \\ 4\zeta_1\zeta_2 \\ 4\zeta_2\zeta_3 \\ 4\zeta_3\zeta_1 \end{Bmatrix}. \tag{6.33}$$

It can be seen that $f_i (i = 1, 2, 3)$ is formed by the product of two parallel lines passing the nodes, not including node i [for example, $f_1 = \zeta_1(2\zeta_1 - 1)$, in which $\zeta_1 = 0$ is the line passing through nodes 2 and 3, and $2\zeta_1 - 1 = 0$ is the line passing through nodes 4 and 6], and that $f_i (i = 4, 5, 6)$ is formed by the product of the two lines that are the sides of the triangle not containing node i. Since the f_i are six linearly independent quadratic functions in terms of ξ and η, (6.32) can form a complete quadratic polynomial in ξ and η. The error in approximating a general function u in the element will be of order $D_\xi^3 u$.

Cubic Interpolation A ten-node interpolation with nodes taken at the three corners, two points on each side, 1/3 of the way from the corners, and at the centroid of the triangle is given by

$$\tilde{u}(\zeta_1, \zeta_2, \zeta_3) = \sum_{i=1}^{10} f_i(\zeta_1, \zeta_2, \zeta_3) u_i = \mathbf{f}^T \mathbf{u}, \tag{6.34}$$

in which

$$\mathbf{f} = \begin{Bmatrix} \frac{1}{2}\zeta_1(3\zeta_1 - 1)(3\zeta_1 - 2) \\ \frac{1}{2}\zeta_2(3\zeta_2 - 1)(3\zeta_2 - 2) \\ \frac{1}{2}\zeta_3(3\zeta_3 - 1)(3\zeta_3 - 2) \\ \frac{9}{2}\zeta_1\zeta_2(3\zeta_1 - 1) \\ \frac{9}{2}\zeta_1\zeta_2(3\zeta_2 - 1) \\ \frac{9}{2}\zeta_2\zeta_3(3\zeta_2 - 1) \\ \frac{9}{2}\zeta_2\zeta_3(3\zeta_3 - 1) \\ \frac{9}{2}\zeta_3\zeta_1(3\zeta_3 - 1) \\ \frac{9}{2}\zeta_3\zeta_1(3\zeta_1 - 1) \\ 27\zeta_1\zeta_2\zeta_3 \end{Bmatrix}. \tag{6.35}$$

The f_i are ten linearly independent cubic functions of ξ and η; therefore (6.34) can be expanded into a complete cubic polynomial in ξ and η. The error in approximating a function u will in general be of order $D_\xi^{(4)}u$.

It should be remarked that, although a continuous function can be formed if the last term of (6.34) (i.e., $f_{10}u_{10}$) is omitted, it will be detrimental to use such a nine-node cubic interpolation in finite-element applications. This is because such an interpolation will always have $u = 0$ at the centroid of the triangle. Thus, using the interpolation (6.35), the interior node is absolutely essential. One may concoct a different form of the nine-node cubic interpolation which will be much better than the case just mentioned. However, any nine-node cubic interpolation cannot be expanded into a complete cubic polynomial in ξ and η, and the order of accuracy will therefore always be less than that of the ten-node interpolation. Thus, to include an interior node in this case is well worthwhile.

6.3.3 Coordinate Transformations

In the finite-element method it is often desirable or necessary to use elements of irregular shape, for example, a general quadrilateral element instead of a rectangular element. Or, in a case where there is a curved boundary, it would be a better geometrical representation of the domain if elements with simple curved boundaries were used. Rather than constructing the interpolation functions for such irregular elements, we can first transform the elements into squares or right triangles, then use the interpolation functions given in this chapter. The transformation is

$$\begin{aligned} x &= x(\xi, \eta), \\ y &= y(\xi, \eta). \end{aligned} \tag{6.36}$$

6.3 Interpolation in Two Dimensions

Differentiation and the integration of the interpolation in ξ- and x-coordinates are related by

$$\begin{Bmatrix} \dfrac{\partial}{\partial \xi} \\ \dfrac{\partial}{\partial \eta} \end{Bmatrix} = \mathbf{J} \begin{Bmatrix} \dfrac{\partial}{\partial x} \\ \dfrac{\partial}{\partial y} \end{Bmatrix}, \tag{6.37}$$

where

$$\mathbf{J} \equiv \text{Jacobian matrix} \equiv \begin{bmatrix} \dfrac{\partial x}{\partial \xi} & \dfrac{\partial y}{\partial \xi} \\ \dfrac{\partial x}{\partial \eta} & \dfrac{\partial y}{\partial \eta} \end{bmatrix}. \tag{6.38}$$

The determinant of $\mathbf{J}$ is called simply the Jacobian:

$$\frac{\partial(x, y)}{\partial(\xi, \eta)} \equiv \det \mathbf{J} \equiv \frac{\partial x}{\partial \xi} \frac{\partial y}{\partial \eta} - \frac{\partial x}{\partial \eta} \frac{\partial y}{\partial \xi} \tag{6.39}$$

In the evaluation of a double integral, we have

$$\iint u(x, y)dx dy = \iint u\bigl(x(\xi, \eta), y(\xi, \eta)\bigr)(\det \mathbf{J})d\xi d\eta. \tag{6.40}$$

Rectangular Element The transformation of a rectangle in the physical coordinates into a square (fig. 6.6) is simply

$$\begin{aligned} x &= \frac{1}{4} \sum_{i=1}^{4} (1 + \xi \xi_i)(1 + \eta \eta_i) x_i, \\ y &= \frac{1}{4} \sum_{i=1}^{4} (1 + \xi \xi_i)(1 + \eta \eta_i) y_i. \end{aligned} \tag{6.41}$$

Then (6.38) and (6.40) become

$$\mathbf{J} = \begin{bmatrix} \dfrac{x_2 - x_1}{2} & \dfrac{y_2 - y_1}{2} \\ \dfrac{x_3 - x_2}{2} & \dfrac{y_3 - y_2}{2} \end{bmatrix}, \tag{6.42}$$

which is constant, and

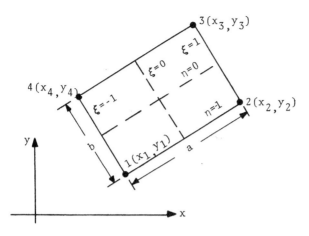

Fig. 6.6 A rectangular element.

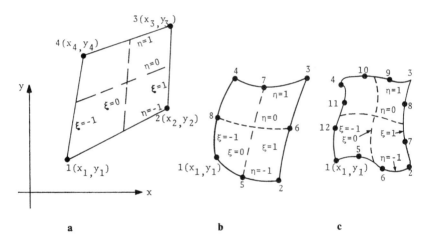

Fig. 6.7 Quadrilateral elements.

$$\iint u(x,y)dxdy = \frac{ab}{4}\int_{-1}^{1}\int_{-1}^{1} u\big(x(\xi,\eta), y(\xi,\eta)\big)d\xi d\eta.$$

Quadrilateral Element Quadrilateral elements such as shown in fig. 6.7 are transformed into a square by means of the transformation

$$x = \sum_{i=1}^{m} f_i(\xi,\eta)x_i,$$
$$y = \sum_{i=1}^{m} f_i(\xi,\eta)y_i,$$
(6.43)

where (x_i, y_i) are the coordinates of the nodes. For the case of fig. 6.7a, where the four sides are straight lines, we have $m = 4$ and f_i is as given in (6.23). For the case of fig. 6.7b, where the sides are expressible as quadratic functions, we have $m = 8$ and f_i as given in (6.25). For the case of fig. 6.7c, where the sides are expressible as cubic functions, we have $m = 12$ and f_i as given in (6.28). Both $\mathbf{J}$ and det $\mathbf{J}$ can be computed accordingly by (6.38) and (6.39). The double integral can be written as

$$\iint u(x,y)dxdy = \int_{-1}^{1}\int_{-1}^{1} u\big(x(\xi,\eta), y(\xi,\eta)\big)(\det \mathbf{J})d\xi d\eta. \quad (6.44)$$

Triangular Element The triangular elements shown in fig. 6.8 can be transformed to the right triangle. For the case of fig. 6.8a, where all the sides are straight lines, the transformation is simply

$$\begin{Bmatrix} 1 \\ x \\ y \end{Bmatrix} = \begin{bmatrix} 1 & 1 & 1 \\ x_1 & x_2 & x_3 \\ y_1 & y_2 & y_3 \end{bmatrix} \begin{Bmatrix} \zeta_1 \\ \zeta_2 \\ \zeta_3 \end{Bmatrix}, \quad (6.45)$$

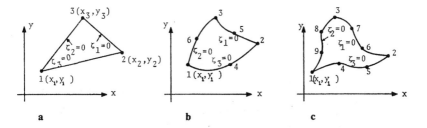

Fig. 6.8 Triangular elements.

or inversely

$$\begin{Bmatrix} \zeta_1 \\ \zeta_2 \\ \zeta_3 \end{Bmatrix} = \frac{1}{2\Delta} \begin{bmatrix} a_1 & b_1 & c_1 \\ a_2 & b_2 & c_2 \\ a_3 & b_3 & c_3 \end{bmatrix} \begin{Bmatrix} 1 \\ x \\ y \end{Bmatrix}, \tag{6.46}$$

where a_i, b_i, c_i, and Δ in terms of x_i, y_i are given in (2.18) and (5.51). The Jacobian matrix is

$$\mathbf{J} = \begin{bmatrix} x_1 - x_3 & y_1 - y_3 \\ x_2 - x_3 & y_2 - y_3 \end{bmatrix}, \tag{6.47}$$

and

$$\det \mathbf{J} = (x_1 - x_3)(y_2 - y_3) - (x_2 - x_3)(y_1 - y_3) = 2\Delta.$$

For the case of fig. 6.8b, where the edges are expressible as quadratic functions in the (x, y)-coordinates, and for the case of fig. 6.8c, where all the edges are expressible as cubic functions in x and y, we have

$$\begin{aligned} x &= \sum_{i=1}^{m} f_i(\zeta_1, \zeta_2, \zeta_3) x_i, \\ y &= \sum_{i=1}^{m} f_i(\zeta_1, \zeta_2, \zeta_3) y_i, \end{aligned} \tag{6.48}$$

where $m = 6$ and 12, respectively, and the f_i are as given in (6.33) and (6.35). Both $\mathbf{J}$ and $\det \mathbf{J}$ can be evaluated by (6.38) and (6.39). The double integral of (6.40) becomes

$$\iint u(x, y) \, dx \, dy = \int_0^1 \int_0^{1-\eta} u(x(\xi, \eta), y(\xi, \eta))(\det \mathbf{J}) \, d\xi \, d\eta. \tag{6.49}$$

Recall that

$$\zeta_1 = \xi, \quad \zeta_2 = \eta, \quad \zeta_3 = 1 - \xi - \eta.$$

When the transformations given in (6.43) and (6.48) are in the same form as the interpolation function, they are often called isoparametric transformations (Ergatoudis, Irons, and Zienkiewicz 1968).

It can be seen that under the transformation the interpolation will still be

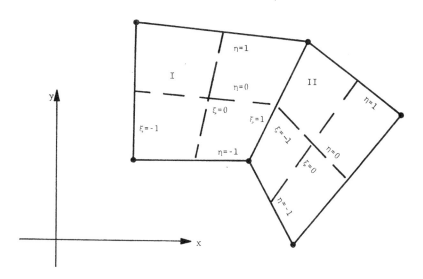

Fig. 6.9 Interelement boundary.

continuous over the physical interelement boundaries. The transformation for the common boundaries of two adjacent elements is the same, even though it is a different transformation for the transformed elements. In other words, we have the same coordinates for the interelement boundaries after transformation. Therefore, the conditions for the continuity of the interpolation are no different from those without transformation. For example, as shown over the common boundary in fig. 6.9, the coordinate is η ranging from -1 to 1 for both elements I and II. If u is a linear function of η along the common boundary, and is compatible at the two common nodes, then u is continuous.

The transformations of (6.41) and (6.46) are linear transformations. That is, the (x, y)-coordinates are linearly related to the (ξ, η)-coordinates. Under such transformation the Jacobian matrix is constant and the order of completeness of the interpolation function does not change. For example, the interpolation given in (6.25) (biquadratic) or (6.32) (quadratic) is second-order complete in (ξ, η)-coordinates; it can be seen that it is also second-order complete in (x, y)-coordinates.

The transformations (6.43) and (6.48) are nonlinear, and the order of completeness of the interpolation function is no longer the same in the two coordinate systems. Generally speaking, the interpolation functions can still represent a linear function in (x, y)-coordinates exactly. Suppose the function u has nodal value $u_i = \alpha_1 + \alpha_2 x_i + \alpha_3 y_i$ at the ith node; then

$$\tilde{u}(x, y) = \sum f_i(\xi, \eta) u_i$$
$$= \alpha_1 \sum f_i(\xi, \eta) + \alpha_2 \sum f_i(\xi, \eta) x_i + \alpha_3 \sum f_i(\xi, \eta) y_i \qquad (6.50)$$
$$= \alpha_1 + \alpha_2 x + \alpha_3 y$$

(see 6.43 or 6.48). However, for a function having nodal values $u_i = y_i^2$ (or $x_i y_i$ or x_i^2) we would have

$$\tilde{u}(x, y) = \sum f_i(\xi, \eta) y_i^2,$$

which in general may not be equal to y^2 exactly—the interpolation function in the (ξ, η)-coordinates cannot exactly represent quadratic terms in (x, y)-coordinates, even though the interpolations such as (6.25), (6.28), (6.33) and, (6.35) are at least second-order complete in ξ and η (i.e., they can represent exactly at least a quadratic function). Because they fail to represent exactly a quadratic term in (x,y)-coordinates, the interpolation in a general quadrilateral or curved element could have lower-order accuracy than that in a rectangular or triangular element.

An eight-node element is shown in fig. 6.10. The transformation between (x, y)- and (ξ, η)-coordinates is simply

$$x = h\xi,$$
$$y = h\left[\eta + \frac{\varepsilon_y}{2} \xi(1 + \eta)\right]. \qquad (6.51)$$

Using (6.24) to approximate a function, say y^2, we have

$$\tilde{u} = \sum_{i=1}^{8} f_i(\xi, \eta) y_i^2.$$

From (6.51) and (6.25),

$$\tilde{u} = \sum_{i=1}^{8} f_i(\xi, \eta) h^2 \left[\eta_i + \frac{\varepsilon_y}{2} \xi_i(1 + \eta_i)\right]^2$$
$$= \sum_{i=1}^{8} f_i(\xi, \eta) h^2 \left(\eta_i + \frac{\varepsilon_y}{2} \xi_i\right)^2 + \sum_{i=1}^{8} 2 f_i(\xi, \eta) h^2 \left(\eta_i + \frac{\varepsilon_y}{2} \xi_i\right) \xi_i \eta_i \frac{\varepsilon_y}{2}$$
$$+ \sum_{i=1}^{8} f_i(\xi, \eta) \frac{h^2 \varepsilon_y^2}{4} \xi_i^2 \eta_i^2.$$

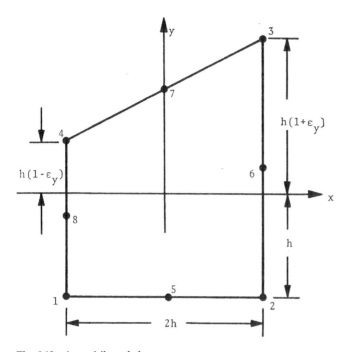

Fig. 6.10 A quadrilateral element.

Using the completeness property (6.26) of the eight-node interpolation, the first two summations in the equation above can be simplified, so that

$$\tilde{u} = h^2\left(\eta + \frac{\varepsilon_y}{2}\xi\right)^2 + \varepsilon_y h^2\left(\eta + \frac{\varepsilon_y}{2}\xi\right)\xi\eta + \sum_{i=1}^{8} f_i(\xi, \eta)\frac{h^2\varepsilon_y^2}{4}\xi_i^2\eta_i^2$$

$$= y^2 + \frac{h^2\varepsilon_y^2}{4}\left(\sum_{i=1}^{8} f_i(\xi, \eta)\xi_i^2\eta_i^2 - \xi^2\eta^2\right).$$

(6.52)

If ε_y is of order one, then $\tilde{u} - y^2$ will be of order h^2—even though from (6.26) $\tilde{u}$ can exactly represent a biquadratic function of ξ and η. Thus one should be careful in using nonlinear transformations for any interpolation that has more than four nodes, because of the additional complexity in computation and the loss of accuracy in the approximation due to the transformation.

6.4 Interpolation in Three Dimensions

6.4.1 Cubic Element

For a cubic element with limits $\xi = \pm 1$, $\eta = \pm 1$, and $\gamma = \pm 1$ and corner

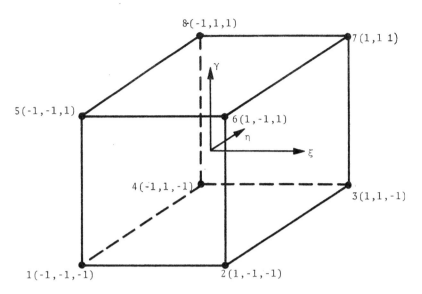

Fig. 6.11 An eight-node cubic element.

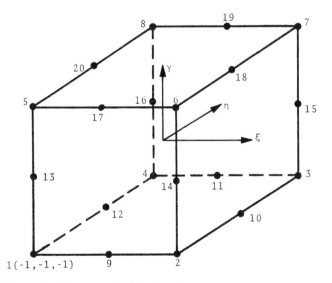

Fig. 6.12 A twenty-node cubic element.

points labeled 1 through 8 as shown in fig. 6.11, the trilinear interpolation function introduced in sec. 5.3.4 can be written as

$$\tilde{u}(\xi, \eta, \gamma) = \sum_{i=1}^{8} f_i(\xi, \eta, \gamma) u_i = \mathbf{f}^T \mathbf{u}, \tag{6.53}$$

in which

$$f_i(\xi, \eta, \gamma) = \tfrac{1}{8}(1 + \xi\xi_i)(1 + \eta\eta_i)(1 + \gamma\gamma_i). \tag{6.54}$$

If the function over a cubic element is to be expressed in terms of the values of the function at the eight corner points and the midpoints of the twelve edges as shown in fig. 6.12, the interpolation can be written as

$$\tilde{u}(\xi, \eta, \gamma) = \sum_{i=1}^{20} f_i(\xi, \eta, \gamma) u_i, \tag{6.55}$$

in which

$$\begin{aligned}
f_i(\xi, \eta, \gamma) &= \tfrac{1}{8}(1 + \xi\xi_i)(1 + \eta\eta_i)(1 + \gamma\gamma_i), \\
&\quad \times (\xi\xi_i + \eta\eta_i + \gamma\gamma_i - 2), & i &= 1, 2, \cdots, 8, \\
&= \tfrac{1}{4}(1 - \xi^2)(1 + \eta\eta_i)(1 + \gamma\gamma_i), & i &= 9, 11, 17, 19, \\
&= \tfrac{1}{4}(1 - \eta^2)(1 + \gamma\gamma_i)(1 + \xi\xi_i), & i &= 10, 12, 18, 20, \\
&= \tfrac{1}{4}(1 - \gamma^2)(1 + \xi\xi_i)(1 + \eta\eta_i), & i &= 13, 14, 15, 16.
\end{aligned} \tag{6.56}$$

In a manner similar to that in the case of the eight-node square (sec. 6.3), it can be shown that (6.55) can be represented by a triquadratic polynomial

$$\begin{aligned}
u(\xi, \eta, \gamma) &= \alpha_1 + \alpha_2 \xi + \alpha_3 \eta + \cdots + \alpha_{10} \gamma^2 \\
&\quad + \alpha_{11} \xi^2 \eta + \alpha_{12} \xi^2 \gamma + \alpha_{13} \eta^2 \gamma + \alpha_{14} \eta^2 \xi + \alpha_{15} \gamma^2 \xi + \alpha_{16} \gamma^2 \eta \\
&\quad + \alpha_{17} \xi \eta \gamma + \alpha_{18} \xi^2 \eta \gamma + \alpha_{19} \eta^2 \gamma \xi + \alpha_{20} \gamma^2 \xi \eta,
\end{aligned} \tag{6.57}$$

which is a complete polynomial of order two.

It is seen that on each face of the cube $\tilde{u}$ is a biquadratic function in the form (6.26); hence, the interpolation function will maintain continuity at the interelement boundary if the eight nodal values are compatible.

6.4.2 Right Tetrahedral Element

As for the right triangle, the interpolation function can be constructed easily for the right tetrahedron (fig. 6.13) by expressing u in terms of $\zeta_1 (= \xi)$, $\zeta_2 (= \eta)$, $\zeta_3 (= \gamma)$ and $\zeta_4 (= 1 - \zeta_1 - \zeta_2 - \zeta_3)$.

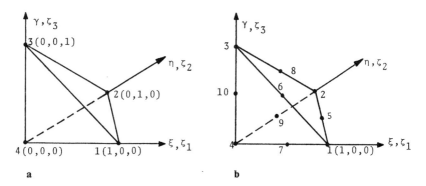

Fig. 6.13 Right tetrahedral elements.

For a tetrahedron with four corner nodes (fig. 6.13a), the interpolation is simply

$$\tilde{u}(\zeta_1, \zeta_2, \zeta_3, \zeta_4) = \sum_{i=1}^{4} \zeta_i u_i = \mathbf{f}^T \mathbf{u}. \tag{6.58}$$

A ten-node interpolation can also be easily constructed over the tetrahedron shown in fig. 6.13b; for this interpolation,

$$\tilde{u} = \sum_{i=1}^{10} f_i(\zeta_1, \zeta_2, \zeta_3, \zeta_4) u_i = \mathbf{f}^T \mathbf{u}, \tag{6.59}$$

where

$$\mathbf{f} = \begin{Bmatrix} \zeta_1(2\zeta_1 - 1) \\ \zeta_2(2\zeta_2 - 1) \\ \zeta_3(2\zeta_3 - 1) \\ \zeta_4(2\zeta_4 - 1) \\ \tfrac{1}{4}\zeta_1\zeta_2 \\ \tfrac{1}{4}\zeta_1\zeta_3 \\ \tfrac{1}{4}\zeta_1\zeta_4 \\ \tfrac{1}{4}\zeta_2\zeta_3 \\ \tfrac{1}{4}\zeta_2\zeta_4 \\ \tfrac{1}{4}\zeta_3\zeta_4 \end{Bmatrix}. \tag{6.60}$$

It can be shown that (6.59) is a complete quadratic polynomial in three-dimensional space.

6.4.3 Coordinate Transformations

In a manner completely analogous to that for the two-dimensional cases discussed in sec. 6.3.3, various shapes of elements in physical coordinates can be transformed to a cube or right tetrahedron by means of

$x = x(\xi, \eta, \gamma)$,
$y = y(\xi, \eta, \gamma)$,
$z = z(\xi, \eta, \gamma)$.

The differentiations are related by

$$\left\{ \begin{array}{c} \dfrac{\partial}{\partial \xi} \\ \dfrac{\partial}{\partial \eta} \\ \dfrac{\partial}{\partial \gamma} \end{array} \right\} = \mathbf{J} \left\{ \begin{array}{c} \dfrac{\partial}{\partial x} \\ \dfrac{\partial}{\partial y} \\ \dfrac{\partial}{\partial z} \end{array} \right\}, \tag{6.61}$$

where

$$\mathbf{J} = \begin{bmatrix} \dfrac{\partial x}{\partial \xi} & \dfrac{\partial y}{\partial \xi} & \dfrac{\partial z}{\partial \xi} \\ \dfrac{\partial x}{\partial \eta} & \dfrac{\partial y}{\partial \eta} & \dfrac{\partial z}{\partial \eta} \\ \dfrac{\partial x}{\partial \gamma} & \dfrac{\partial y}{\partial \gamma} & \dfrac{\partial z}{\partial \gamma} \end{bmatrix}. \tag{6.62}$$

The integrations are related by

$$\iiint u(x, y, z)\,dx\,dy\,dz = \iiint u \det \mathbf{J} \; d\xi\,d\eta\,d\gamma. \tag{6.63}$$

Rectangular Parallelepiped A rectangular block in (x, y, z)-coordinates with corners at (x_i, y_i, z_i), $i = 1, 2, \cdots, 8$, is transformed into a cube with limits $\xi = \pm 1$, $\eta = \pm 1$ and $\gamma = \pm 1$ by

$$\begin{aligned} x &= \frac{1}{8} \sum_{i=1}^{8} [1 + \xi\xi_i + \eta\eta_i + \gamma\gamma_i]x_i, \\ y &= \frac{1}{8} \sum_{i=1}^{8} [1 + \xi\xi_i + \eta\eta_i + \gamma\gamma_i]y_i, \\ z &= \frac{1}{8} \sum_{i=1}^{8} [1 + \xi\xi_i + \eta\eta_i + \gamma\gamma_i]z_i, \end{aligned} \tag{6.64}$$

in which the relations between **x** and **ξ** are linear.

Hexahedron A general six-faced element is transformed to a cube by

$$x = \sum_{i=1}^{m} f_i(\xi, \eta, \gamma) x_i,$$
$$y = \sum_{i=1}^{m} f_i(\xi, \eta, \gamma) y_i, \qquad (6.65)$$
$$z = \sum_{i=1}^{m} f_i(\xi, \eta, \gamma) z_i.$$

For the element with each face bounded by straight lines (fig. 6.14a), we have $m = 8$ and f_i as given in (6.54). For an element with each face bounded by curves expressible as quadratic functions in x, y, z, (fig. 6.14b) we have $m = 20$ and f_i as given in (6.56). In both cases, the relations between **x** and **ξ** are nonlinear. The integrations are related by (6.63) with the appropriate limits:

$$\iiint u(x, y, z)dxdydz = \int_{-1}^{1}\int_{-1}^{1}\int_{-1}^{1} u(x, y, z)(\det \mathbf{J})d\xi d\eta d\gamma. \qquad (6.66)$$

Tetrahedron A tetrahedron with vertices at (x_i, y_i, z_i), $i = 1, 2, 3, 4$ (fig. 6.15a), is transformed to the right tetrahedron of fig. 6.13a by the linear transforma-

$$\begin{Bmatrix} 1 \\ x \\ y \\ z \end{Bmatrix} = \begin{bmatrix} 1 & 1 & 1 & 1 \\ x_1 & x_2 & x_3 & x_4 \\ y_1 & y_2 & y_3 & y_4 \\ z_1 & z_2 & z_3 & z_4 \end{bmatrix} \begin{Bmatrix} \zeta_1 \\ \zeta_2 \\ \zeta_3 \\ \zeta_4 \end{Bmatrix}, \qquad (6.67)$$

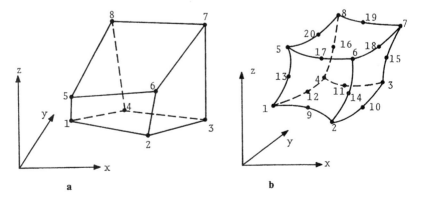

Fig. 6.14 Hexahedral elements.

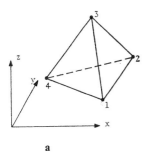 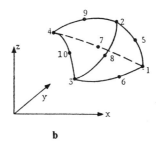

Fig. 6.15 Tetrahedral elements.

tion or inversely,

$$\begin{Bmatrix} \zeta_1 \\ \zeta_2 \\ \zeta_3 \\ \zeta_4 \end{Bmatrix} = \frac{1}{6V} \begin{bmatrix} a_1 & b_1 & c_1 & d_1 \\ a_2 & b_2 & c_2 & d_2 \\ a_3 & b_3 & c_3 & d_3 \\ a_4 & b_4 & c_4 & d_4 \end{bmatrix} \begin{Bmatrix} 1 \\ x \\ y \\ z \end{Bmatrix}, \tag{6.68}$$

where a_i, b_i, c_i, d_i, and V are given in terms of the coordinates of the vertices x_i, y_i, and z_i.

For the tetrahedron shown in fig. 6.15b, where the faces are bounded by curves expressible as quadratic functions, the transformation is

$$\begin{aligned} x &= \sum_{i=1}^{10} f_i(\zeta_1, \zeta_2, \zeta_3, \zeta_4) x_i, \\ y &= \sum_{i=1}^{10} f_i(\zeta_1, \zeta_2, \zeta_3, \zeta_4) y_i, \\ z &= \sum_{i=1}^{10} f_i(\zeta_1, \zeta_2, \zeta_3, \zeta_4) z_i, \end{aligned} \tag{6.69}$$

where the f_i are as given in (6.60). The relationship between the (x, y, z)- and (ξ, η, γ)-coordinates is nonlinear. The integrations are related by

$$\iiint u(x, y, z) dx dy dz = \int_0^1 \int_0^{1-\gamma} \int_0^{1-\eta-\gamma} u(x, y, z)(\det \mathbf{J}) d\xi d\eta d\gamma. \tag{6.70}$$

6.5 Numerical Integration

If $u(\mathbf{x})$ is approximated by the interpolation functions given in (6.1), then one can approximate the integral of $u(\mathbf{x})$ by

$$\int u(x)dx = \sum_{i=1}^{m} W_i q_i + E, \qquad (6.71)$$

where

$$W_i = \int f_i(x)dx$$

is called the *weighting factor*, and the error E of the integral,

$$E = \int [u(x) - \tilde{u}(x)] \, dx, \qquad (6.72)$$

can be estimated.

In the ordinary rules for numerical integration—generally called *quadrature formulas*—the q_is are the values of u at some discrete points in the domain. These points may or may not represent nodes. Two quadrature formulas for one-dimensional space that have been found useful in the finite-element formulation are *Newton-Cotes* quadrature and *Gaussian* quadrature. The former is based on the zeroth order Hermite interpolation (i.e., Lagrange's interpolation), while the latter is based on the first order Hermite Interpolation.

6.5.1 Newton-Cotes Quadrature

If a function $u(\xi)$ is approximated by $\tilde{u}(\xi)$ according to (6.5), then the integral of $u(\xi)$ may be approximated by

$$\int_{-1}^{1} \tilde{u}(\xi)d\xi = \sum_{i=1}^{m} W_i u(\xi_i), \qquad (6.73)$$

where

$$W_i = \int_{-1}^{1} f_i(\xi)d\xi.$$

In Newton-Cotes quadrature the m discrete points along ξ are equally spaced and the integration limits are the discrete points at the two ends. Therefore, it is sometimes called closed-end integration. For $m = 2$ we have

$$\int_{-1}^{1} u(\xi)d\xi = u(-1) + u(1) - \tfrac{2}{3} u''(\zeta), \qquad (6.74)$$

where ζ is some value between -1 and $+1$. Thus the term $-2/3\, u''(\zeta)$ gives the order of magnitude of the numerical integration error. This integration scheme is clearly nothing more than the so-called trapezoidal rule: For $m = 3$,

$$\int_{-1}^{1} u(\xi)d\xi = \tfrac{1}{3}[u(-1) + 4u(0) + u(1)] - \tfrac{1}{90} u^{(4)}(\zeta); \qquad (6.75)$$

for $m = 4$,

$$\int_{-1}^{1} u(\xi)d\xi = \tfrac{1}{4}[u(-1) + 3u(-\tfrac{1}{3}) + 3u(\tfrac{1}{3}) + u(1)] - \frac{2}{405} u^{(4)}(\zeta); \qquad (6.76)$$

and for $m = 5$,

$$\int_{-1}^{1} u(\xi)d\xi = \tfrac{1}{45}[7u(-1) + 32u(-\tfrac{1}{2}) + 12u(0) + 32u(\tfrac{1}{2}) + 7u(1)]$$
$$- \frac{1}{15120} u^{6}(\zeta). \qquad (6.77)$$

It is well known that in using the Newton-Cotes quadrature it is advantageous to adopt *odd* numbers of discrete points. For example, in the above the two cases $m = 3$ (Simpson's rule) and $m = 4$ yield integration errors of the same order.

6.5.2 Gaussian Quadrature

If first-order m-point Hermite interpolation is used, then

$$\int_{-1}^{1} u(\xi)d\xi = \sum_{i=1}^{m} W_i u(\xi_i) + \sum_{i=1}^{m} W'_i u'(\xi_i) + E, \qquad (6.78)$$

where

$$W_i = \int_{-1}^{1} H_{0i}^{(1)}(\xi)d\xi, \qquad W'_i = \int_{-1}^{1} H_{1i}^{(1)}(\xi)d\xi,$$

and E is the error.

If the points $\xi_1, \cdots, \xi_m$ are so chosen that the weighting factors W'_i all vanish, then the resulting integration formulas, which are called the Gaussian quadratures, are of the same form as the Newton-Cotes quadratures. Such quadratures will integrate a polynomial of degree $2m - 1$ correctly. For example, when $m = 1$,

$$\int_{-1}^{1} u(\xi)d\xi = 2u(0) + \frac{u''(\zeta)}{3}.$$

The stations for the Gaussian quadrature and the corresponding weighting factors for $m = 2, \cdots, 8$ are given in table 6.1. The construction of this table

Table 6.1 Stations and weighting factors for Gaussian quadrature,

$$\int_{-1}^{1} u(\xi)d\xi \approx \sum_{i=1}^{m} W_i u(\xi_i).$$

$\pm \xi_i$ (Stations)				W_i (Weighting Factors)		
			$m = 2$			
0.57735	02691	89626		1.00000	00000	00000
			$m = 3$			
0.00000	00000	00000		0.88888	88888	88889
0.77459	66692	41483		0.55555	55555	55556
			$m = 4$			
0.33998	10435	84856		0.65214	51548	62546
0.86113	63115	94053		0.34785	48451	37454
			$m = 5$			
0.00000	00000	00000		0.56888	88888	88889
0.53846	93101	05683		0.47862	86704	99366
0.90617	98459	38664		0.23692	68850	56189
			$m = 6$			
0.23861	91860	83197		0.46791	39345	72691
0.66120	93864	66265		0.36076	15730	48139
0.93246	95142	03152		0.17132	44923	79170
			$m = 7$			
0.00000	00000	00000		0.41795	91836	73469
0.40584	51513	77397		0.38183	00505	05119
0.74153	11855	99394		0.27970	53914	89277
0.94910	79123	42759		0.12948	49661	68870
			$m = 8$			
0.18343	46424	95650		0.36268	37833	78362
0.52553	24099	16329		0.31370	66458	77887
0.79666	64774	13627		0.22238	10344	53374
0.96028	98564	97536		0.10122	85362	90376

Source: P. Davis and P. Rabinowitz. Abscissas and Weights for Gaussian Quadratures of High Order. *J. Res. Nat. Bur. Stand.* 56 (1956): RP2645.

can be found in Davis and Rabinowitz (1956). Note that, for the same m, the Gaussian quadrature is much more accurate than the Newton-Cotes quadrature. It is seen that the error in using three integration stations in Gaussian quadrature is the same as that in using five integration stations in Newton-Cotes quadrature. Thus Gaussian quadrature is comparatively a more efficient integration scheme.

6.5.3 Numerical Integration in Two Dimensions

If the domain of integration is rectangular, Newton-Cotes or Gaussian quadratures can be immediately adopted:

$$\int_{-1}^{1}\int_{-1}^{1} u(\xi,\eta)d\xi d\eta = \int_{-1}^{1}\sum_{i=1}^{m} u(\xi_i,\eta)W_i d\eta = \sum_{i=1}^{m}\sum_{j=1}^{m} u(\xi_i,\eta_j)W_i W_j. \qquad (6.79)$$

For Gaussian quadrature, the integration of a polynomial $\sum \alpha_{ij}\xi^i\eta^j$, $i,j \leq 2m - 1$ will be done correctly. For example, fig. 6.16 illustrates a nine-point Gaussian quadrature over a square area. Here i and j are from 1 to 3; the factors W_i and W_j are obtained from table 6.1.

Integration over a general quadrilateral area can be easily derived by using (6.44) with the quadrilateral element in (x, y)-coordinates being transformed into a square element in (ξ, η)-coordinates (6.43). Furthermore, a triangle can be treated as a degenerate quadrilateral in either of two ways as shown in fig.

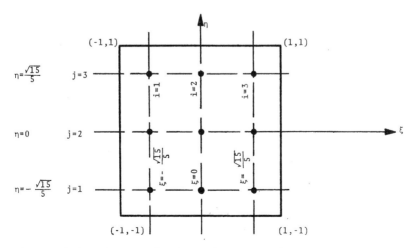

Fig. 6.16 A nine-point square Gaussian quadrature.

6.17; hence, an integral over a triangular area can be obtained by Newton-Cotes or Gaussian quadrature. However, it is not advisable to use integration formulas involving many integration stations in a degenerate case because singularities will occur in the transformation—and this will lead to the loss of accuracy.

The integration formula over an arbitrary triangle can be expressed more conveniently by using triangular coordinates. Table 6.2 lists the stations, the weighting factors, and the order of magnitude of the integration error for an integration scheme known as Hammer's formula (Hammer and Stroud 1958; Hammer, Marlowe, and Stroud 1956).

A useful formula, derived from (6.49), for integration over a triangular area in terms of triangular coordinates is

$$\int_\Delta \zeta_1^\alpha \zeta_2^\beta \zeta_3^\gamma \, dx dy = \det \mathbf{J} \int_0^1 \int_0^{1-\eta} \xi^\alpha \eta^\beta (1 - \xi - \eta)^\gamma d\xi d\eta$$
$$= 2\Delta \frac{\alpha! \beta! \gamma!}{(\alpha + \beta + \gamma + 2)!}. \qquad (6.80)$$

Table 6.2 Stations, weighting factors, and error for Hammer's formula.

Total Number of Integration Points m	Integration Point i	Position in Triangular Coordinates ξ_1, ξ_2, ξ_3			Weighting Factor (W_i/Triangle Area)	Error[a]
1	1	.3333333	.3333333	.3333333	1.0000000	$O(h^2)$
3	1	.5000000	.5000000	.0000000	.3333333	
	2	.0000000	.5000000	.5000000	.3333333	$O(h^3)$
	3	.5000000	.0000000	.5000000	.3333333	
4	1	.3333333	.3333333	.3333333	.5625000	
	2	.7333333	.1333333	.1333333	.5208333	$O(h^4)$
	3	.1333333	.7333333	.1333333	.5208333	
	4	.1333333	.1333333	.7333333	.5208333	
7	1	.3333333	.3333333	.3333333	.2250000	
	2	.79742699	.10128651	.10128651	.12593918	
	3	.10128651	.79742699	.10128651	.12593918	
	4	.10128651	.10128651	.79742699	.12593918	$O(h^6)$
	5	.05971587	.47014206	.47014206	.13239415	
	6	.47014206	.05971587	.47014206	.13239415	
	7	.47014206	.47014206	.05971587	.13239415	

[a] Here $O(h^i)$ means that the numerical integration is exact if the order of the integrated function is up to the $(i - 1)$th power of $\xi_1, \xi_2,$ and ξ_3.

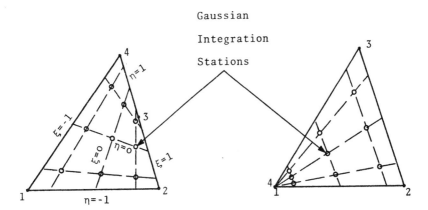

Fig. 6.17 Triangles as degenerate quadrilaterals.

For example,

$$\int_\Delta \zeta_1\zeta_2\zeta_3\, dx dy = 2\Delta \frac{1!\,1!\,1!}{5!}$$

$$= \frac{\Delta}{60}.$$

6.5.4 Numerical Integration in Three Dimensions

If the domain is a cube, similar to (6.79), Newton-Cotes or Gaussian quadrature can be written down exactly as in (6.79):

$$\int_{-1}^{1}\int_{-1}^{1}\int_{-1}^{1} u(\xi,\eta,\gamma)d\xi d\eta d\gamma = \sum_{i=1}^{m}\sum_{j=1}^{m}\sum_{k=1}^{m} W_i W_j W_k u(\xi_i,\eta_j,\gamma_k). \tag{6.81}$$

For Gaussian quadrature the location of ξs, ηs, γs, and the corresponding Ws are given in Table 6.1, and (6.81) will integrate a polynomial $\sum \alpha_{ijk}\xi^i\eta^j\gamma^k$ ($i,j,k \leq 2m-1$) correctly. More efficient integration rules can be achieved by designing the integration to integrate a complete polynomial $\sum \alpha_{ijk}\xi^i\eta^j\gamma^k$, where $i+j+k$ is less than or equal to a fixed integer (Irons 1971), and the integration takes the form

$$\int_{-1}^{1}\int_{-1}^{1}\int_{-1}^{1} u(\xi,\eta,\gamma)d\xi d\eta d\gamma = A_1 u(0,0,0) \quad \text{(1 term)}$$
$$+ B_6\{u(-b,0,0) + u(b,0,0) + u(0,-b,0) + \cdots \quad \text{(6 terms)}\}$$

$$+ C_8\{u(-c, -c, -c) + u(c, -c, -c) + u(c, c, -c) + \cdots \quad (6.82)$$
$$+ u(-c, -c, c) + \cdots \quad (8 \text{ terms})\}$$
$$+ D_{12}\{u(-d, -d, 0) + u(d, -d, 0) + u(-d, 0, -d)$$
$$+ u(d, 0, -d) + u(0, -d, -d) + \cdots \quad (12 \text{ terms})\};$$

here the weighting factors A_1, B_6, $\cdots$, and the coordinates b, c, and d are those listed in Table 6.3. The comparison of the error of the integration with that of Gaussian quadrature is given in Table 6.4 (Irons 1971).

The integration formula for a hexahedron can be obtained by using (6.66). For integration over a tetrahedron, the polynomial terms may be integrated by

$$\int_V \zeta_1^\alpha \zeta_2^\beta \zeta_3^\theta \zeta_4^\nu \, dxdydz = 6V \int_0^1 \int_0^{1-\gamma} \int_0^{1-\eta-\gamma} \xi^\alpha \eta^\beta \gamma^\theta (1 - \xi - \eta - \gamma)^\nu d\xi d\eta d\gamma$$
$$= 6V \frac{\alpha! \, \beta! \, \theta! \, \nu!}{(\alpha + \beta + \theta + \nu + 2)!}. \quad (6.83)$$

Table 6.3 Weighting factors and coordinates for (6.82).

Number of Integration Points m	Weighting Factors[a]	Coordinates	Error
1	$A_1 = 8$		$0\,(h^2)$
6	$B_6 = 8/6$	$b = 1$	$0\,(h^4)$
14	$B_6 = 0.886426593$	$b = 0.7958622426$	$0\,(h^6)$
	$C_8 = 0.335180055$	$c = 0.758786911$	
27	$A_1 = 0.788073483$		$0\,(h^8)$
	$B_6 = 0.499369002$	$b = 0.848418011$	
	$C_8 = 0.478508449$	$c = 0.652816472$	
	$D_{12} = 0.032303742$	$d = 1.106412899$	

[a]Unlisted weighting factors are zero.

Table 6.4 Errors in the numerical integration (6.82) and in Gaussian quadrature. (G in the first column denotes Gaussian quadrature.)

Number of integration points	Quartic Terms		Sixth-Degree Terms			Eighth-Degree Terms			
	x^4	x^2y^2	x^6	x^4y^2	$x^2y^2z^2$	x^8	x^6y^2	x^4y^4	$x^4y^2z^2$
6	1.1	−0.89							
8G	−0.71	0	−0.85	−0.24	0				
14	0	0	−0.18	−0.02	0.22				
27G	0	0	0	0	0	0.09	0.04	0.10	−0.05
27G	0	0	−0.18	0	0	−0.31	−0.06	0	0
64G	0	0	0	0	0	−0.05	0	0	0

For example,

$$\int_V \zeta_1\zeta_2\zeta_3\zeta_4 dx dy dz = \frac{6V}{7!} = \frac{V}{840}.$$

6.6 Higher-Order Two-Dimensional Elements

As can be seen in secs. 6.2–6.4, the order of a complete polynomial which can be exactly represented by the interpolation functions is directly related to the number of nodal points of the elements. The four-node [(6.22) and fig. 6.2] or three-node [(6.31) and fig. 6.5a] interpolation $\tilde{u}$ can represent exactly any linear function of ξ and η. The eight-node [(6.24) and fig. 6.3)] or the six-node [(6.32) and fig. 6.5b] interpolation can exactly represent any quadratic function of ξ and η. That is to say, the highest order of accuracy in the representation of a function by the $\tilde{u}$ is directly related to the number of nodal points used in the interpolation. (The accuracy is also affected by the transformation, as shown by way of the example in sec. 6.3.) The procedures for constructing the element matrices in high-order interpolation is similar to that outlined in sec. 2.6. In general, the displacement vector for a plane-stress or -strain problem can be written in the form

$$\mathbf{u} = \begin{Bmatrix} u \\ v \end{Bmatrix} = \begin{bmatrix} \mathbf{f}^T & 0 \\ 0 & \mathbf{f}^T \end{bmatrix} \begin{Bmatrix} \mathbf{q}_u \\ \mathbf{q}_v \end{Bmatrix} = \mathbf{D}^* \begin{Bmatrix} \mathbf{q}_u \\ \mathbf{q}_v \end{Bmatrix}, \tag{6.84a}$$

where $\mathbf{f}$ is the interpolation function given in sec. 6.3, and $\mathbf{q}_u$ and $\mathbf{q}_v$ are the vectors for the nodal values of u and v, respectively. Using strain-displacement and stress-strain relations, we can evaluate

$$\mathbf{e}_{3\times 1} = \begin{Bmatrix} e_x \\ e_y \\ e_{xy} \end{Bmatrix} = \mathbf{E}^* \begin{Bmatrix} \mathbf{q}_u \\ \mathbf{q}_v \end{Bmatrix},$$

$$\boldsymbol{\sigma} = \mathbf{Ce} = \mathbf{CE}^* \begin{Bmatrix} \mathbf{q}_u \\ \mathbf{q}_v \end{Bmatrix},$$
(6.84b)

Where $\mathbf{C}$ is the elastic-constant matrix defined in sec. 5.1 and

$$\mathbf{E}^* = \begin{bmatrix} \dfrac{\partial \mathbf{f}^T}{\partial x} & 0 \\ 0 & \dfrac{\partial \mathbf{f}^T}{\partial y} \\ \dfrac{\partial \mathbf{f}^T}{\partial y} & \dfrac{\partial \mathbf{f}^T}{\partial x} \end{bmatrix}. \tag{6.85}$$

The interpolation functions are in terms of ξ and η (or ζ_1, ζ_2, and ζ_3 in triangular coordinates). From (6.37), we have

$$\left\{ \begin{array}{c} \dfrac{\partial \mathbf{f}^T}{\partial x} \\ \dfrac{\partial \mathbf{f}^T}{\partial y} \end{array} \right\} = \mathbf{J}^{-1} \left\{ \begin{array}{c} \dfrac{\partial \mathbf{f}^T}{\partial \xi} \\ \dfrac{\partial \mathbf{f}^T}{\partial \eta} \end{array} \right\}, \tag{6.86}$$

where

$$\mathbf{J}^{-1} = \left[\begin{array}{cc} \dfrac{\partial y}{\partial \eta} & -\dfrac{\partial y}{\partial \xi} \\ -\dfrac{\partial x}{\partial \eta} & \dfrac{\partial x}{\partial \xi} \end{array} \right] \dfrac{1}{\det \mathbf{J}}. \tag{6.87}$$

The element matrices associated with the generalized-coordinate vector $\mathbf{q}^* = \{{\mathbf{q}_r^*}\}$ are evaluated according to (5.46) and (5.47):

$$\mathbf{k}_n^* = \iint t\mathbf{E}^{*T}\mathbf{C}\mathbf{E}^* \det \mathbf{J}\, d\xi d\eta, \tag{6.88}$$

$$\mathbf{Q}_n^* = \iint t\mathbf{D}^{*T}\mathbf{F} \det \mathbf{J}\, d\xi d\eta + \int_{(\partial A_e)_n} t\mathbf{D}^{*T}\bar{\mathbf{T}} \dfrac{ds}{d\zeta} d\zeta, \tag{6.89}$$

where the limits of integration will depend on the shape of the elements, as will be discussed later, and where

$$ds = [(dx)^2 + (dy)^2]^{1/2}, \tag{6.90}$$

which can be evaluated from (6.43) or (6.48) for different cases; ζ is the arc length measured in (ξ, η)-coordinates. Very often, $ds/d\zeta$ is taken to be a constant (though this should be recognized as an approximation). The integrations in (6.88) and (6.89) are generally carried out numerically.

The element-stiffness matrix $\mathbf{k}_n$ and the loading vector $\mathbf{Q}_n$ associated with the generalized coordinate vector

$$\mathbf{q}^T = \{u_1 \quad v_1 \quad u_2 \quad v_2 \cdots\} \tag{6.91}$$

can be obtained by rearranging the components of $\mathbf{k}_n^*$ and $\mathbf{Q}_n^*$. Let the components of $\mathbf{q}$ and $\mathbf{q}^*$ be related by

$$q_i = q_{I(i)}^*, \tag{6.92}$$

where

$$I(2i - 1) = i,$$
$$I(2i) = i + \text{NNPE}, \tag{6.93}$$

in which NNPE is the number of nodes of the element and i ranges from 1 to NNPE. Then the components of $\mathbf{k}_n$ and $\mathbf{Q}_n$ are

$$k_{ij} = k^*_{I(i)I(j)}, \tag{6.94}$$
$$Q_i = Q^*_{I(i)}, \quad i,j = 1, 2, \cdots, \text{NNPE}.$$

Depending on the actual shape of the element, the transformation from (x, y)-coordinates to (ξ, η)-coordinates is given in (6.43) or (6.48). The order of the transformation, that is, the number of nodal points required for the transformation of the coordinates, does *not* have to be the same as the order of the interpolation function. In practice, one uses less or the same order for the transformation as that for interpolation. For example, one may have an element in the shape of fig. 6.6 or 6.7a, for which only the coordinates of the four corner points are needed for the transformation; but the nodal points for the interpolation functions are the corners and the points on each side, for which the interpolation functions are given by (6.25) for the eight-node element and (6.28) for the twelve-node element. Although in principle one could have a higher-order transformation than the interpolation, in view of the error introduced by the transformation, one should not use this stratagem in practice. For example, if one has an element of the shape of fig. 6.7c, and if one is going to use eight-node (6.25) or even four-node (6.23) interpolation, one should approximate the element by the shape of fig. 6.7b or a. This will not cause any loss of accuracy in representing a function u in the (x, y)-coordinates and will save a lot of time in evaluating $\mathbf{J}^{-1}$, det $\mathbf{J}$, and $ds/d\zeta$ in (6.86) through (6.90).

6.6.1 Quadrilateral Element

In the evaluation of the element matrices, the integrations in (6.88) and (6.89) become

$$\mathbf{k}^*_n = \int_{-1}^{1}\int_{-1}^{1} t\,(\det \mathbf{J})\,\mathbf{E}^{*\mathrm{T}}\mathbf{C}\mathbf{E}^* d\xi d\eta,$$
$$\mathbf{Q}^*_n = \int_{-1}^{1}\int_{-1}^{1} t\,(\det \mathbf{J})\,\mathbf{D}^{*\mathrm{T}}\mathbf{\bar{F}} d\xi d\eta + \int_{(\partial A_\sigma)_n} t\,\mathbf{D}^{*\mathrm{T}}\mathbf{\bar{T}}\frac{ds}{d\zeta} d\zeta. \tag{6.95}$$

The line integral over $(\partial A_\sigma)_n$ is zero if there are no tractions prescribed over the element boundaries and the line integral has integration limits $-1, +1$, with ζ the absolute value of ξ or η (depending upon which side of the element the tractions are prescribed). To integrate (6.95) numerically, we use (6.79) for the area integral and (6.78) for the line integral:

$$\mathbf{k}_n^* = \sum_{i,l} (t\, \mathbf{E}^{*T}\mathbf{C}\mathbf{E}^* \det \mathbf{J})_{il} W_i W_l,$$
$$\mathbf{Q}_n^* = \sum_{i,l} (t\, \mathbf{D}^{*T}\bar{\mathbf{F}} \det \mathbf{J})_{il} W_i W_l + \sum_i \left(t \frac{ds}{d\zeta} \mathbf{D}^{*T}\bar{\mathbf{T}} \bigg|_{\text{side }j} \right)_i W_i, \tag{6.96}$$

in which the range of the summations is from one to the number of integration stations, $(\ )_{il}$ denotes the value of the quantity in parentheses at the integration stations (ξ_i and η_l), and the W_i are the corresponding weighting factors. In the second summation in (6.96), $(\ )|_{\text{side }j}$ denotes that side j has prescribed tractions and the quantity is being evaluated on side j. The location of the integration points and the weighting factors for various orders of accuracy are given in table 6.1.

In the case that the element is a rectangle, with homogeneous material within the element, both $\mathbf{k}_n^*$ and $\mathbf{Q}_n^*$ can be integrated out explicitly. The results are given in app. B.

6.6.2 Triangular Element

For triangular elements, (6.88) and (6.89) become

$$\mathbf{k}_n^* = \int_0^1 \int_0^{1-\eta} t\, (\det \mathbf{J})\, \mathbf{E}^{*T}\mathbf{C}\mathbf{E}^* d\xi d\eta, \tag{6.97}$$

$$\mathbf{Q}_n^* = \int_0^1 \int_0^{1-\eta} t\, (\det \mathbf{J})\, \mathbf{D}^{*T}\bar{\mathbf{F}} d\xi d\eta + \int_{(\partial A_\sigma)_n} t \frac{ds}{d\zeta} \mathbf{D}^{*T}\bar{\mathbf{T}} d\zeta. \tag{6.98}$$

In evaluating $\mathbf{E}^*$ and $\det \mathbf{J}$, one may directly use the expression given in (6.86) and (6.87). However, by noting that $\zeta_1 = \xi$, $\zeta_2 = \eta$, and $\zeta_3 = 1 - \xi - \eta$, we have

$$\left\{ \begin{array}{c} \dfrac{\partial \mathbf{f}^T}{\partial x} \\ \dfrac{\partial \mathbf{f}^T}{\partial y} \end{array} \right\} = \mathbf{J}^{-1} \left\{ \begin{array}{cc} \dfrac{\partial \mathbf{f}^T}{\partial \zeta_1} - \dfrac{\partial \mathbf{f}^T}{\partial \zeta_3} \\ \dfrac{\partial \mathbf{f}^T}{\partial \zeta_2} - \dfrac{\partial \mathbf{f}^T}{\partial \zeta_3} \end{array} \right\}, \tag{6.99}$$

in which

$$\mathbf{J}^{-1} = \begin{bmatrix} \dfrac{\partial \xi}{\partial x} & \dfrac{\partial \eta}{\partial x} \\ \dfrac{\partial \xi}{\partial y} & \dfrac{\partial \eta}{\partial y} \end{bmatrix}^{-1}, \qquad (6.100)$$

or

$$\mathbf{J}^{-1} = \begin{bmatrix} \dfrac{\partial y}{\partial \zeta_2} - \dfrac{\partial y}{\partial \zeta_3} & \dfrac{\partial y}{\partial \zeta_3} - \dfrac{\partial y}{\partial \zeta_1} \\ \dfrac{\partial x}{\partial \zeta_3} - \dfrac{\partial x}{\partial \zeta_2} & \dfrac{\partial x}{\partial \zeta_1} - \dfrac{\partial x}{\partial \zeta_3} \end{bmatrix} \dfrac{1}{\det \mathbf{J}}, \qquad (6.101)$$

and

$$\det \mathbf{J} = \left(\dfrac{\partial x}{\partial \zeta_1} - \dfrac{\partial x}{\partial \zeta_3} \right)\left(\dfrac{\partial y}{\partial \zeta_2} - \dfrac{\partial y}{\partial \zeta_3} \right) - \left(\dfrac{\partial x}{\partial \zeta_2} - \dfrac{\partial x}{\partial \zeta_3} \right)\left(\dfrac{\partial y}{\partial \zeta_1} - \dfrac{\partial y}{\partial \zeta_3} \right). \qquad (6.102)$$

The matrix $\mathbf{J}^{-1}$ can be evaluated directly from the transformation of (x, y)-into $(\zeta_1, \zeta_2, \zeta_3)$-coordinates.

To evaluate $\mathbf{k}_n^*$ and $\mathbf{Q}_n^*$ numerically, we have

$$\mathbf{k}_n^* = \sum_i (t\, \mathbf{E}^{*T}\mathbf{C}\mathbf{E}^* \det \mathbf{J})_i W_i, \qquad (6.103)$$

$$\mathbf{Q}_n^* = \sum_i (t\, \mathbf{D}^{*T}\bar{\mathbf{F}} \det \mathbf{J})_i W_i + \sum_i' \left(t\dfrac{ds}{d\zeta} \mathbf{D}^{*T}\bar{\mathbf{T}} \bigg|_{\text{side}\,j} \right)_i \dfrac{W_i'}{2}, \qquad (6.104)$$

in which $\sum_i$ denotes summing over all the integration stations of the triangle, $(\ )_i$ denotes the value of the quantity in parentheses at integration stations, and the W_i are the corresponding weighting factors of table 6.2. $\sum_i'$ sums over the integration stations for the line integral with the weighting factors W_i' of table 6.1. The factor $1/2$ in (6.104) is due to the fact that the weighting factors in table 6.1 are based on the integration from -1 to $+1$, an interval of length 2.

In the case where the triangle has all the edges straight, integration of $\mathbf{k}_n^*$ and $\mathbf{Q}_n^*$ in (6.97) and (6.98) can be carried out analytically without much trouble. Explicit expressions for $\mathbf{k}_n^*$ and $\mathbf{Q}_n^*$ are given in app. C for the case of homogeneous material within the element.

6.7 Higher-Order Three-Dimensional Elements

The procedures for constructing three-dimensional elements are very much the same as those for two dimensions. The displacement, strain, and stress vectors are written as

$$\mathbf{u} = \begin{Bmatrix} u \\ v \\ w \end{Bmatrix} = \mathbf{D}^* \begin{Bmatrix} \mathbf{q}_u \\ \mathbf{q}_v \\ \mathbf{q}_w \end{Bmatrix}, \tag{6.105}$$

$$\mathbf{e} = \begin{Bmatrix} e_x \\ e_y \\ \vdots \\ e_{yz} \end{Bmatrix} = \mathbf{E}^* \begin{Bmatrix} \mathbf{q}_u \\ \mathbf{q}_v \\ \mathbf{q}_w \end{Bmatrix}, \tag{6.106}$$

$$\boldsymbol{\sigma} = \begin{Bmatrix} \sigma_x \\ \sigma_y \\ \vdots \\ \sigma_{yz} \end{Bmatrix} = \mathbf{CE}^* \begin{Bmatrix} \mathbf{q}_u \\ \mathbf{q}_v \\ \mathbf{q}_w \end{Bmatrix}, \tag{6.107}$$

in which $\mathbf{C}$ is the elastic-coefficient matrix given in sec. 5.1, and

$$\mathbf{D}^* = \begin{bmatrix} \mathbf{f}^T & & 0 \\ & \mathbf{f}^T & \\ 0 & & \mathbf{f}^T \end{bmatrix}, \tag{6.108}$$

$$\mathbf{E}^* = \begin{bmatrix} \dfrac{\partial \mathbf{f}^T}{\partial x} & 0 & 0 \\ 0 & \dfrac{\partial \mathbf{f}^T}{\partial y} & 0 \\ 0 & 0 & \dfrac{\partial \mathbf{f}^T}{\partial z} \\ \dfrac{\partial \mathbf{f}^T}{\partial y} & \dfrac{\partial \mathbf{f}^T}{\partial x} & 0 \\ \dfrac{\partial \mathbf{f}^T}{\partial z} & 0 & \dfrac{\partial \mathbf{f}^T}{\partial x} \\ 0 & \dfrac{\partial \mathbf{f}^T}{\partial z} & \dfrac{\partial \mathbf{f}^T}{\partial y} \end{bmatrix}, \tag{6.109}$$

and

$$\begin{Bmatrix} \dfrac{\partial \mathbf{f}^T}{\partial x} \\ \dfrac{\partial \mathbf{f}^T}{\partial y} \\ \dfrac{\partial \mathbf{f}^T}{\partial z} \end{Bmatrix} = \mathbf{J}^{-1} \begin{Bmatrix} \dfrac{\partial \mathbf{f}^T}{\partial \xi} \\ \dfrac{\partial \mathbf{f}^T}{\partial \eta} \\ \dfrac{\partial \mathbf{f}^T}{\partial \gamma} \end{Bmatrix}, \tag{6.110}$$

where

$$\mathbf{J} = \begin{bmatrix} \dfrac{\partial x}{\partial \xi} & \dfrac{\partial y}{\partial \xi} & \dfrac{\partial z}{\partial \xi} \\ \dfrac{\partial x}{\partial \eta} & \dfrac{\partial y}{\partial \eta} & \dfrac{\partial z}{\partial \eta} \\ \dfrac{\partial x}{\partial \gamma} & \dfrac{\partial y}{\partial \gamma} & \dfrac{\partial z}{\partial \gamma} \end{bmatrix}. \tag{6.111}$$

6.7.1 Hexahedral Element

The interpolation functions are given in (6.54) and (6.56). The element matrices are similar to those of (6.95):

$$\mathbf{k}_n^* = \int_{-1}^{1}\int_{-1}^{1}\int_{-1}^{1} \mathbf{E}^{*\mathrm{T}}\mathbf{C}\mathbf{E}^* \det \mathbf{J}\, d\xi d\eta d\gamma, \tag{6.112}$$

$$\mathbf{Q}_n^* = \int_{-1}^{1}\int_{-1}^{1}\int_{-1}^{1} \mathbf{D}^{*\mathrm{T}}\bar{\mathbf{F}} \det \mathbf{J}\, d\xi d\eta d\gamma + \int_{-1}^{1}\int_{-1}^{1} \mathbf{D}^{*\mathrm{T}}\bar{\mathbf{T}} \left.\frac{dS_j}{d\Pi}\right|_{S_j} d\alpha d\beta, \tag{6.113}$$

where S_j is the surface j of the element where the surface tractions are prescribed, $d\alpha d\beta$ denotes $d\eta d\gamma$ for the faces with $\xi = \pm 1$, or $d\xi d\gamma$ for the faces $\eta = \pm 1$, or $d\xi d\eta$ for the faces with $\gamma = \pm 1$, and $dS_j/d\Pi$ is evaluated from (6.65), the transformation of (x, y, z)- to (ξ, η, γ)-coordinates:

$$\frac{dS_j}{d\Pi} = \left\{\left(\frac{\partial y}{\partial \alpha}\frac{\partial z}{\partial \beta} - \frac{\partial z}{\partial \alpha}\frac{\partial y}{\partial \beta}\right)^2 + \left(\frac{\partial z}{\partial \alpha}\frac{\partial x}{\partial \beta} - \frac{\partial x}{\partial \alpha}\frac{\partial z}{\partial \beta}\right)^2 \right. \\ \left. + \left(\frac{\partial x}{\partial \alpha}\frac{\partial y}{\partial \beta} - \frac{\partial y}{\partial \alpha}\frac{\partial x}{\partial \beta}\right)^2\right\}^{1/2}, \tag{6.114}$$

with x, y, z evaluated on the face j.

To evaluate (6.112) and (6.113) numerically, we may use (6.81) for the volume integral and (6.79) for the surface integral:

$$\mathbf{k}_n^* = \sum_{i,k,l} (\mathbf{E}^{*\mathrm{T}}\mathbf{C}\mathbf{E}^* \det \mathbf{J})_{ikl} W_i W_k W_l,$$

$$\mathbf{Q}_n^* = \sum_{i,k,l} (\mathbf{D}^{*\mathrm{T}}\bar{\mathbf{F}} \det \mathbf{J})_{ikl} W_i W_k W_l + \sum_{i,k} \left(\mathbf{D}^{*\mathrm{T}}\bar{\mathbf{T}}\frac{dS_j}{d\Pi}\bigg|_{S_j}\right)_{ik} W_i W_k, \tag{6.115}$$

where $(\ )_{ikl}$ denotes the value of the enclosed quantity at the integration stations ξ_i, η_j, and γ_k, and the W_i are the corresponding weighting factors. The location of the integration points and the W_i by Gaussian quadrature for various orders of accuracy are given in table 6.1.

It is perhaps more efficient to use the integration rules given in (6.82) for the volume integral. For example, if the rule for 14 integration points is chosen, we have

$$\mathbf{k}_n^* = B_6\{(\det \mathbf{J} \cdot \mathbf{E}^{*T}\mathbf{CE}^*)_{(-b,0,0)} + (\det \mathbf{J} \cdot \mathbf{E}^{*T}\mathbf{CE})_{(b,0,0)} + \cdots \text{ (6 terms)}\}$$
$$+ C_8\{(\det \mathbf{J} \cdot \mathbf{E}^{*T}\mathbf{CE}^*)_{(-c,-c,-c)}$$
$$+ (\det \mathbf{J} \cdot \mathbf{E}^{*T}\mathbf{CE}^*)_{(-c,-c,c)} + \cdots \text{ (8 terms)}\}, \qquad (6.116)$$

where B_6, C_8, b, and c are given in table 6.3. It should be noted that the above integration has the same order of accuracy as that of Gaussian quadrature with $m = 3$, which has 27 integration points.

6.7.2 Tetrahedral Element

The interpolation function f_i is given in (6.58) and (6.68). Since $\mathbf{f}$ is expressed in terms of $\zeta_1 (= \xi)$, $\zeta_2 (= \eta)$, $\zeta_3 (= \gamma)$ and $\zeta_4 (= 1 - \xi - \eta - \gamma)$ it is more convenient to express (6.110) and (6.111) in terms of these variables,

$$\left\{ \begin{array}{c} \dfrac{\partial \mathbf{f}^T}{\partial x} \\ \dfrac{\partial \mathbf{f}^T}{\partial y} \\ \dfrac{\partial \mathbf{f}^T}{\partial z} \end{array} \right\} = \mathbf{J}^{-1} \left\{ \begin{array}{c} \dfrac{\partial \mathbf{f}^T}{\partial \zeta_1} - \dfrac{\partial \mathbf{f}^T}{\partial \zeta_4} \\ \dfrac{\partial \mathbf{f}^T}{\partial \zeta_2} - \dfrac{\partial \mathbf{f}^T}{\partial \zeta_4} \\ \dfrac{\partial \mathbf{f}^T}{\partial \zeta_3} - \dfrac{\partial \mathbf{f}^T}{\partial \zeta_4} \end{array} \right\}, \qquad (6.117)$$

where

$$\mathbf{J} = \begin{bmatrix} \dfrac{\partial x}{\partial \zeta_1} - \dfrac{\partial x}{\partial \zeta_4} & \dfrac{\partial y}{\partial \zeta_1} - \dfrac{\partial y}{\partial \zeta_4} & \dfrac{\partial z}{\partial \zeta_1} - \dfrac{\partial z}{\partial \zeta_4} \\ \dfrac{\partial x}{\partial \zeta_2} - \dfrac{\partial x}{\partial \zeta_4} & \dfrac{\partial y}{\partial \zeta_2} - \dfrac{\partial y}{\partial \zeta_4} & \dfrac{\partial z}{\partial \zeta_2} - \dfrac{\partial z}{\partial \zeta_4} \\ \dfrac{\partial x}{\partial \zeta_3} - \dfrac{\partial x}{\partial \zeta_4} & \dfrac{\partial y}{\partial \zeta_3} - \dfrac{\partial y}{\partial \zeta_4} & \dfrac{\partial z}{\partial \zeta_3} - \dfrac{\partial z}{\partial \zeta_4} \end{bmatrix}. \qquad (6.118)$$

Then, as in (6.97) and (6.98), we have

$$\mathbf{k}_n^* = \int_0^1 \int_0^{1-\gamma} \int_0^{1-\eta-\gamma} (\mathbf{E}^{*T}\mathbf{CE}^* \det \mathbf{J}) d\xi d\eta d\gamma, \qquad (6.119)$$

$$Q_n^* = \int_0^1 \int_0^{1-\gamma} \int_0^{1-\eta-\gamma} D^{*T}\bar{F} \det J \, d\xi d\eta d\gamma + \int_0^1 \int_0^{1-\beta} D^{*T}\bar{T} \frac{dS_j}{dII} d\alpha d\beta, \qquad (6.120)$$

in which S_j is the surface j of the element, with $\zeta_j = 0$ where the tractions are prescribed, and where α and β are coordinates on the jth face, with dS_i/dII as defined in (6.114).

For numerical integration,

$$k_n^* = \sum_i (E^{*T}CE^* \det J)_i W_i, \qquad (6.121)$$

$$Q_n^* = \sum_i (D^{*T}\bar{F} \det J)_i W_i + \sum_i' \left(D^{*T}\bar{T} \frac{dS_j}{dII}\bigg|_{S_j}\right) W_i', \qquad (6.122)$$

in which $(\ )_i$ denotes the value of the function at the ith integration station, the W_i are corresponding weight factors for the volume integral, and the W_i' are the weighting factors for the surface integral. The location of integration stations and weighting factors are given in table 6.5 for the volume integrals and in table 6.2 for the surface integrals. Table 6.5 gives the same data for Hammer's formula.

Table 6.5. Stations, weighting factors, and error for Hammer's formula for interpolation over a tetrahedron V.

Total Number of Integration Points m	Integration Point i	Position in Volume Coordinates $\zeta_1, \zeta_2, \zeta_3, \zeta_4$				Weight Factor W_i/V	Error[a]
1	1	1/4,	1/4,	1/4,	1/4	1.0000	$O(h^2)$
4	1	α,	β,	β,	β[b]	.2500	$O(h^3)$
	2	β,	α,	β,	β	.2500	
	3	β,	β,	α,	β	.2500	
	4	β,	β,	β,	α	.2500	
5	1	1/4,	1/4,	1/4,	1/4	$-.8000$	$O(h^4)$
	2	1/3,	1/6,	1/6,	1/6	.0450	
	3	1/6,	1/3,	1/6,	1/6	.0450	
	4	1/6,	1/6,	1/3,	1/6	.0450	
	5	1/6,	1/6,	1/6,	1/3	.0450	

[a] Here $O(h^i)$ means that the numerical integration is exact if the order of the integrated functions is up to the $(i-1)$ power of $\zeta_1, \cdots, \zeta_4$.
[b] $\alpha = 0.58541020$, $\beta = 0.13819660$.

6.7.3 Concerning Matrix Multiplication

It is quite costly to generate a higher-order three-dimensional element. The matrix multiplication of

$$\mathbf{E}^{*\mathrm{T}}_{3m\times 6} \mathbf{C}_{6\times 6} \mathbf{E}^{*}_{6\times 3m} = (E_{ji}C_{jk}E_{kl})$$

is very time-consuming, because it requires

$$\frac{2 \times 6 \times 6 \times 3m(3m+1)}{2} = 324m\left(m + \frac{1}{3}\right) \approx 324m^2$$

multiplications, where m is the number of nodes of the element (with 3 degrees of freedom per node). For $m = 20$, the value will be roughly 130,000, after the symmetry of $\mathbf{E}^{*\mathrm{T}}\mathbf{C}\mathbf{E}^{*}$ has been accounted for. If there are, say, $n = 27$ integration points, so that the foregoing matrix will have to be evaluated 27 times, then the total number of multiplications will be 4×10^6.

The cost can be reduced by carrying out the multiplication in two steps, by evaluating $\mathbf{C}\,\mathbf{E}^{*}$ first, then evaluating $\mathbf{E}^{*\mathrm{T}}(\mathbf{C}\,\mathbf{E}^{*})$. The number of multiplications will be

$$6 \times 6 \times 3m + \frac{6 \times 3m(3m+1)}{2} = 27m\left(m + \frac{13}{3}\right) \approx 27m^2.$$

Still further reduction in the number of multiplications can be achieved. Since $\mathbf{C}$ is a symmetric positive definite matrix, it can be factored into

$$(\det \mathbf{J})\,\mathbf{C} = \mathbf{L}^{\mathrm{T}}\mathbf{L}$$

by the Cholesky method; $\mathbf{L}$ is an upper triangular matrix, so that

$$(\det \mathbf{J})\mathbf{E}^{*\mathrm{T}}\mathbf{C}\mathbf{E}^{*} = (\mathbf{L}\mathbf{E}^{*})^{\mathrm{T}}(\mathbf{L}\mathbf{E}^{*}),$$

where (using 6.109)

$$\mathbf{L}\mathbf{E}^{*}_{6\times 3m} = \begin{bmatrix} L_{11}\frac{\partial \mathbf{f}^{\mathrm{T}}}{\partial x} + L_{14}\frac{\partial \mathbf{f}^{\mathrm{T}}}{\partial y} + L_{15}\frac{\partial \mathbf{f}^{\mathrm{T}}}{\partial z} & L_{14}\frac{\partial \mathbf{f}^{\mathrm{T}}}{\partial x} + L_{12}\frac{\partial \mathbf{f}^{\mathrm{T}}}{\partial y} + L_{16}\frac{\partial \mathbf{f}^{\mathrm{T}}}{\partial z} \\ L_{24}\frac{\partial \mathbf{f}^{\mathrm{T}}}{\partial y} + L_{25}\frac{\partial \mathbf{f}^{\mathrm{T}}}{\partial z} & L_{24}\frac{\partial \mathbf{f}^{\mathrm{T}}}{\partial x} + L_{22}\frac{\partial \mathbf{f}^{\mathrm{T}}}{\partial y} + L_{26}\frac{\partial \mathbf{f}^{\mathrm{T}}}{\partial z} \\ L_{34}\frac{\partial \mathbf{f}^{\mathrm{T}}}{\partial y} + L_{35}\frac{\partial \mathbf{f}^{\mathrm{T}}}{\partial z} & L_{34}\frac{\partial \mathbf{f}^{\mathrm{T}}}{\partial x} \qquad\qquad + L_{36}\frac{\partial \mathbf{f}^{\mathrm{T}}}{\partial z} \\ L_{44}\frac{\partial \mathbf{f}^{\mathrm{T}}}{\partial y} + L_{45}\frac{\partial \mathbf{f}^{\mathrm{T}}}{\partial z} & L_{44}\frac{\partial \mathbf{f}^{\mathrm{T}}}{\partial x} \qquad\qquad + L_{46}\frac{\partial \mathbf{f}^{\mathrm{T}}}{\partial z} \\ L_{55}\frac{\partial \mathbf{f}^{\mathrm{T}}}{\partial z} & L_{56}\frac{\partial \mathbf{f}^{\mathrm{T}}}{\partial z} \\ 0 & L_{66}\frac{\partial \mathbf{f}^{\mathrm{T}}}{\partial z} \end{bmatrix}$$

$$\left. \begin{array}{l} L_{15}\dfrac{\partial \mathbf{f}^T}{\partial x} + L_{16}\dfrac{\partial \mathbf{f}^T}{\partial y} + L_{13}\dfrac{\partial \mathbf{f}^T}{\partial z} \\[4pt] L_{25}\dfrac{\partial \mathbf{f}^T}{\partial x} + L_{26}\dfrac{\partial \mathbf{f}^T}{\partial y} + L_{23}\dfrac{\partial \mathbf{f}^T}{\partial z} \\[4pt] L_{35}\dfrac{\partial \mathbf{f}^T}{\partial x} + L_{36}\dfrac{\partial \mathbf{f}^T}{\partial y} + L_{33}\dfrac{\partial \mathbf{f}^T}{\partial z} \\[4pt] L_{45}\dfrac{\partial \mathbf{f}^T}{\partial x} + L_{46}\dfrac{\partial \mathbf{f}^T}{\partial y} \\[4pt] L_{55}\dfrac{\partial \mathbf{f}^T}{\partial x} + L_{56}\dfrac{\partial \mathbf{f}^T}{\partial y} \\[4pt] L_{66}\dfrac{\partial \mathbf{f}^T}{\partial y} \end{array} \right]$$

Noting that $\partial \mathbf{f}^T/\partial x$ is a $1 \times m$ row vector, we see that it requires $36m$ multiplications to evaluate $\mathbf{L}\,\mathbf{E}^*$. It requires $6 \times 3m(3m + 1)/2$ multiplications to evaluate $(\mathbf{L}\,\mathbf{E}^*)^T\,\mathbf{L}\,\mathbf{E}^*$, so that the total number of multiplications is

$$27m\left(m + \frac{5}{3}\right).$$

If the material is isotropic, only L_{11}, L_{12}, L_{13}, L_{22}, L_{23}, L_{33}, L_{44}, L_{55}, and L_{66} are nonzero. Then the number of multiplications for $\mathbf{L}\,\mathbf{E}^*$ is only $12m$, and for $(\mathbf{L}\,\mathbf{E}^*)^T(\mathbf{L}\,\mathbf{E}^*)$ it is

$$3\frac{m(m+1)}{2} + 2m^2 + 2m^2 + 4\frac{m(m+1)}{2} + 3m^2 + 5\frac{m(m+1)}{2}$$
$$= 13m^2 + 6m.$$

Then the total number is

$$13m\left(m + \frac{18}{13}\right).$$

The following procedure is used to evaluate $(\mathbf{LE}^*)^T(\mathbf{LE}^*)$ for an isotropic material. Let

$$\mathbf{LE}^* = (A_{ij})$$

Then for $j = 1, 2, \cdots, m$, $\quad i = j, j+1, \cdots, m$,

$$k_{ij} = A_{1i}A_{1j} + A_{4i}A_{4j} + A_{5i}A_{5j},$$
$$k_{i+m,\,j+m} = A_{1,\,i+m}A_{1,\,j+m} + A_{2,\,i+m}A_{2,\,j+m} + A_{4,\,i+m}A_{4,\,j+m} + A_{6,\,i+m}A_{6,\,j+m},$$

$$k_{i+2m,\,j+2m} = A_{1,\,i+2m}A_{1,\,j+2m} + A_{2,\,i+2m}A_{2,\,j+2m} + A_{3,\,i+2m}A_{3,\,j+2m};$$
$$+ A_{5,\,i+2m}A_{5,\,j+2m} + A_{6,\,i+2m}A_{6,\,j+2m}$$

and for $i, j = 1, 2, 3, \cdots, m$,

$$k_{i+m,\,j} = A_{1,\,i+m}A_{1,\,j} + A_{4,\,i+m}A_{4j},$$
$$k_{i+2m,\,j} = A_{1,\,i+2m}A_{1,\,j} + A_{5,\,i+2m}A_{5j},$$
$$k_{i+2m,\,j+m} = A_{1,\,i+2m}A_{1,\,j+m} + A_{2\,i+2m}A_{2,\,j+m} + A_{6,\,i+2m}A_{6,\,j+m}.$$

6.7.4 The Element Matrices

In order to associate the element matrices $\mathbf{k}_n$ and $\mathbf{Q}_n$ with the generalized-coordinate vector

$$\mathbf{q}^T = \{u_1 \quad v_1 \quad w_1 \quad u_2 \quad v_2 \quad w_2 \cdots\}, \tag{6.123}$$

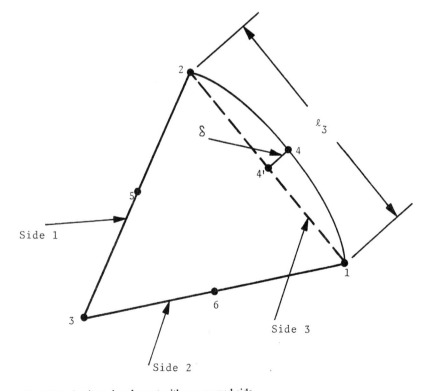

Fig. 6.18 A triangular element with one curved side.

a rearrangement of the elements of $\mathbf{k}_n^*$ and $\mathbf{Q}_n^*$ must be performed. Since

$$q_i = q_{I(i)}^*, \tag{6.124}$$

where

$$\begin{aligned} I(3i-2) &= i, \\ I(3i-1) &= i + \text{NNPE}, \\ I(3i) &= i + \text{NNPE} + \text{NNPE}, \end{aligned} \tag{6.125}$$

where NNPE is the number of nodes per element, then

$$\begin{aligned} k_{ij} &= k_{I(i)I(j)}^*, \\ Q_i &= Q_{I(i)}^*, \quad i,j = 1, 2, \cdots, \text{NNPE}. \end{aligned} \tag{6.126}$$

6.8 Another Way to Account for a Curved Boundary

It is natural to use curved elements for curved boundaries, especially when higher-order elements are used. To construct the interpolation functions, one often maps the physical space of the element to ξ-coordinates (secs. 6.6 and 6.7). The transformation from $\mathbf{x}$-coordinates to ξ-coordinates is nonlinear. However, as pointed out at the end of sec. 6.3, nonlinear transformations could reduce the order of accuracy of the interpolation in the physical space and thus reduce the accuracy of the high-order elements. In the following we shall discuss a way to construct a curved element for the boundary without having to use a nonlinear transformation.

Consider in fig. 6.18 the 6-node triangular element with two straight edges and one curved edge. The curved edge coincides with the boundaries of the problem being considered. The nodes are 1, 2, 3, 4, 5, and 6. We construct the interpolation using nodes 1, 2, 3, 4', 5, 6:

$$\mathbf{u} = \begin{Bmatrix} u \\ v \end{Bmatrix} = \begin{bmatrix} \mathbf{f}^T & 0 \\ 0 & \mathbf{f}^T \end{bmatrix} \begin{Bmatrix} \mathbf{q}'_u \\ \mathbf{q}'_v \end{Bmatrix} = \mathbf{D}\mathbf{q}', \tag{6.127}$$

where

$$\begin{aligned} \mathbf{q}'_u &= \{u_1 \quad u_2 \quad u_3 \quad u_4' \quad u_5 \quad u_6\}, \\ \mathbf{q}'_v &= \{v_1 \quad v_2 \quad v_3 \quad v_4' \quad v_5 \quad v_6\}, \end{aligned} \tag{6.128}$$

and where $\mathbf{f}^T$ is as given in (6.33). Then the stiffness matrix for the element associated with $\mathbf{q}'$ is

$$\mathbf{q}'^T \mathbf{k}'_n \mathbf{q}' = \int_\triangle t\mathbf{e}^T \mathbf{C}\mathbf{e} dx dy + \int_\frown t\mathbf{e}^T \mathbf{C}\mathbf{e} dx dy, \tag{6.129}$$

where $\triangle$ is the area of triangle 123 and $\frown$ is the area between the curve and chord 12. From the first integral, we have

$$\mathbf{q}' \mathbf{k}^*_n \mathbf{q}' = \int_\triangle t\mathbf{e}^T \mathbf{C}\mathbf{e} dx dy, \tag{6.130}$$

where $\mathbf{k}^*_n$ is the same as that for the element with nodes 1, 2, 3, 4', 5, 6 [eqs. (C. 2), (C. 14) and (C. 15) in app. C]. Since $\triangle$ is usually much smaller than $\frown$, the second integral is approximately $\mathbf{q}'^T \mathbf{k}''_n \mathbf{q}'$, or

$$\mathbf{q}'^T \mathbf{k}''_n \mathbf{q}' = \int_\frown t\mathbf{e}^T \mathbf{C}\mathbf{e} dx dy = \frac{2}{3} l_3 \delta \mathbf{e}^T \mathbf{C}\mathbf{e} \Big|_{\text{node}4'}, \tag{6.131}$$

or

$$\mathbf{k}''_n = \frac{2}{3} \delta l_3 \mathbf{E}^{*T} \mathbf{C} \mathbf{E}^* \Big|_{\substack{\zeta_1=\zeta_2=1/2 \\ \zeta_3=0}}, \tag{6.132}$$

where δ and l_3 are as given in fig. 6.18. Then we have

$$\mathbf{k}'_n = \mathbf{k}^*_n + \mathbf{k}''_n. \tag{6.134}$$

If the displacement vector at node 4 is constrained, we must make a transformation

$$\begin{aligned}\mathbf{q}'_u &= \mathbf{s}\mathbf{q}_u, \\ \mathbf{q}'_v &= \mathbf{s}\mathbf{q}_v,\end{aligned} \tag{6.135}$$

with

$$\mathbf{s} = \begin{bmatrix} 1 & 0 & 0 & 0 & 0 & 0 \\ 0 & 1 & 0 & 0 & 0 & 0 \\ 0 & 0 & 1 & 0 & 0 & 0 \\ -f_1/f_4 & -f_2/f_4 & -f_3/f_4 & 1/f_4 & -f_5/f_4 & -f_6/f_4 \\ 0 & 0 & 0 & 0 & 1 & 0 \\ 0 & 0 & 0 & 0 & 0 & 1 \end{bmatrix}, \tag{6.136}$$

where the f_i are as given in (6.33) and are evaluated at node 4. The element-stiffness matrix associated with

$$\mathbf{q}^* = \begin{Bmatrix} \mathbf{q}_u \\ \mathbf{q}_v \end{Bmatrix}$$

is

$$\begin{bmatrix} \mathbf{s} & 0 \\ 0 & \mathbf{s} \end{bmatrix}^{\mathrm{T}} \mathbf{k}'_n \begin{bmatrix} \mathbf{s} & 0 \\ 0 & \mathbf{s} \end{bmatrix}. \tag{6.137}$$

In order to obtain the element-stiffness matrix associated with

$$\mathbf{q}^{\mathrm{T}} = \{u_1 \quad v_1 \quad u_2 \quad v_2 \quad \cdots \quad u_6 \quad v_6\},$$

a further transformation using (6.94) must be performed.

This technique can be generalized to account for curved boundaries in three-dimensional problems.

7 Bending of Beams and Plates

In previous chapters we have discussed second-order ordinary and partial differential equations. Functionals for the associated variational statement of a given problem were indicated, and appropriate interpolation functions were developed so that compatibility and completeness requirements were met, thereby leading to converging solutions for the finite-element equations. An opportunity to deal with fourth-order equations is provided in the study of the bending of beams and plates. We will see that in the case of plates the problem of choosing appropriate interpolation functions to satisfy continuity between elements is a relatively difficult task. In the case of beam bending, which represents a one-dimensional problem, this difficulty is avoided. We therefore begin by considering a fourth-order ordinary differential equation encountered in Bernoulli-Euler beam theory. This will allow us to point out several new features intrinsic to such problems which were not discussed in the earlier chapters.

7.1 One-Dimensional Problems: Bending of Beams

Consider the simple fourth-order equation in the region $a \leq x \leq b$:

$$EI \frac{d^4 w}{dx^4} = p(x), \tag{7.1}$$

where EI is the beam flexural stiffness. The associated variational statement can be derived as discussed in ch. 1 and app. D, and is given by $\delta \Pi = 0$ where Π is

$$\Pi = \int_a^b \left[\frac{EI}{2} \left(\frac{d^2 w}{dx^2} \right)^2 - pw \right] dx. \tag{7.2}$$

The rigid (geometric) and natural boundary conditions can be determined by performing the variations and integrating by parts. We have

$$\begin{aligned}
\delta \Pi &= \int_a^b \left[EI \frac{d^2 w}{dx^2} \delta\left(\frac{d^2 w}{dx^2}\right) - p \delta w \right] dx = 0, \\
&= EI \frac{d^2 w}{dx^2} \delta\left(\frac{dw}{dx}\right) \Big|_a^b - EI \frac{d^3 w}{dx^3} \delta w \Big|_a^b + \int_a^b \left(EI \frac{d^4 w}{dx^4} - p \right) \delta w \, dx = 0.
\end{aligned} \tag{7.3}$$

According to (7.3) possible boundary conditions at $x = a, b$ are the rigid (geometric) conditions $w = w' = 0$ and the natural conditions $w'' = w''' = 0$, where primes denote differentiation with respect to x. Now, if nonzero values are prescribed for w'' and w''' at, say, $x = b$, so that $EIw''(b) = \alpha$ and $EIw'''(b) = \beta$, (7.2) should be altered to give

$$\Pi = \int_a^b \left[\frac{EI}{2}(w'')^2 - pw\right]dx + \beta w(b) - \alpha w'(b). \tag{7.4}$$

In a finite-element formulation the assumed interpolation functions used to represent w should be such that w and w' are continuous. This assures that Π is defined and that, as discussed in ch. 1, we can write Π as a sum of contributions from the elements into which the region is divided. Therefore, using (7.2), we have

$$\Pi = \sum_{i=1}^{n-1} \int_{x_i}^{x_{i+1}} \left[\frac{EI}{2}(w'')^2 - pw\right]dx, \tag{7.5a}$$

where n denotes the number of nodes (i.e., there are $n - 1$ elements). Equation (7.5a) can also be written in matrix form with w of each element expressed in terms of generalized coordinates and interpolation functions, as

$$\Pi = \sum_{i=1}^{n-1} \left(\frac{1}{2}\mathbf{q}^T\mathbf{k}\mathbf{q} - \mathbf{q}^T\mathbf{Q}\right), \tag{7.5b}$$

where $\mathbf{k}$ is the element-stiffness matrix.

Interelement compatibility is no problem for one-dimensional elements since there are no sides, and matching occurs only at interelement nodes. Since both w and w' are required to be continuous, it is possible to achieve interelement compatibility by using the values of both w and its derivative at each node as the generalized displacements or degrees of freedom. This means that if the values for deflection w and slope w' are the same for adjacent elements, then w and w' are continuous at that node. Completeness requires that the interpolating polynomial assumed for w contain all the terms of order lower than cubic, and that, as the element size approaches zero, it contain constant curvature (w'') states. Therefore, both compatibility and completeness are easily achieved in this case with a cubic interpolation which contains

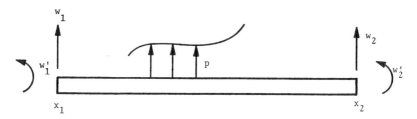

Fig. 7.1 A convenient element for beam-bending problems.

the four constants necessary to handle the four degrees of freedom (w and w' at both ends) for the element.

In order to obtain the element properties, we consider the typical element shown in fig. 7.1, where 1 and 2 are local numbers. Consider the cubic polynomial for w,

$$w(x) = \alpha_1 + \alpha_2 x + \alpha_3 x^2 + \alpha_4 x^3. \tag{7.6}$$

If $w(x)$ and its derivative are evaluated at the nodes x_1 and x_2, we obtain

$$\begin{aligned} w_1 &= \alpha_1 + \alpha_2 x_1 + \alpha_3 x_1^2 + \alpha_4 x_1^3, \\ w_1' &= \alpha_2 + 2\alpha_3 x_1 + 3\alpha_4 x_1^2, \\ w_2 &= \alpha_1 + \alpha_2 x_2 + \alpha_3 x_2^2 + \alpha_4 x_2^3, \\ w_2' &= \alpha_2 + 2\alpha_3 x_2 + 3\alpha_4 x_2^2. \end{aligned} \tag{7.7}$$

Equation (7.7) can be used to solve for the α_i ($i = 1, 2, 3, 4$) in terms of the four nodal values w_i and $w_i'(i = 1, 2)$. If we define the generalized nodal displacement vector $\mathbf{q}$ by

$$\mathbf{q}^T = \begin{bmatrix} w_1 & w_1' & w_2 & w_2' \end{bmatrix} = \begin{bmatrix} q_1 & q_2 & q_3 & q_4 \end{bmatrix}, \tag{7.8}$$

then $w(x)$ can be written in a form where appropriate interpolation functions are exhibited. And in this case the functions turn out to be first-order Hermetian interpolation functions, from which we obtain

$$w = \sum_{i=1}^{4} q_i f_i(x) \tag{7.9}$$

where

7.1 One-Dimensional Problems: Bending of Beams

$$f_1(x) = H_{01}^{(1)}(x) = 1 - 3\left(\frac{x-x_1}{\varepsilon}\right)^2 + 2\left(\frac{x-x_1}{\varepsilon}\right)^3,$$

$$f_2(x) = H_{11}^{(1)}(x) = (x-x_1)\left(\frac{x-x_1}{\varepsilon} - 1\right)^2,$$

$$f_3(x) = H_{02}^{(1)}(x) = \left(\frac{x-x_1}{\varepsilon}\right)^2\left[3 - 2\left(\frac{x-x_1}{\varepsilon}\right)\right],$$

$$f_4(x) = H_{12}^{(1)}(x) = \frac{(x-x_1)^2}{\varepsilon}\left(\frac{x-x_1}{\varepsilon} - 1\right).$$

(7.10)

The quantity $\varepsilon = x_2 - x_1$ is the size of the element. In terms of nodal quantities, (7.9) can be written as

$$w(x) = H_{01}^{(1)}w_1 + H_{11}^{(1)}w_1' + H_{02}^{(1)}w_2 + H_{12}^{(1)}w_2'. \tag{7.11}$$

When (7.9) is substituted into (7.5a) and the procedure given in sec. 1.9 is followed, one obtains results similar in form to (1.40) and (1.47). We write

$$\Pi = \sum_{i=1}^{n-1}\left\{\int_{x_i}^{x_{i+1}} \frac{EI}{2}(w'')^2 dx - \int_{x_i}^{x_{i+1}} pw\, dx\right\} = \sum_{i=1}^{n-1}\left(\frac{1}{2}\mathbf{q}^T\mathbf{k}\mathbf{q} - \mathbf{q}^T\mathbf{F}\right), \tag{7.12}$$

where $\mathbf{k}$ is the element-stiffness matrix and $\mathbf{F}$ is the element-loading vector. In this case the 4×4 matrix $\mathbf{k}$ is given by

$$\mathbf{k} = \frac{EI}{\varepsilon^3}\begin{bmatrix} 12 & & & \text{sym} \\ 6\varepsilon & 4\varepsilon^2 & & \\ -12 & -6\varepsilon & 12 & \\ 6\varepsilon & 2\varepsilon^2 & -6\varepsilon & 4\varepsilon^2 \end{bmatrix}. \tag{7.13}$$

The assembly process and solution of the finite-element equations again follows the procedure developed in previous chapters. It should be recalled (see ch. 1) an essential difference between second-order and fourth-order ordinary differential-equation problems is the kind of interpolation functions which must be used in each case. For the second-order equation in ch. 1, linear interpolation was sufficient. For the case considered here, cubic interpolation is necessary for convergence (i.e., in order to satisfy both completeness and compatibility as indicated in secs. 1.7 and 1.8).

The development implied by (7.12) and (7.13) is certainly applicable to nonhomogeneous or nonuniform beams where EI differs for each element. The results can also be generalized to a beam on an elastic foundation; the governing equation is in this case

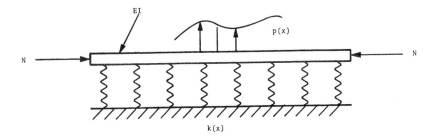

Fig. 7.2 A beam on an elastic foundation, and with axial loading.

$$\frac{d^2}{dx^2}\left(EI\frac{d^2w}{dx^2}\right) + k(x)w = p(x). \tag{7.14}$$

In such a case, (7.12) must be altered to include the foundation modulus $k(x)$, so that we have

$$\Pi = \sum_{i=1}^{n-1}\left\{\int_{x_i}^{x_{i+1}}\left[\frac{EI}{2}(w'')^2 - pw + \frac{k}{2}w^2\right]dx\right\}. \tag{7.15}$$

In order to include the effect of an axial load N (see fig. 7.2), where the governing equation is

$$(EIw'')'' + (Nw')' + k(x)w = p(x), \tag{7.16}$$

there must be added within the summation of (7.15), the integral

$$\int_{x_i}^{x_{i+1}} -\tfrac{1}{2}N(w')^2 dx. \tag{7.17}$$

The element-stiffness matrix **k** for the problem illustrated in fig. 7.2 can also be considered to consist of three parts, so that we can write

$$\mathbf{k} = \mathbf{k}_1 + \mathbf{k}_2 + \mathbf{k}_3. \tag{7.18}$$

The matrix $\mathbf{k}_1$, involving EI, the matrix $\mathbf{k}_2$, containing the elastic foundation modulus **k**, and the matrix, $\mathbf{k}_3$, containing the axial load N, are defined by the relations

$$\tfrac{1}{2}\mathbf{q}^T\mathbf{k}_1\mathbf{q} = \int \frac{EI}{2}(w'')^2 dx, \tag{7.19a}$$

$$\frac{1}{2}\mathbf{q}^T\mathbf{k}_2\mathbf{q} = \int \frac{k}{2} w^2 dx, \tag{7.19b}$$

$$\frac{1}{2}\mathbf{q}^T\mathbf{k}_3\mathbf{q} = \int -\frac{N}{2}(w')^2 dx, \tag{7.19c}$$

where the integration is carried out over the element and where $\mathbf{q}$ is defined by (7.8). Incidentally, instead of energy, virtual-work terms can be used on the left-hand side of eqs. (7.19): for, say, (7.19b), we would write

$$\delta\mathbf{q}^T\mathbf{k}_2\mathbf{q} = \int kw\delta w \, dx. \tag{7.20}$$

This latter approach is useful when dealing with nonconservative force fields, where the corresponding element matrices are seen as contributions of specific integrals involving the virtual work of those forces. For example, in the case of flutter problems (Rossettos and Tong 1974), proper virtual-work terms can yield appropriate aerodynamic-damping and force matrices (Olsen 1970).

The inclusion of the axial load N also leads to buckling eigenvalue problems where N plays the role of the eigenvalue. Additional detail for such problems is given by Cook (1974). The matrices $\mathbf{k}_2$ (for constant modulus k) and $\mathbf{k}_3$ (for constant N) for the problem of fig. 7.2 are given by

$$\mathbf{k}_2 = \frac{k\varepsilon}{420}\begin{bmatrix} 156 & & & \text{sym} \\ 22\varepsilon & 4\varepsilon^2 & & \\ 54 & 13\varepsilon & 156 & \\ -13\varepsilon & -3\varepsilon^2 & -22\varepsilon & 4\varepsilon^2 \end{bmatrix}, \tag{7.21}$$

and

$$\mathbf{k}_3 = -\frac{N}{30\varepsilon}\begin{bmatrix} 36 & & & \text{sym} \\ 3\varepsilon & 4\varepsilon^2 & & \\ -36 & -3\varepsilon & 36 & \\ 3\varepsilon & -\varepsilon^2 & -3\varepsilon & 4\varepsilon^2 \end{bmatrix}. \tag{7.22}$$

We conclude this section by discussing the conditions under which it is possible to obtain exact solutions to certain one-dimensional problems, such as beam bending, by the finite-element method (Tong 1969). First, note that a proper finite-element model can always be derived from a variational principle (Pian and Tong 1969). It turns out that with only one independent variable, if the homogeneous solutions of the Euler differential equations of a positive-definite functional are used as the interpolation functions, regardless of the number of elements used and the form of the particular solution, the values of the generalized coordinates at the nodal points given by the finite-element method based on this functional are the exact values for the problem. The proof given in Tong (1969) is in terms of a single independent variable, but it can be extended to many dependent variables.

7.2 Two-Dimensional Problems: Bending of Plates

In the one-dimensional bending of beams, neighboring beam elements are joined only at end-nodal points, so that the satisfaction of compatibility is easily achieved. In two-dimensional problems such as the bending of plates, however, neighboring elements are joined not only at certain discrete end- (or corner-) nodal points, but also along the interelement boundaries. Furthermore, in plate bending one must deal with a fourth-order differential equation. It will be seen now that the problem of choosing appropriate interpolation functions to satisfy the compatibility* and completeness requirements for an element is more complex. First, the assumed polynomial for the lateral displacement of the plate, w, and its first derivatives must be continuous within the element. Also, w and its derivative normal to the interface between neighboring elements must be continuous along the interelement boundary, and should be uniquely determined at the interelement boundary by the nodal values of the generalized displacements associated with that boundary. Finally, for completeness, all terms of order less than cubic and constant-curvature states (the curvatures are $w_{,xx}$, $w_{,yy}$ and the twist is $w_{,xy}$) as the elements approach zero size should be contained in any polynomial interpolation used to represent the displacement w.

Plate-bending finite elements that satisfy the just described requirements on the interpolation functions are often called conforming elements. Otherwise, they are nonconforming elements. The solution using conforming elements will assure the monotonic convergence for the associated functional. In what follows, we will see that it is impossible to achieve compatibility by using simple interpolation functions when only w and its slopes are prescribed at the nodes. One of the early plate elements will be discussed next, since it not only serves the purpose of explaining in detail how to develop a plate-bending element, but also helps to indicate why compatibility requirements cannot be achieved in a simple manner. This early "non-

---

*Compatibility is only required for monotonic convergence of the functional, which is positive definite. It has been shown that if the interpolation satisfies the conditions of the Irons patch test (Strang and Fix 1973) (it need not be compatible everywhere), the solution will converge.

conforming" rectangular-plate element, although not satisfying compatibility completely, turns out to be a relatively useful aid in the study of deflection behavior, and also has good convergence characteristics (completeness is satisfied). The satisfaction of compatibility is only essential for monotonic convergence of the strain energy as smaller elements are taken. It turns out that relaxing the compatibility requirement makes such a displacement model less stiff, which fortuitously aids the accuracy.

The thin-plate assumptions are reviewed first so that a variational statement consistent with these assumptions can be derived. Then the usual procedure of domain subdivision and local approximate representation of the displacement by appropriate interpolation is carried out to obtain the relevant element matrices. We consider a thin plate with its midsurface in the (x,y)-plane, and assume the following: (1) midsurface strains are negligible; (2) a normal to the undeformed midsurface is assumed to remain normal to the deformed midsurface; this means that the transverse shear strains e_{xz} and e_{yz} in eqs. (5.18) vanish; and (3) the normal stress σ_z is negligible and is taken to be 0. With these assumptions one can write the displacements u and v in the x and y directions, respectively, as

$$u = z\frac{\partial w}{\partial x}, \qquad v = z\frac{\partial w}{\partial y}. \tag{7.23}$$

The isotropic stress-strain relations (5.22) also reduce in this case to

$$e_x = \frac{1}{E}(\sigma_x - \nu\sigma_y), \qquad e_y = \frac{1}{E}(\sigma_y - \nu\sigma_x), \qquad e_{xy} = \frac{1}{G}\sigma_{xy}. \tag{7.24}$$

These assumptions and the resulting eqs. (7.23) and (7.24) are often referred to collectively as the Kirchhoff hypothesis. As a consequence of the assumptions, the deformed state of the plate can be described completely in terms of the transverse displacement w, which is a function of x and y. The strain energy of deformation, therefore, can also be written solely in terms of w, and, after some manipulation, the final result can be written as

$$U = \frac{1}{2}\iint_A \frac{Eh^3}{12(1-\nu^2)}\left[(w_{,xx} + w_{,yy})^2 - 2(1-\nu)(w_{,xx}w_{,yy} - w_{,xy}^2)\right]dxdy, \tag{7.25}$$

where the quantity $Eh^3/12(1-\nu^2)$ is called the plate flexural stiffness, where h is the thickness of the plate, and where the integration is carried out over the area of the plate.

For a given plate under a lateral load, an appropriate variational statement is $\delta \Pi = 0$ with the functional Π given by

$$\Pi = U - \iint_A pw\,dx\,dy. \tag{7.26}$$

The corresponding Euler equation is the single equilibrium equation for the plate written in terms of w:

$$\nabla^2\left(\frac{Eh^3}{12(1-\nu^2)}\nabla^2 w\right) = p(x, y), \tag{7.27}$$

where $p(x, y)$ is the lateral loading function, and ∇^2 is the harmonic operator. Note the resemblance between (7.27) and (7.1), where, for the beam, EI is the beam flexural stiffness, d^4/dx^4 replaces the biharmonic operator $\nabla^2\nabla^2$, and there is only the one independent variable x.

We can now proceed to develop the relevant element matrices as before by using the variational statement associated with the functional given in (7.26). Consider the rectangular plate element in the (x, y)-plane shown in fig. 7.3, and take the three nodal displacements w, $w_{,x}$, and $w_{,y}$ as the generalized nodal coordinates, so that with four nodes the element has a total of 12 degrees of freedom. The nodal vector $\mathbf{q}$ for the element is therefore written (in matrix form, to emphasize the number of components) as

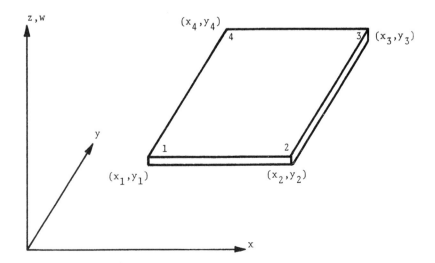

Fig. 7.3 A rectangular element.

7.2 Two-Dimensional Problems: Bending of Plates

$$\mathbf{q}^T = [w_1 \quad w_{,x1} \quad w_{,y1} \quad w_2 \quad w_{,x2} \quad w_{,y2} \quad \cdots \quad w_{,y4}]_{1\times 12}. \tag{7.28}$$

In fig. 7.3 we let $x_1 = x_4$, $x_2 = x_3$, $y_1 = y_2$ and $y_3 = y_4$ for convenience. We describe the displacement w within the element in terms of 12 parameters by means of the polynomial

$$w(x,y) = \alpha_1 + \alpha_2 x + \alpha_3 y + \alpha_4 x^2 + \alpha_5 xy + \alpha_6 y^2 + \alpha_7 x^3 + \alpha_8 x^2 y \\ + \alpha_9 xy^2 + \alpha_{10} y^3 + \alpha_{11} x^3 y + \alpha_{12} xy^3 = \mathbf{P}\boldsymbol{\alpha}, \tag{7.29}$$

where

$$\mathbf{P} = [1 \quad x \quad y \quad x^2 \quad \cdots \quad xy^3]_{1\times 12},$$
$$\boldsymbol{\alpha}^T = [\alpha_1 \quad \alpha_2 \quad \alpha_3 \quad \cdots \quad \alpha_{12}]_{1\times 12}.$$

Note that the polynomial in (7.29) contains only two fourth-order terms and that along lines $x =$ constant or $y =$ constant, the displacement w represents a cubic curve which is uniquely defined by four constants. Since the interelement boundaries do in fact represent such lines, the two end values of w and its tangential derivatives at the ends of a given boundary (i.e., on the boundary $x =$ constant the tangential derivatives are $w_{,y}$ taken at both ends, etc.) will define w along that boundary uniquely. As a result, if two neighboring elements have the same values for the nodal coordinates, continuity of w will be assured all along the interface of such elements. It can be seen from (7.29) that the derivative of w normal to an interface will also vary as a cubic, but only two values of normal derivative are defined at the ends of an interelement boundary. This cubic is therefore not defined uniquely, and a discontinuity in the normal derivative (or normal slope) can be expected to occur in general along the interface, so that the interpolation given in (7.29) is not a compatible one.

The difficulty just described can be overcome by using more degrees of freedom for each node. In particular, if one adds the twist $w_{,xy}$ as an additional degree of freedom at each node (thereby producing an element of 16 degrees of freedom, with four degrees of freedom per node), the uniqueness of normal slope along the interelement boundaries, and therefore compatibility, can be established. In such a case the four terms

$$\alpha_{13} x^2 y^2 + \alpha_{14} x^3 y^2 + \alpha_{15} x^2 y^3 + \alpha_{16} x^3 y^3 \tag{7.30}$$

should be added to the interpolating polynomial in (7.29), thereby providing the higher-order matching required between adjacent elements. This produces a compatible rectangular element (Bogner, Fox, and Schmit 1965) which also turns out to have excellent convergence properties.

To continue with the development of the noncompatible rectangular element, we next make use of the variational statement associated with (7.26). There are two approaches that one can take at this point. In the first, one can relate the constants α_1 to α_{12} to the generalized nodal coordinates q_1 to q_{12} by writing down the twelve simultaneous equations for the values of w and its derivatives at the nodes (analogous to eqs. (7.7) for the beam-bending problem). These equations can be written in matrix form as

$$A\alpha = q, \tag{7.31}$$

which can be inverted to yield the αs in terms of the qs:

$$\alpha = A^{-1}q = Tq. \tag{7.32}$$

Using (7.29) with (7.32) one can write $w(x, y)$ in a form analogous to (7.9) for beam bending. In this case one obtains

$$w(x, y) = \sum_{i=1}^{2} \sum_{j=1}^{2} [f_{ij}(\xi, \eta) w(\xi_i, \eta_j) + (f_x)_{ij}(\xi, \eta) w_{,x}(\xi_i, \eta_j) \\ + (f_y)_{ij}(\xi, \eta) w_{,y}(\xi_i, \eta_j)], \tag{7.33}$$

in which (see fig. 7.4)

$$\xi = \frac{x}{a}, \quad \eta = \frac{y}{b}, \quad \xi_1 = \eta_1 = -1, \quad \xi_2 = \eta_2 = +1, \tag{7.34}$$

and

$$f_{ij}(\xi, \eta) = \frac{\xi_i \eta_j}{8}(2 + \xi\xi_i + \eta\eta_j - \xi^2 - \eta^2)(\xi + \xi_i)(\eta + \eta_j),$$

$$(f_x)_{ij}(\xi, \eta) = \frac{\xi_i \eta_j}{8}(\xi^2 - 1)(\xi + \xi_i)(\eta + \eta_j)a, \tag{7.35}$$

$$(f_y)_{ij}(\xi, \eta) = \frac{\xi_i \eta_j}{8}(\eta^2 - 1)(\xi + \xi_i)(\eta + \eta_j)b.$$

Equation (7.33) can be written in matrix form as

$$w = f^T q, \tag{7.36}$$

where

$$\mathbf{f}^T = \{f_{11}\ (f_x)_{11}\ (f_y)_{11}\ f_{21}\ (f_x)_{21}\ (f_y)_{21}\ f_{22}\ (f_x)_{22}\ (f_y)_{22}\ f_{12}\ (f_x)_{12}\ (f_y)_{12}\},$$
$$\mathbf{q}^T = \{w(\xi_1, \eta_1)\ w_{,x}(\xi_1, \eta_1)\ w_{,y}(\xi_1, \eta_1)\ \cdots\ w_{,x}(\xi_1, \eta_2)\ w_{,y}(\xi_1, \eta_2)\}.$$

If we now substitute (7.33) or (7.36) into (7.26), Π can be written as

$$\Pi = \tfrac{1}{2}\mathbf{q}^T\mathbf{k}\mathbf{q} - \mathbf{q}^T\mathbf{F} \tag{7.37}$$

for a single element, where $\mathbf{k}$ and $\mathbf{F}$ are the element-stiffness matrix and element-loading vector, respectively. They are given by

$$\mathbf{k} = \iint \mathbf{f}''^T \mathbf{D} \mathbf{f}'' \, dxdy \tag{7.38a}$$

and

$$\mathbf{F} = \iint p\mathbf{f}\, dxdy, \tag{7.38b}$$

where

$$\mathbf{D} = \frac{Eh^3}{12(1-\nu^2)} \begin{bmatrix} 1 & \nu & 0 \\ \nu & 1 & 0 \\ 0 & 0 & \dfrac{1-\nu}{2} \end{bmatrix}. \tag{7.39}$$

The vector $\mathbf{f}''$ which appears in (7.38a) arises from the second derivatives in the energy expression (7.25), and is given by

$$\mathbf{f}'' = \begin{Bmatrix} \partial^2 \mathbf{f}^T/\partial x^2 \\ \partial^2 \mathbf{f}^T/\partial y^2 \\ 2\partial^2 \mathbf{f}^T/\partial x \partial y \end{Bmatrix}. \tag{7.40}$$

The vanishing of the first variation, $\delta\Pi = 0$, leads to the 12 equations $\partial\Pi/\partial q_i = 0$ ($i = 1, 2, \cdots, 12$). These equations are the equilibrium conditions for the element:

$$\mathbf{k}\mathbf{q} = \mathbf{F}, \tag{7.41}$$

which relates the nodal generalized coordinates $\mathbf{q}$ to the corresponding generalized forces.

In a second approach, instead of using (7.32) to express the αs in terms of the qs, we can evaluate the element-stiffness matrix directly in terms of the αs, and then utilize the transformation matrix **T** of (7.32). That is, by substituting (7.29) directly into the energy expression (7.25), one obtains

$$U = \tfrac{1}{2}\boldsymbol{\alpha}^T \mathbf{k}' \boldsymbol{\alpha}. \tag{7.42}$$

We can then use (7.32) to write

$$U = \tfrac{1}{2}\mathbf{q}^T \mathbf{T}^T \mathbf{k}' \mathbf{T} \mathbf{q}, \tag{7.43}$$

from which the element-stiffness matrix **k** is seen to be

$$\mathbf{k} = \mathbf{T}^T \mathbf{k}' \mathbf{T}. \tag{7.44}$$

This second approach simplifies evaluating the derivatives $w_{,xx}$, $w_{,yy}$, etc., since $w_{,xx} = 2\alpha_4 + 6\alpha_7 x + \cdots$, $w_{,yy} = 2\alpha_6 + \cdots$, but it leads to the necessity of using the transformation indicated in (7.44). It should be pointed out, however, that many of the operations such as the inversion in (7.32) can be done analytically, and the integrations in (7.38a,b) can be done numerically (although in cases where Eh^3 is constant, integration can also be done analytically and is straightforward). Once the appropriate element matrices are determined one can use the assembly and solution techniques already de-

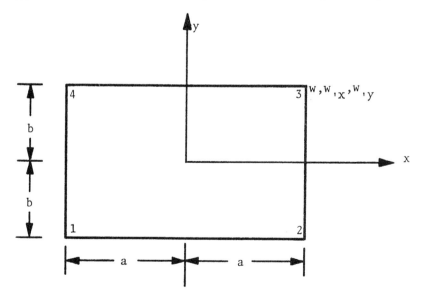

Fig. 7.4 Rectangular-element geometry.

scribed in previous chapters, so the details need not be repeated at this point.

For the rectangular geometry shown in fig. 7.4, the element-stiffness matrix associated with the generalized coordinates $\mathbf{q} = \{w_1, w_2, w_3, w_4, w_{,x1}, w_{,x2}, w_{,x3}, w_{,x4}, w_{,y1}, w_{,y2}, w_{,y3}, w_{,y4}\}$ can be written as $\mathbf{k} = \mathbf{B}^T \bar{\mathbf{k}} \mathbf{B}$ where $\mathbf{B}$ is given by

$$\mathbf{B} = \begin{bmatrix} \mathbf{A} & 0 & 0 \\ 0 & \mathbf{A} & 0 \\ 0 & 0 & \mathbf{A} \end{bmatrix}; \quad \mathbf{A} = \begin{bmatrix} 1 & -1 & 1 & -1 \\ -1 & 1 & 1 & -1 \\ 1 & 1 & 1 & 1 \\ -1 & -1 & 1 & 1 \end{bmatrix}. \quad (7.45)$$

The nonzero $\bar{k}_{ij}$s are written out explicitly in (7.46), as given in Holand and Bell (1972). All other $\bar{k}_{ij}$s are zero.

$$\bar{k}_{22} = \frac{3b}{a^3}, \quad \bar{k}_{25} = -\frac{3b}{a^2}, \quad \bar{k}_{2,12} = -\frac{\nu}{a},$$

$$\bar{k}_{33} = \frac{3a}{b^3}, \quad \bar{k}_{38} = -\frac{\nu}{b}, \quad \bar{k}_{39} = -\frac{3a}{b^2},$$

$$\bar{k}_{44} = \frac{1}{5a^3b^3}(5a^4 + 5b^4 + 14a^2b^2 - 4\nu a^2b^2),$$

$$\bar{k}_{47} = -\frac{1}{5a^2b}(5b^2 + 2a^2 + 3\nu a^2),$$

$$\bar{k}_{4,10} = -\frac{1}{5ab^2}(5a^2 + 2b^2 + 3\nu b^2), \quad (7.46)$$

$$\bar{k}_{55} = \frac{3b}{a}, \quad \bar{k}_{5,12} = \nu, \quad \bar{k}_{66} = \frac{b}{a}, \quad \bar{k}_{6,11} = \nu,$$

$$\bar{k}_{77} = \frac{1}{5ab}(5b^2 + 2a^2 - 2\nu a^2), \quad \bar{k}_{7,10} = \nu,$$

$$\bar{k}_{88} = \frac{1}{3ab}(2a^2 + b^2 - 2\nu a^2), \quad \bar{k}_{89} = \nu,$$

$$\bar{k}_{99} = \frac{3a}{b}, \quad \bar{k}_{10,10} = \frac{1}{5ab}(5a^2 + 2b^2 - 2\nu b^2),$$

$$\bar{k}_{11,11} = \frac{a}{b}, \quad \bar{k}_{12,12} = \frac{1}{3ab}(2b^2 + a^2 - 2\nu b^2).$$

This element-stiffness matrix and those of other types of elements are given in Clough and Tocher (1965) where numerical examples indicating the behavior of several elements are also presented.

In the search to resolve the interelement compatibility difficulties for plate bending, various conforming plate elements based upon the displacement model have been developed. The different elements that have evolved are distinguished by the type of interpolation functions used, the generalized coordinates defined at nodes, the number of nodes, and the type of element shape. Compatible higher-order triangular elements (Bell 1969; Cowper et al. 1969) have been developed by using an incomplete fifth-degree polynomial for w, and using the six quantities w, $w_{,x}$, $w_{,y}$, $w_{,xy}$, $w_{,xx}$, and $w_{,yy}$ as the generalized nodal coordinates at the vertices of the triangle. These 18-degree-of-freedom elements are quite accurate but they are, of course computationally expensive. Also, many structural applications have natural discontinuities such as abrupt changes in thickness or composition, discrete stiffening elements, and members meeting at angles. In such cases we have regions where continuity should not be enforced. Therefore in such applications (where the finite-element method is particularly suited and often relied on), these higher-order elements lead to awkward computational strategies. Several plate elements have been compared by Abel and Desai (1972) for accuracy against computational expense; as it turns out, if one weighs the expense against the incremental increase in accuracy, simpler elements appear to do much better than higher-order elements. It should be emphasized, of course, that in specific applications such as problems where accurate stress (i.e., bending moments, which are proportional to the higher derivatives of w) prediction is important, especially in regions of high stress gradients, the use of higher-order elements is often necessary.

A 21-degree-of-freedom triangular element is described in Holand and Bell (1972) where the three additional degrees of freedom that are added to the higher-order element described in the previous paragraph are the normal derivatives, denoted by $w_{,n}$, at each of the three nodes located at the midpoints of the three sides of the triangle. The interpolating polynomial is complete of fifth degree. Numerical results (*ibid.*) are obtained for deflections, and bending and twisting moments for variously supported rectangular and equilateral triangular plates under uniformly distributed and hydrostatic loads. A comparison is made between 9-, 15-, 18-, and 21-degree-of-freedom triangular elements, together with the mesh sizes required to yield equivalent results, and these are compared with analytical solutions. (Note that the 9-degree-of-freedom nonconforming element uses w, $w_{,x}$, and $w_{,y}$ as the nodal values at the 3 vertices, and the 15-degree-of-freedom element adds the nodal values w and $w_{,n}$ to the three midpoints of the sides of the triangle.) Marked improvement in results is seen to occur as one goes from the 9- to the 15-degree-of-freedom element, and even more significant improvement occurs as one goes from the 15- to the 21- degree-of-freedom element. The various element-stiffness matrices are derived and written out explicitly in Holand and Bell.

In addition to the refined triangular elements just described, a conforming and complete bending arbitrary quadrilateral element has been developed by Clough and Felippa (1968). In this element the six degrees of freedom w, $w_{,x}$, $w_{,y}$, $w_{,xy}$, $w_{,xx}$, and $w_{,yy}$ are again used at each node. This element therefore represents a 24-degree-of-freedom element in contrast to the 18 and 16 degrees of freedom for the early conformable triangular and rectangular elements, respectively.

To include transverse shear deformation, the rotation of normals θ_x and θ_y about the x and y axes are not necessarily equated to the slopes $w_{,x}$ and $w_{,y}$ respectively, since the transverse shear strains are proportional to the quantities $w_{,x} - \theta_x$ and $w_{,y} - \theta_y$. This fact must, therefore, be reflected in the choice of nodal coordinates for the element. A plate element that includes transverse shear strains has been presented by Cook (1974), Pawsey and Clough (1971), Zienkiewicz, Taylor, and Too (1971), and Ahmad, Irons, and Zienkiewicz (1970). The element is an isoparametric plate element with eight nodes; four at the corners and four at the midpoints of the sides. The nodal degrees of freedom are the displacement w and the rotations θ_x and θ_y, yielding a 24-degree-of-freedom element.

The discussion to this point and the ability to overcome the difficulties of satisfying compatibility with the resulting conforming plate elements has focused on using assumed displacement fields in order to interpolate the nodal values into the element. Another approach is based on the hybrid stress method (Pian 1964) which is to be discussed in the following chapter; the approach leads to one of the more successful plate elements. The success of this approach in resolving the compatibility dilemma lies in its flexibility at the element-formulation stage, where both stress and displacement fields are assumed to advantage over different portions of the element. Hybrid stress plate elements have yielded excellent results for problems involving stress distribution, deflections and vibrations (Mau, Pian, and Tong 1973; Pian and Mau 1972; Rossettos, Tong, and Perl 1972).

Problems

1 Consider the beam-bending boundary-value problem consisting of eq. (7.1) and the boundary conditions $w = w' = 0$ at each end of a beam $0 \leq x \leq L$. Divide the beam into three elements, use cubic interpolations, and

set up the appropriate finite-element equations. Properly constrain the assembled system to obtain a set of four algebraic equations. Solve for the nodal values and compare with the exact solution. (For convenience, set the quantities $EI = L = p = 1$.) With cubic interpolation the results should compare exactly. Why? (See Tong 1969.)

2 Use quadratic interpolation and w as the generalized coordinate at each node to solve prob. 1. For an element with nodes i and $i + 1$, use interpolation.

$$w = w_i\left(1 - \frac{x - x_i}{\varepsilon_i}\right)^2 + w_{i+1}\left(\frac{x - x_i}{\varepsilon_i}\right) + \alpha\frac{(x_{i+1} - x)(x - x_i)}{\varepsilon_i^2},$$

where $\varepsilon_i = x_{i+i} - x_i$. The parameter α can first be determined in terms of w_i and w_{i+1} by minimizing π_i with respect to α, where

$$\pi_i = \int_{x_i}^{x_{i+1}} \left[\frac{EI}{2}(w'')^2 - pw\right]dx,$$

and then the usual procedure can be followed. Divide the region into three and six equal elements, and in each case solve for the nodal values. Compare the results with prob. 1. In what ways do the approaches in probs. 1 and 2 differ? Why does the present approach not converge as $\varepsilon_i \to 0$?

3 (a) Carry out the necessary details, using the interpolation function of sec. 7.1, to obtain the element matrices **k** and **F** implied in eq. (7.12). For **F** let $p(x) = p =$ constant. (b) In a similar manner derive the element matrices given in eqs. (7.21) and (7.22).

4 (a) Consider the problem illustrated in fig. 7.2 (eq. 7.16) with $p(x) = 0$, and where the beam of length L is simply supported at each end so that w and w'' vanish there. Divide the beam into three elements, use cubic interpolation, and with constant values for $k(x)$ and N derive the constrained set of finite-element equations. (b) Find the lowest eigenvalue for N in the case where $k = 0$. (For convenience, set $EI = L = 1$ and use a computer; the exact eigenvalue is π^2.)

5 Consider the functional π for an element, consisting of the integral in eq. (7.15), with the integral in eq. (7.17) added. Take appropriate variations, integrate by parts, and seek conditions to make $\delta\pi = 0$, thereby showing that the resulting Euler equation is given by eq. (7.16).

6 Evaluate the element-loading vector **F** in eq. (7.12) using cubic interpolation and (a) $p(x) = p_0 =$ constant; (b) $p(x) = p_0 + p_1 x$; (c) $p(x) = p_0 + p_1 x$

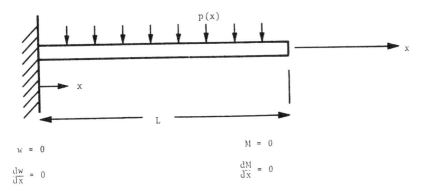

Fig. 7.5

$+ p_2 x^2$. How is the total load on the element weighted at each of the nodes for the three cases? (The function $p(x)$ can be regarded as a loading function.)

7 In the finite-element method for beams one can use both the displacement w, and the moment M ($M = EIw''$) as unknown nodal coordinates. The functional for constructing the finite-element equations is related to the so-called Reissner principle and is

$$\Pi_R = \int_0^L \left[\frac{\partial M}{\partial x} \frac{\partial w}{\partial x} + \frac{M^2}{2EI} + p(x)w \right] dx.$$

(a) Construct the element matrices using the nodal values of M and w as the generalized coordinates (assume EI and p are constant for each element).
(b) What are the natural and rigid boundary conditions for the general problem? Some definitions are indicated in the example of the cantilever shown in fig. 7.5.
(c) Solve the cantilever problem shown in fig 7.5 with $EI = p = L = 1$. Compare the results with those of the exact solution ($d^2M/dx = 1$, $M = d^2w/dx^2$). Use 2 elements.

8 Use the ideas in sec. 7.2 and develop a procedure for finding a polynomial interpolation function that will describe a compatible triangular bending element. Select six nodes; three at the corners and three at the midpoints of the sides of the triangle. Use appropriate generalized coordinates at the nodes, and discuss the reasons for the steps you take.

9 Integrate eq. (7.38a) to obtain several elements of the stiffness matrix. In particular determine k_{25}, k_{33}, k_{38}, k_{47}, and k_{55}.

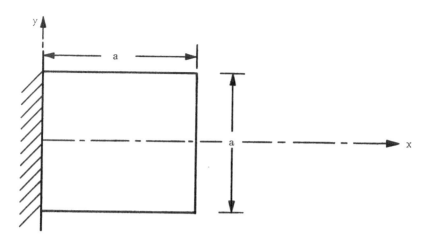

Fig. 7.6

10 Consider the deflection of the cantilevered square plate shown in fig 7.6 under a uniform load $p(x) = p$. The governing equation is eq. (7.27), and at the edge $y = 0$ the rigid boundary conditions are $w = w_{,x} = 0$, (also $w_{,y} = 0$). Divide the plate into four elements, determine the finite-element equations, and constrain the system appropriately. (For convenience let $Eh^3/12(1 - \nu^2) = 1$, $p = 1$, and $a = 1$.) Use the rectangular element-stiffness matrix given in sec. 7.2 [eq. (7.38a)] and the element vector $\mathbf{F}$ given by eq. (7.38b). The interpolation polynomial is that of eq. (7.29). Write a simple computer program to assemble the element matrices and to constrain the overall system of algebraic equations. Solve for the deflection w at the nodes.

11 (a) Develop the necessary relations required to determine the element-stiffness matrix for a 9-degree-of-freedom triangular bending element by taking w, $w_{,x}$ and $w_{,y}$ as the generalized coordinates at each node and assuming the third-degree polynomial

$$w(x, y) = \alpha_1 + \alpha_2 x + \alpha_3 y + \alpha_4 x^2 + \alpha_5 xy + \alpha_6 y^2 + \alpha_7 x^3 \\ + \alpha_8(x^2 y + xy^2) + \alpha_9 y^3.$$

Will this yield a conforming element? Explain.

(b) Using area coordinates as discussed in ch. 6 express $w(x, y)$ as $w = \mathbf{q}^T \mathbf{f}$, where

$$\mathbf{q}^T = \{w_1 \ w_{,x1} \ w_{,y1} \ w_2 \ w_{,x2} \ w_{,y2} \ w_3 \ w_{,x3} \ w_{,y3}\}, \qquad \mathbf{f} = \begin{Bmatrix} f_1 \\ f_2 \\ \vdots \\ f_9 \end{Bmatrix}.$$

Find the interpolation functions $f_1, f_2, \cdots f_9$. Since there is freedom of choice in selecting terms, there are only nine generalized coordinates in the foregoing expressions. For convenience, the tenth term of the third degree polynomial may be dropped.

Show that the following useful relations between Cartesian and area coordinates hold:

$$\begin{Bmatrix} x \\ y \\ 1 \end{Bmatrix} = \begin{bmatrix} x_1 & x_2 & x_3 \\ y_1 & y_2 & y_3 \\ 1 & 1 & 1 \end{bmatrix} \begin{Bmatrix} \zeta_1 \\ \zeta_2 \\ \zeta_3 \end{Bmatrix}, \qquad \begin{Bmatrix} \dfrac{\partial}{\partial x} \\ \dfrac{\partial}{\partial y} \end{Bmatrix} = \frac{1}{2A} \begin{bmatrix} y_{23} & y_{31} \\ x_{32} & x_{13} \end{bmatrix} \begin{Bmatrix} \dfrac{\partial}{\partial \zeta_1} \\ \dfrac{\partial}{\partial \zeta_2} \end{Bmatrix},$$

where $y_{23} = y_2 - y_3$, etc., and A = area of triangle.

(c) Set up the element-stiffness matrix $\mathbf{k}$ as an integral in the same manner as was done for (7.38a), and define the various matrices in the integrand. It will be useful to derive and use the following relation for the curvatures:

$$\begin{Bmatrix} w_{,xx} \\ w_{,yy} \\ 2w_{,xy} \end{Bmatrix} = \frac{1}{4A^2} \begin{bmatrix} y_{23}^2 & y_{31}^2 & 2y_{31}y_{23} \\ x_{32}^2 & x_{13}^2 & 2x_{13}x_{32} \\ 2x_{32}y_{23} & 2x_{13}y_{31} & 2(x_{13}y_{23} + x_{32}y_{31}) \end{bmatrix} \begin{Bmatrix} w_{,\zeta_1\zeta_1} \\ w_{,\zeta_2\zeta_2} \\ w_{,\zeta_1\zeta_2} \end{Bmatrix}.$$

8 Hybrid Methods

8.1 Introduction

Pian (1964) first introduced the concept of separately assuming displacements and stresses over different parts of the continuum; he named it the hybrid model. Subsequently, Tong and Pian (1969) rationalized this approach by means of variational formulations; it was then generalized into different finite-element models (Tong 1970) with various applications (Chen and Mei 1974; Mau and Tong 1974; Rossettos and Tong 1974; Mau, Pian, and Tong 1973; Tong, Pian, and Lasry 1973; Tong and Fung 1971). A general review of the method has been given by Pian and Tong (1972).

The hybrid model of the finite-element method can be derived in a straightforward manner by following the development in sec. 1.3. This model utilizes more fully the power of the finite-element method by making use of different types and orders of local approximations for each element; this model also solves some of the difficulties of enforcing compatibility conditions in the displacement model.

In this chapter, we shall follow the discussion of the hybrid approach given by Pian and Tong (1970), by examining the harmonic and biharmonic equations.

8.2 Foundations

For simplicity we shall treat the two-dimensional Laplace equation. Consider a domain A with a boundary ∂A. The governing equation is

$$\nabla^2 \phi = 0, \tag{8.1}$$

in A; the boundary conditions are

$$\partial \phi / \partial \nu = \bar{\phi}_\nu \quad \text{on} \quad \partial A_\nu, \tag{8.2}$$
$$\phi = \bar{\phi} \quad \text{on} \quad \partial A_\phi, \tag{8.3}$$

where ∂ / ∂_ν is the normal derivative and where $\bar{\phi}_\nu$ and $\bar{\phi}$ are given quantities over the boundaries ∂A_ν and ∂A_ϕ, respectively, in which $\partial A_\nu + \partial A_\phi = \partial A$.

This problem can be formulated in terms of a variational statement, as discussed in ch. 2. In this case the appropriate functional is

$$\Pi = \tfrac{1}{2} \int_A (\nabla\phi)^2 dA - \int_{\partial A_\nu} \phi \bar{\phi}_\nu \, ds, \tag{8.4}$$

with the constraint that ϕ satisfies (8.3) over ∂A_ϕ.

This same problem can be formulated in other ways. For example, we may introduce two additional unknowns

$$u = \partial\phi/\partial x \quad \text{and} \quad v = \partial\phi/\partial y. \tag{8.5}$$

Then (8.1) becomes

$$\nabla \cdot \mathbf{V} = \partial u/\partial x + \partial v/\partial y = 0, \tag{8.6}$$

and (8.2), (8.3) become

$$\mathbf{V} \cdot \boldsymbol{\nu} = \bar{\phi}_\nu \quad \text{on} \quad \partial A_\nu, \tag{8.7}$$
$$\phi = \bar{\phi} \quad \text{on} \quad \partial A_\phi, \tag{8.8}$$

where $\mathbf{V} = \{u, v\}^T$ and where $\boldsymbol{\nu}$ is the unit outward normal. The corresponding functional is

$$\Pi = \int_A [\mathbf{V} \cdot \nabla\phi - \tfrac{1}{2}(u^2 + v^2)] \, dA - \int_{\partial A_\nu} \phi \bar{\phi}_\nu \, ds. \tag{8.9a}$$

The solution of (8.5) through (8.8) is given by the stationary condition of Π with respect to ϕ and $\mathbf{V}$ in which (8.8) represents the rigid boundary condition for ϕ.

By using the divergence theorem on the first term in the first integral of (8.9a) and the relation (8.8), Π can be written as

$$\Pi = -\int_A [\phi \nabla \cdot \mathbf{V} + \tfrac{1}{2}(u^2 + v^2)] \, dA + \int_{\partial A_\nu} \phi(\mathbf{V} \cdot \boldsymbol{\nu} - \bar{\phi}_\nu) ds + \int_{\partial A_\phi} \bar{\phi} \mathbf{V} \cdot \boldsymbol{\nu} \, ds. \tag{8.9b}$$

This is the form most convenient for deriving the hybrid equations. We first divide the domain A into a finite number of elements and write Π as

$$\Pi = \sum_{\substack{\text{all} \\ \text{elements}}} \Big\{ -\int_{A_m} [\phi \nabla \cdot \mathbf{V} + \tfrac{1}{2}(u^2 + v^2)] \, dA + \int_{(\partial A_\nu)_m} \phi(\mathbf{V} \cdot \boldsymbol{\nu} - \bar{\phi}_\nu) ds$$
$$+ \int_{(\partial A_\phi)_m} \bar{\phi} \mathbf{V} \cdot \boldsymbol{\nu} \, ds \Big\}, \tag{8.10}$$

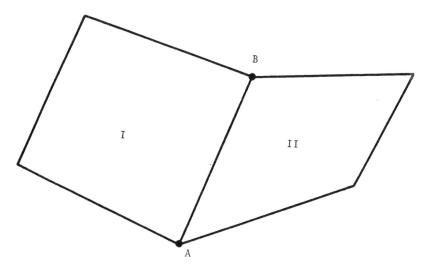

Fig. 8.1 Two adjacent elements.

where m denotes the mth element and $(\partial A_\nu)_m$ and $(\partial A_\phi)_m$ are the parts of the element boundary ∂A_m in coincidence with ∂A_ν and ∂A_ϕ, respectively. The integral over $(\partial A_\nu)_m$ [or $(\partial A_\phi)_m$] vanishes if no part of ∂A_m is on ∂A_ν (or ∂A_ϕ). Since the functional Π in (8.10) is the sum of the integrals over each individual element, in order that it be the same as that of (8.9b) it is necessary to place the constraint condition on **V** that the normal component of **V** is to be continuous across all interelement boundaries. In other words, for the two adjacent elements I and II as shown in fig. 8.1, we must have

$$(\mathbf{V}\cdot\boldsymbol{\nu})_\mathrm{I} + (\mathbf{V}\cdot\boldsymbol{\nu})_\mathrm{II} = 0$$

along the common boundary between points A and B of the two elements. To relax this constraint condition, we modify the functional in (8.10) by introducing the Lagrangian multiplier $\tilde{\phi}$, which is an unknown function along all interelement boundaries and is equal to ϕ on ∂A_ν and to $\bar{\phi}$ on ∂A_ϕ. We next define

$$\begin{aligned}\Pi' &= \int_{\substack{\text{all}\\ \text{interelement}\\ \text{boundaries}}} \tilde{\phi}[(\mathbf{V}\cdot\boldsymbol{\nu})_\mathrm{I} + (\mathbf{V}\cdot\boldsymbol{\nu})_\mathrm{II}]\,ds \\ &= \sum_{\substack{\text{all}\\ \text{elements}}} \left[\int_{\partial A_m}\tilde{\phi}\mathbf{V}\cdot\boldsymbol{\nu}\,ds - \int_{(\partial A_\nu)_m}\phi\mathbf{V}\cdot\boldsymbol{\nu}\,ds - \int_{(\partial A_\phi)_m}\bar{\phi}\mathbf{V}\cdot\boldsymbol{\nu}\,ds\right].\end{aligned} \quad (8.11)$$

It is clear that Π' vanishes if $\mathbf{V} \cdot \boldsymbol{\nu}$ is continuous across all interelement boundaries, and that $\tilde{\phi} = \phi$ on ∂A_ν and $\tilde{\phi} = \bar{\phi}$ on ∂A_ϕ. By adding (8.11) to (8.10) we have

$$\Pi_h = \Pi + \Pi' = \sum_m \pi_{hm}, \qquad (8.12)$$

in which $\sum_m$ indicates the sum over all elements, and where

$$\pi_{hm} = \int_{\partial A_m} \tilde{\phi} \mathbf{V} \cdot \boldsymbol{\nu} ds - \int_{A_m} [\phi \nabla \cdot \mathbf{V} + \tfrac{1}{2}(u^2 + v^2)] dA - \int_{(\partial A_\nu)_m} \tilde{\phi} \bar{\phi}_\nu ds. \qquad (8.13)$$

In the functional Π_h of (8.12), $\tilde{\phi}$ is a function defined on all interelement boundaries and on the boundary ∂A, while ϕ and $\mathbf{V}$ are functions defined within each element. The only constraint condition for $\tilde{\phi}$ is that

$$\tilde{\phi} = \bar{\phi} \quad \text{on} \quad \partial A_\phi; \qquad (8.14)$$

there is no constraint condition for ϕ or $\mathbf{V}$. The first variation of Π_h with respect to $\tilde{\phi}, \phi,$ and $\mathbf{V}$ is

$$\delta\Pi_h = \sum_m \left(\int_{\partial A_m} \mathbf{V} \cdot \boldsymbol{\nu} \delta\tilde{\phi} ds - \int_{(\partial A_\nu)_m} \bar{\phi}_\nu \delta\tilde{\phi} ds \right) \\ + \sum_m \left[\int_{\partial A_m} \tilde{\phi} \delta\mathbf{V} \cdot \boldsymbol{\nu} ds - \int_{A_m} (\phi \nabla \cdot \delta\mathbf{V} + \mathbf{V} \cdot \delta\mathbf{V}) dA - \int_{A_m} \nabla \cdot \mathbf{V} \delta\phi dA \right]. \qquad (8.15)$$

The vanishing of $\delta\Pi_h$ for arbitrary $\delta\tilde{\phi}, \delta\phi$, and $\delta\mathbf{V}$ (with $\delta\tilde{\phi} = 0$ on ∂A_ϕ) implies that each of the sums in (8.15) vanish. Also, since there are no constraints on $\delta\phi$ and $\delta\mathbf{V}$ for different elements, the vanishing of the second term in (8.15) implies that each term (all quantities in brackets) in the sum vanishes separately. Therefore $\delta\Pi_h = 0$ implies

$$\sum_m \int_{\partial A_m} \mathbf{V} \cdot \boldsymbol{\nu} \delta\tilde{\phi} ds - \int_{(\partial A_\nu)_m} \bar{\phi}_\nu \delta\tilde{\phi} ds = 0 \qquad (8.16)$$

with $\delta\tilde{\phi} = 0$ on $(\partial A_\phi)_m$, and

$$\int_{\partial A_m} \tilde{\phi} \boldsymbol{\nu} \cdot \delta\mathbf{V} ds - \int_{A_m} (\phi \nabla \cdot \delta\mathbf{V} + \mathbf{V} \cdot \delta\mathbf{V}) dA - \int_{A_m} \nabla \cdot \mathbf{V} \delta\phi dA = 0 \qquad (8.17)$$

for all m.

In this problem, both ϕ and V within A_m can be determined from the requirements that $V \cdot \nu$ be continuous across all the interelement boundaries and that it satisfies (8.7). Equation (8.17) will require (8.5) and (8.6) to be satisfied within each element A_m, and $\tilde{\phi} = \phi$ over all element boundaries ∂A_m.

It is clear that from (8.17), the solution for ϕ and V within A_m can be first completely determined in terms of the boundary value $\tilde{\phi}$ on ∂A_m and then $\tilde{\phi}$ itself can be determined from (8.16) (with V in terms of $\tilde{\phi}$). Recall from the discussion in sec. 1.3 that the unknown function $\tilde{\phi}$ is the so-called imaginary boundary value of each element. The solution within the element is determined from the stationary condition for π_{hm} with respect to ϕ and V; that is, from (8.17).

8.3 Hybrid Element-Stiffness Matrix

The finite-element solution can be constructed by using the stationary condition for Π_h of (8.12). We assume $\tilde{\phi}$ in terms of known functions with unknown parameters on all interelement boundaries and on the boundary of A, and assume ϕ and V within each element. However, since there are no constraint conditions on ϕ and V, they can be independently assumed for different elements. For instance, we may assume ϕ and V in terms of polynomials with unknown coefficients in one element and in terms of trigonometric functions in another. The unknowns of ϕ and V of each element can be first expressed in terms of those of $\tilde{\phi}$ from (8.17), the stationary conditions of π_{hm}; then the unknown parameters of $\tilde{\phi}$ can be obtained from (8.16).

The hybrid element-stiffness matrix is to be obtained from π_{hm} by expressing the unknowns of ϕ and V of one element in terms of those of $\tilde{\phi}$ of the element boundary. It is noted that π_{hm} as defined in (8.13) is a general form for the hybrid finite-element model. In this section we shall consider a more restricted type of hybrid model. Instead of arbitrarily assuming V within A_m, we shall restrict ourselves to selecting that V which satisfies (8.6) identically,

$$\nabla \cdot V = \frac{\partial u}{\partial x} + \frac{\partial v}{\partial y} = 0. \tag{8.18}$$

In this case (8.13) becomes

8.3 Hybrid Element-Stiffness Matrix

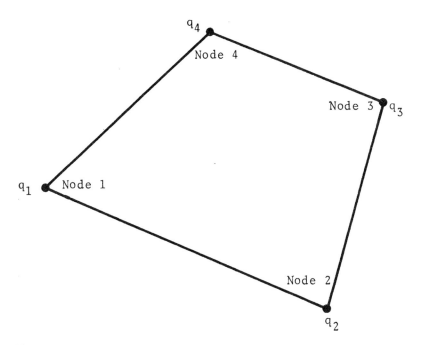

Fig. 8.2 General quadrilateral element.

$$\pi_{hm} = \int_{\partial A_m} \tilde{\phi} \mathbf{V} \cdot \boldsymbol{\nu} ds - \tfrac{1}{2} \int_{A_m} (u^2 + v^2) dA - \int_{(\partial A_\nu)_m} \tilde{\phi} \bar{\phi}_\nu ds. \tag{8.19}$$

Note that now the quantity ϕ no longer appears explicitly. In order to construct the finite-element equations, all we need to do is separately assume $\tilde{\phi}$ over the element boundary ∂A_m, and assume $\mathbf{V}$ within the element A_m so that it satisfies (8.18). The finite-element model based on the functional (8.19) for an element is called the hybrid stress model by Pian.

We shall assume $\tilde{\phi}$ in terms of its nodal values along the element boundaries; these nodal values will be the unknown parameters or generalized coordinates for $\tilde{\phi}$. For example, for a quadrilateral element* as shown in fig. 8.2, we assume that $\tilde{\phi}$ is equal to q_1, q_2, q_3, q_4 at the corner nodes 1, 2, 3, 4, and that it varies linearly between any two nodes. Between nodes 1 and 2,

$$\tilde{\phi} = q_1\left(1 - \frac{s}{s_{12}}\right) + q_2 \frac{s}{s_{12}}, \tag{8.20}$$

---
*The reason we do not consider the three-node triangular element is that it will result in the same element-stiffness matrix derived in ch. 2.

where s is the distance coordinate from point 1 to point 2, while s_{12} is the actual distance between the two points. Similarly, the expression for $\tilde{\phi}$ between any other two nodes can be obtained by cyclic permutation of 1, 2, 3, 4. In matrix form we have

$$\tilde{\phi} = \mathbf{L}\mathbf{q} = \mathbf{L}\begin{Bmatrix} q_1 \\ q_2 \\ q_3 \\ q_4 \end{Bmatrix}, \tag{8.21}$$

where

$$\begin{aligned}
\mathbf{L} &= \left\{ 1 - \frac{s}{s_{12}} \quad \frac{s}{s_{12}} \quad 0 \quad 0 \right\} & \text{between 1 and 2;} \\
&= \left\{ 0 \quad 1 - \frac{s}{s_{23}} \quad \frac{s}{s_{23}} \quad 0 \right\} & \text{between 2 and 3;} \\
&= \left\{ 0 \quad 0 \quad 1 - \frac{s}{s_{34}} \quad \frac{s}{s_{34}} \right\} & \text{between 3 and 4;} \\
&= \left\{ \frac{s}{s_{41}} \quad 0 \quad 0 \quad 1 - \frac{s}{s_{41}} \right\} & \text{between 4 and 1.}
\end{aligned}$$

Within the elements, we assume

$$\begin{aligned}
u &= \beta_1 + \beta_3 y + \beta_5 x, \\
v &= \beta_2 + \beta_4 x - \beta_5 y,
\end{aligned} \tag{8.22}$$

or, in matrix form,

$$\mathbf{V} = \begin{Bmatrix} u \\ v \end{Bmatrix} = \mathbf{P}\boldsymbol{\beta} = \begin{bmatrix} 1 & 0 & y & 0 & x \\ 0 & 1 & 0 & x & -y \end{bmatrix} \begin{Bmatrix} \beta_1 \\ \beta_2 \\ \vdots \\ \beta_5 \end{Bmatrix}, \tag{8.23}$$

where the βs are unknown constants. It can be seen that the expressions given in (8.22) identically satisfy (8.18) for any β. In general, one may include higher-order terms in x and y for u and v in the expressions (8.22). However, it turns out that for the quadrilateral element with $\tilde{\phi}$ given in the linear form of (8.22), the number of terms used in (8.22) is sufficient (Tong and Pian 1969). Using (8.23), we can also write $\mathbf{V} \cdot \boldsymbol{\nu}$ in matrix form

$$\mathbf{V} \cdot \boldsymbol{\nu} = \boldsymbol{\nu}^\mathrm{T} \mathbf{P} \boldsymbol{\beta}, \tag{8.24}$$

where

$$\boldsymbol{\nu} = \begin{Bmatrix} \nu_x \\ \nu_y \end{Bmatrix}. \tag{8.25}$$

For the element boundary between nodes i and j, we have

$$\nu_x = \frac{y_i - y_j}{s_{ij}}, \quad \nu_y = \frac{x_j - x_i}{s_{ij}},$$
$$s_{ij} = [(x_i - x_j)^2 + (y_i - y_j)^2]^{1/2}, \tag{8.26}$$

where x_i and y_i are the spatial coordinates of node i. Substitution of (8.21), (8.23), and (8.24) into (8.19) yields

$$\pi_{hm} = \boldsymbol{\beta}^T \mathbf{G} \mathbf{q} - \tfrac{1}{2} \boldsymbol{\beta}^T \mathbf{H} \boldsymbol{\beta} - \mathbf{q}^T \mathbf{Q}', \tag{8.27}$$

where

$$\mathbf{H} = \int_{A_m} \mathbf{P}^T \mathbf{P} \, dA, \quad \mathbf{G} = \int_{\partial A_m} \mathbf{P}^T \boldsymbol{\nu} \mathbf{L} \, ds, \quad \mathbf{Q}' = \int_{(\partial A_\nu)_m} \bar{\phi}_\nu \mathbf{L}^T \, ds. \tag{8.28}$$

The integration of $\mathbf{H}$ can be carried out numerically. It is also relatively easy to integrate $\mathbf{H}$ analytically by using the triangular coordinates discussed in sec. 6.3. If principal axes, with the centroid of the element chosen as the origin, are used, and if $\mathbf{P}$ is defined as in (8.23), $\mathbf{H}$ is a diagonal matrix. For example, in the case of a rectangular element ($0 \leq x \leq a$, $0 \leq y \leq b$) we have

$$\mathbf{H} = \begin{bmatrix} ab & & & & 0 \\ & ab & & & \\ & & ab^3/12 & & \\ & & & a^3b/12 & \\ 0 & & & & \dfrac{a^3b + ab^3}{12} \end{bmatrix}. \tag{8.29}$$

The evaluation of $\mathbf{G}$ is carried out by integration from one side of the element to another,

$$\int_{\partial A_m} \bar{\phi} \mathbf{V} \cdot \boldsymbol{\nu} \, ds = \int_1^2 \cdots + \int_2^3 \cdots + \int_3^4 \cdots + \int_4^1, \tag{8.30}$$

where $\int_i^j$ denotes integration from node i to node j. For example,

$$\int_1^2 = \int_0^{s_{12}} [(\beta_1 + \beta_3 y + \beta_5 x)\nu_x + (\beta_2 + \beta_4 x - \beta_5 y)\nu_y] \times \left[q_1\left(1 - \frac{s}{s_{12}}\right) + q_2\left(\frac{s}{s_{12}}\right) \right] ds, \tag{8.31}$$

where

$$s_{12} = [(x_1 - x_2)^2 + (y_1 - y_2)^2]^{1/2},$$
$$x = x_1\left(1 - \frac{s}{s_{12}}\right) + x_2 \frac{s}{s_{12}}, \qquad \nu_x = \frac{y_1 - y_2}{s_{12}},$$
$$y = y_1\left(1 - \frac{s}{s_{12}}\right) + y_2 \frac{s}{s_{12}}, \qquad \nu_y = \frac{x_2 - x_1}{s_{12}}.$$

Carrying out the integration, we find that

$$\int_1^2 = [\beta_1, \beta_2, \cdots \beta_5] \mathbf{G}_{12} \begin{Bmatrix} q_1 \\ q_2 \end{Bmatrix}, \tag{8.32}$$

where $\mathbf{G}_{12}$ is a 5×2 matrix, and that

$$\mathbf{G}_{12}^T = \begin{bmatrix} \dfrac{\nu_x}{2} & \dfrac{\nu_y}{2} & \left(\dfrac{y_1}{3} + \dfrac{y_2}{6}\right)\nu_x & \left(\dfrac{x_1}{3} + \dfrac{x_2}{6}\right)\nu_y \\ \dfrac{\nu_x}{2} & \dfrac{\nu_y}{2} & \left(\dfrac{y_1}{6} + \dfrac{y_2}{3}\right)\nu_x & \left(\dfrac{x_1}{6} + \dfrac{x_2}{3}\right)\nu_y \\ & & \left(\dfrac{x_1}{3} + \dfrac{x_2}{6}\right)\nu_x - \left(\dfrac{y_1}{3} + \dfrac{y_2}{6}\right)\nu_y \\ & & \left(\dfrac{x_1}{6} + \dfrac{x_2}{3}\right)\nu_x - \left(\dfrac{y_1}{6} + \dfrac{y_2}{3}\right)\nu_y \end{bmatrix} s_{12}. \tag{8.33}$$

Similarly, integration of the other sides can be obtained from the cyclic permutation of 1, 2, 3, and 4 in the subscripts of $\mathbf{G}_{12}$. The matrix $\mathbf{G}$ is obtained by summing up the contributions from all the sides. If we write $\mathbf{G}_{ij}$ in the form

$$\mathbf{G}_{ij} = \{\mathbf{g}, \mathbf{h}\}_{ij}$$

for the side with nodes i and j, where $\mathbf{g}$ and $\mathbf{h}$ are the first and the second column of $\mathbf{G}_{ij}$ respectively, and are associated with the generalized coordinates q_i and q_j, then $\mathbf{G}$ can be written out explicitly as

$$\mathbf{G} = \{\mathbf{g}_{12} + \mathbf{h}_{41} \quad \mathbf{h}_{12} + \mathbf{g}_{23} \quad \mathbf{h}_{23} + \mathbf{g}_{34} \quad \mathbf{h}_{34} + \mathbf{g}_{41}\}. \tag{8.34}$$

The unknowns β for each element are determined from the stationary condition of π_{hm} in (8.27) with respect to β, since in general β for each element can be independently assumed. The stationary condition of π_{hm} leads to

$$\mathbf{H}\beta - \mathbf{G}\mathbf{q} = 0$$

or

$$\beta = \mathbf{H}^{-1}\mathbf{G}\mathbf{q}. \tag{8.35}$$

Substituting into (8.27), we obtain

$$\pi_{hm} = \tfrac{1}{2}\mathbf{q}^T\mathbf{k}\mathbf{q} - \mathbf{q}^T\mathbf{Q}', \tag{8.36}$$

with the hybrid element-stiffness matrix $\mathbf{k}$ given by

$$\mathbf{k} = \mathbf{G}^T\mathbf{H}^{-1}\mathbf{G}. \tag{8.37}$$

Equation (8.12) can now be written as

$$\Pi_h = \sum_{\substack{\text{all} \\ \text{elements}}} \tfrac{1}{2}\mathbf{q}^T\mathbf{k}\mathbf{q} - \mathbf{q}^T\mathbf{Q}'. \tag{8.38}$$

The form of (8.38) is identical to the displacement model discussed in ch. 2, where only the nodal values of ϕ were used as generalized coordinates. The final finite-element equations can be obtained in a routine manner by assembling the element matrices $\mathbf{k}$ and $\mathbf{Q}'$, and by imposing the proper boundary constraints on the $\mathbf{q}$s.

8.4 Poisson's Equation

If the governing equation is Poisson's equation

$$\nabla^2 \phi = f(x, y), \tag{8.39}$$

the formulation using (8.19) is still valid except that (8.18) is replaced by

$$\nabla \cdot \mathbf{V} = \frac{\partial u}{\partial x} + \frac{\partial v}{\partial y} = f(x, y). \tag{8.40}$$

In constructing the element matrices, we assume

$$V = P\beta + P_F\beta_F, \tag{8.41}$$

where $P\beta$ can be in the same form as (8.23), satisfying the homogeneous divergence equation, while $P_F\beta_F$ is a particular solution of (8.40) with β_F consisting of known constants related to the form of $f(x, y)$. It is general practice to approximate $f(x, y)$ as a constant or a linear function of x and y within the element. For example, if $f(x, y) = f_o$ is a constant, we have

$$P_F\beta_F = \begin{Bmatrix} f_o x \\ 0 \end{Bmatrix}.$$

A substitution of (8.41) into (8.19) yields

$$\pi_{hm} = \beta^T G q + \beta_F^T G_F q - \tfrac{1}{2}\beta^T H\beta - \beta^T H_F\beta_F + C - q^T Q', \tag{8.42}$$

where G, H, and Q' are given in (8.28) and

$$G_F = \int_{\partial A_m} P_F^T \nu L \, ds, \qquad H_F = \int_{A_m} P^T P_F \, dA, \tag{8.43}$$

$$C = -\tfrac{1}{2}\beta_F^T \left(\int_{A_m} P_F^T P_F \, dA \right) \beta_F.$$

The stationary condition of π_{hm} with q fixed gives

$$H\beta = Gq - H_F\beta_F$$

or

$$\beta = H^{-1}(Gq - H_F\beta_F). \tag{8.44}$$

Substituting (8.44) into (8.42), we obtain

$$\pi_{hm} = \tfrac{1}{2} q^T k q - q^T Q'' + C'_m, \tag{8.45}$$

where k is the same as in (8.37), and

$$Q'' = Q' - G_F^T \beta_F + G^T H^{-1} H_F \beta_F. \tag{8.46}$$

The particular solution $P_F\beta_F$ of (8.40) is not unique, and Q'' may depend on its choice. However, if both P and P_F involve polynomials of the same degree, and **P** includes the complete polynomial of that degree, then **Q″** is independent of the choice of the particular solution (Tong and Pian 1969). For example, **P** in (8.23) includes all polynomials in x and y of first degree, (i.e., all linear equations) that satisfy (8.18). If $\mathbf{P}_F$ is also only linear in x and y, then **Q** is independent of the choice of $\mathbf{P}_F \beta_F$. The quantity C'_m is a constant, and has no effect on the final solution. Equation (8.45) can be substituted into (8.12) to construct the final finite-element solution.

As another generalization, if the governing equation for ϕ is

$$\frac{\partial}{\partial x}\left(b_{11}\frac{\partial \phi}{\partial x}\right) + \frac{\partial}{\partial x}\left(b_{12}\frac{\partial \phi}{\partial x}\right) + \frac{\partial}{\partial y}\left(b_{12}\frac{\partial \phi}{\partial x}\right) + \frac{\partial}{\partial y}\left(b_{22}\frac{\partial \phi}{\partial x}\right) = 0, \tag{8.47}$$

the previous formulation is still valid except that the exact relations between u, v and ϕ are now

$$\begin{aligned} u &= b_{11}\frac{\partial \phi}{\partial x} + b_{12}\frac{\partial \phi}{\partial y}, \\ v &= b_{12}\frac{\partial \phi}{\partial x} + b_{22}\frac{\partial \phi}{\partial y}, \end{aligned} \tag{8.48}$$

and (8.19) is in the form

$$\pi_{hm} = \int_{\partial A_m} \tilde{\phi} \mathbf{V} \cdot \boldsymbol{\nu} ds - \tfrac{1}{2}\int_{A_m} (c_{11}u^2 + 2c_{12}uv + c_{22}v^2)\,dA - \int_{(\partial A_v)_m} \tilde{\phi}_\nu \phi\, ds, \tag{8.49}$$

where the c_{ij} are elements of the inverse of the matrix having b_{ij} as its elements. The element matrices can now be determined using (8.49) in the same manner and with the same $\tilde{\phi}$ and **V** as discussed in the last section.

8.5 Biharmonic Equation

In this section, we shall use subscript notation where x_1 and x_2 are to denote x and y respectively. In the history of the finite-element method, the compatibility condition was a major hurdle in the development of a compatible triangular or general quadrilateral element for a problem governed by fourth-order partial differential equations. The hybrid approach is one of the best ways to resolve the difficulty of obtaining compatible functions.

Consider the biharmonic equation over a domain A,

$$\nabla^2 \nabla^2 w = p(x_1, x_2), \tag{8.50}$$

with the boundary conditions on ∂A:

$$w = \bar{w}, \tag{8.51}$$

$$\frac{\partial w}{\partial \nu} = \bar{w}_\nu \tag{8.52}$$

in which

$$\nabla^2 = \frac{\partial^2}{\partial x_1^2} + \frac{\partial^2}{\partial x_2^2}. \tag{8.53}$$

In much the same way as for the harmonic equation this problem can also be formulated in different ways. In the present case we introduce the three additional dependent variables

$$M_{11} = -\frac{\partial^2 w}{\partial x_1^2}, \quad M_{22} = -\frac{\partial^2 w}{\partial x_2^2}, \quad M_{12} = -\frac{\partial^2 w}{\partial x_1 \partial x_2}, \tag{8.54}$$

which are called the stress couples, or moments. Using (8.54), we can write (8.50) in the form

$$\frac{\partial^2 M_{11}}{\partial x_1^2} + 2\frac{\partial^2 M_{12}}{\partial x_1 \partial x_2} + \frac{\partial^2 M_{22}}{\partial x_2^2} = -p(x_1, x_2). \tag{8.55}$$

The solution of (8.50) is the same as that of (8.54) and (8.55) with the same boundary conditions.

Similarly to the treatment of the harmonic equation, a general hybrid functional can be written for this problem in the form

$$\Pi_h = \sum_{\substack{\text{all} \\ \text{elements}}} \pi_{hm},$$

in which

$$\pi_{hm} = \int_{\partial A_m} (-M_{\alpha\beta}\nu_\alpha \tilde{w}_{,\beta} + Q_\alpha \nu_\alpha \tilde{w})\, ds - \int_{A_m} (M_{\alpha\beta,\alpha\beta} w + pw + \tfrac{1}{2} M_{\alpha\beta} M_{\alpha\beta})\, dA, \tag{8.56}$$

where repeated indices denote summation from 1 to 2, ν_α is the direction cosine between the boundary normal and the x_α-axis, and

$$Q_\alpha = M_{\alpha\beta,\beta}. \tag{8.57}$$

The constraint conditions for $\tilde{w}$ are

$$\tilde{w} = \bar{w} \quad \text{and} \quad \frac{\partial \tilde{w}}{\partial \nu} = \bar{w}_\nu \quad \text{on} \quad \partial A. \tag{8.58}$$

The Euler equations of (8.56) can be shown to be (8.54), (8.55), and, in addition, the continuity of $M_{\alpha\beta}\nu_\alpha\nu_\beta$ and $Q_\alpha\nu_\alpha$ across the interelement boundaries, and

$$w = \tilde{w} \quad \text{and} \quad \frac{\partial w}{\partial \nu} = \frac{\partial \tilde{w}}{\partial \nu} \tag{8.59}$$

over the element boundaries. Since $\tilde{w}$ and $\partial \tilde{w}/\partial \nu$ are the same for two adjacent elements over their common boundaries, we have the continuity of w and $\partial w/\partial \nu$ across the interelement boundaries.

By requiring different combinations of Euler equations to be satisfied, we can derive different finite-element models. In particular, we shall consider here the so-called hybrid stress model; for this model, we require that (8.55) be satisfied exactly. In this case, the functional in (8.56) is reduced to

$$\pi_{hm} = -\int_{\partial A_m} (M_{\alpha\beta}\nu_\alpha\tilde{w}_{,\beta} - Q_\alpha\nu_\alpha\tilde{w})ds - \tfrac{1}{2}\int_{A_m} M_{\alpha\beta}M_{\alpha\beta}dA. \tag{8.60}$$

In the above functional, the Ms for different elements can be independently assumed, and $\tilde{w}$, $\tilde{w}_{,\beta}$ are only assumed over the interelement boundaries and over the boundary of the domain. Since we do not have to assume w (or $\tilde{w}$) within the element, compatibility difficulties associated with assuming w are avoided.

8.6 Element-Stiffness Matrix for the Biharmonic Equation

In formulating the finite-element solution, we assume the stress vector in the form

$$\boldsymbol{\sigma} = \begin{Bmatrix} M_{11} \\ M_{22} \\ M_{12} \end{Bmatrix} = \mathbf{P}(x_1, x_2)\boldsymbol{\beta} + \mathbf{P}_F(x_1, x_2)\boldsymbol{\beta}_F \tag{8.61}$$

for the mth element in terms of unknown parameters $\boldsymbol{\beta}$. In (8.61), $\mathbf{P}$ is chosen to satisfy the homogeneous part of (8.55) for arbitrary $\boldsymbol{\beta}$. For example, when all the quadratic and lower-order terms for the stress couples are included in (8.61), there are 17 independent βs and the matrix $\mathbf{P}$ of (8.61) is given by

$$\mathbf{P} = \begin{bmatrix} 1 & 0 & 0 & x_1 & 0 & 0 & x_2 & 0 & 0 & x_1^2 & 0 & 0 & x_2^2 & 0 & 0 & x_1 x_2 & 0 \\ 0 & 1 & 0 & 0 & x_1 & 0 & 0 & x_2 & 0 & 0 & x_1^2 & 0 & 0 & x_2^2 & 0 & 0 & x_1 x_2 \\ 0 & 0 & 1 & 0 & 0 & x_1 & 0 & 0 & x_2 & -x_1 x_2 & 0 & x_1^2 & 0 & -x_1 x_2 & x_2^2 & 0 & 0 \end{bmatrix}.$$
(8.62)

If only the linear and the constant terms are included, there will be 9 independent βs, and $\mathbf{P}$ will be given by the first nine columns of (8.62). The second term in (8.61) may be any particular solution of (8.55). For example, for an element where $p(x_1, x_2) = -P_0 = $ constant, one can take

$$\mathbf{P}_F \beta_F = \begin{Bmatrix} x_1^2/4 \\ x_2^2/4 \\ 0 \end{Bmatrix} P_0, \quad \text{or} \quad \begin{Bmatrix} x_1^2/2 \\ 0 \\ 0 \end{Bmatrix} P_0, \quad \text{or} \quad \begin{Bmatrix} 0 \\ 0 \\ x_1 x_2/2 \end{Bmatrix} P_0. \tag{8.63}$$

The finite-element analysis using $\mathbf{P}$ of (8.62) will yield identical generalized nodal forces using any of the particular solutions in (8.63). This is because (8.62) includes all the quadratic terms of the homogeneous solutions of (8.55) (Tong and Pian 1969). From $\boldsymbol{\sigma}$ of (8.61), we can also define the matrix

$$\mathbf{T} = \begin{Bmatrix} Q_\alpha \nu_\alpha \\ -M_{1\alpha} \nu_\alpha \\ -M_{2\alpha} \nu_\alpha \end{Bmatrix} = \mathbf{NP}\beta + \mathbf{NP}_F \beta_F \tag{8.64}$$

over the boundaries of the element, in which

$$\mathbf{N} = \begin{bmatrix} \nu_1 \dfrac{\partial}{\partial x_1} & \nu_2 \dfrac{\partial}{\partial x_2} & \nu_1 \dfrac{\partial}{\partial x_2} + \nu_2 \dfrac{\partial}{\partial x_1} \\ -\nu_1 & 0 & -\nu_2 \\ 0 & -\nu_2 & -\nu_1 \end{bmatrix}. \tag{8.65}$$

We also have to assume $\tilde{w}$ and its derivatives $\tilde{w}_{,\alpha}$ ($\alpha = 1, 2$) along the boundaries of the elements. For convenience we shall drop the tilde over w in the remainder of this section. Let us consider a typical edge with normal direction ν and tangential direction s (see fig. 8.3). Our objective is to describe the function w and its derivatives $w_{,\alpha}$ ($\alpha = 1, 2$) along this edge in terms of the values w and $w_{,\alpha}$ at the two end nodes in such a manner that when the corresponding nodal values of two neighboring elements coincide the function w and its first derivatives are continuous along this interelement boundary. This condition may be stated in another way: when for two neighboring elements the corresponding nodal values of w and the derivatives $w_{,s}$ and $w_{,\nu}$, respec-

8.6 Element-Stiffness Matrix for the Biharmonic Equation

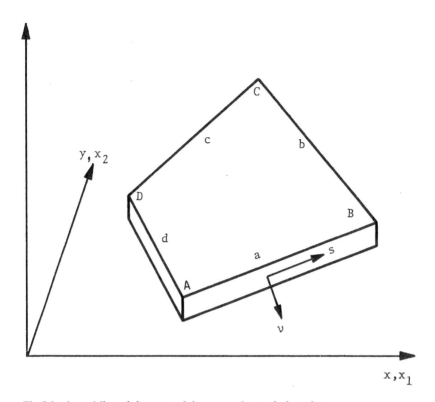

Fig. 8.3 A quadrilateral element, and sign convention on the boundary.

tively, along the tangential and normal directions of that boundary coincide, the function w as well as the normal slope $w_{,\nu}$ is continuous along the interelement boundary. The interpolation functions which satisfy the compatibility condition are, of course,

$$w(s) = H_{01}^{(1)} w_1 + H_{02}^{(1)} w_2 + H_{11}^{(1)} (w_{,s})_1 + H_{12}^{(1)} (w_{,s})_2 \tag{8.66}$$

and

$$w_{,\nu}(s) = H_{01}^{(0)} (w_{,\nu})_1 + H_{02}^{(0)} (w_{,\nu})_2, \tag{8.67}$$

where the subscripts 1 and 2 are used for the two ends of the edge under consideration, and s is measured from node 1. The Hermite functions (ch.6) used in the above equations are

$$\begin{aligned} H_{01}^{(0)}(s) &= 1 - \xi, \\ H_{02}^{(0)}(s) &= \xi, \end{aligned} \tag{8.68}$$

and

$$\begin{aligned}
H_{01}^{(1)}(s) &= 1 - 3\xi^2 + 2\xi^3, \\
H_{02}^{(1)}(s) &= 3\xi^2 - 2\xi^3, \\
H_{11}^{(1)}(s) &= l[\xi - 2\xi^2 + \xi^3], \\
H_{12}^{(1)}(s) &= l[\xi^3 - \xi^2],
\end{aligned} \qquad (8.69)$$

where $\xi = s/l$ and l is the length of the edge. The derivative $w_{,s}$ can be obtained from (8.66),

$$w_{,s}(s) = \frac{dH_{01}^{(1)}}{ds} w_1 + \frac{dH_{02}^{(1)}}{ds} w_2 + \frac{dH_{11}^{(1)}}{ds}(w_{,s})_1 + \frac{dH_{12}^{(1)}}{ds}(w_{,s})_2. \qquad (8.70)$$

In order to express $w(s)$ and $w_{,\alpha}(s)$ in terms of the nodal values of w and $w_{,\alpha}$ it is only necessary to express $w_{,\alpha}(s)$ by

$$\begin{aligned}
w_{,1}(s) &= -\nu_2 w_{,s}(s) + \nu_1 w_{,\nu}(s), \\
w_{,2}(s) &= \nu_1 w_{,s}(s) + \nu_2 w_{,\nu}(s),
\end{aligned} \qquad (8.71)$$

where $w_{,s}(s)$ and $w_{,\nu}(s)$ are given by (8.70) and (8.67), and then to substitute

$$\begin{aligned}
w_{,\nu} &= w_{,1}\nu_1 + w_{,2}\nu_2, \\
w_{,s} &= -w_{,1}\nu_2 + w_{,2}\nu_1
\end{aligned} \qquad (8.72)$$

for the nodal values of $w_{,s}$ and w.

Thus for a typical edge the boundary-displacement matrix is

$$\mathbf{u} = \{w \quad w_{,1} \quad w_{,2}\}^{\mathrm{T}} = \mathbf{Lq}, \qquad (8.73)$$

where $\mathbf{L}$ is defined as follows. Let

$$\mathbf{L}_1 = \begin{bmatrix} 1 & 0 & 0 \\ 0 & -\nu_2 & \nu_1 \\ 0 & \nu_1 & \nu_2 \end{bmatrix},$$

$$\mathbf{L}_2 = \begin{bmatrix} H_{01}^{(1)} & H_{11}^{(1)} & 0 \\ \dfrac{dH_{01}^{(1)}}{ds} & \dfrac{dH_{11}^{(1)}}{ds} & 0 \\ 0 & 0 & H_{01}^{(0)} \end{bmatrix}, \qquad (8.74)$$

$$\mathbf{L}_3 = \begin{bmatrix} H_{02}^{(1)} & H_{12}^{(1)} & 0 \\ \dfrac{dH_{02}^{(1)}}{ds} & \dfrac{dH_{12}^{(1)}}{ds} & 0 \\ 0 & 0 & H_{02}^{(0)} \end{bmatrix},$$

8.6 Element-Stiffness Matrix for the Biharmonic Equation

and

$$\begin{aligned} \mathbf{L}' &= \mathbf{L}_1\mathbf{L}_2\mathbf{L}_1, \\ \mathbf{L}'' &= \mathbf{L}_1\mathbf{L}_3\mathbf{L}_1. \end{aligned} \tag{8.75}$$

The matrix $\mathbf{L}$ can be expressed in terms of $\mathbf{L}'$ and $\mathbf{L}''$. These relations will depend on the particular side of the element. We use a quadrilateral element (fig. 8.3) as an example. The four sides are labeled a, b, c, and d, and the four corners, A, B, C, and D. Let the column matrix $\mathbf{q}$ be

$$\mathbf{q} = \{w_A \quad (w_{,1})_A \quad (w_{,2})_A \quad w_B \quad (w_{,1})_B \quad (w_{,2})_B \quad w_C \quad (w_{,1})_C \quad (w_{,2})_C \\ w_D \quad (w_{,1})_D \quad (w_{,2})_D\}^{\mathrm{T}}. \tag{8.76}$$

Then

$$\begin{aligned} \mathbf{L} &= \{\mathbf{L}'_a \quad \mathbf{L}''_a \quad 0 \quad 0 \ \} \quad \text{on side} \quad a, \\ &= \{0 \quad \mathbf{L}'_b \quad \mathbf{L}''_b \quad 0 \ \} \quad \text{on side} \quad b, \\ &= \{0 \quad 0 \quad \mathbf{L}'_c \quad \mathbf{L}''_c\} \quad \text{on side} \quad c, \\ &= \{\mathbf{L}'_d \quad 0 \quad 0 \quad \mathbf{L}''_d\} \quad \text{on side} \quad d. \end{aligned} \tag{8.77}$$

The subscripts a, b, c, and d mean that $\mathbf{L}'$ and $\mathbf{L}''$ are evaluated on edges a, b, c, and d. A substitution of (8.61) and (8.64) into (8.56) yields

$$\begin{aligned} \pi_{hm} &= \int_{\partial A_m} \mathbf{T}\mathbf{u}\, ds - \tfrac{1}{2}\int_{A_m} \boldsymbol{\sigma}^{\mathrm{T}}\boldsymbol{\sigma}\, dA \\ &= \boldsymbol{\beta}^{\mathrm{T}}\mathbf{G}\mathbf{q} + \boldsymbol{\beta}_F^{\mathrm{T}}\mathbf{G}_F\mathbf{q} - \tfrac{1}{2}\boldsymbol{\beta}^{\mathrm{T}}\mathbf{H}\boldsymbol{\beta} - \boldsymbol{\beta}^{\mathrm{T}}\mathbf{H}_F\boldsymbol{\beta}_F + C, \end{aligned} \tag{8.78}$$

in which

$$\begin{aligned} \mathbf{G} &= \int_{\partial A_m} (\mathbf{NP})^{\mathrm{T}}\mathbf{L}\, ds, \quad \mathbf{G}_F = \int_{\partial A_m} (\mathbf{NP}_F)^{\mathrm{T}}\mathbf{L}\, ds, \\ \mathbf{H} &= \int_{A_m} \mathbf{P}^{\mathrm{T}}\mathbf{P}\, dA, \quad \mathbf{H}_F = \int_{A_m} \mathbf{P}^{\mathrm{T}}\mathbf{P}_F\, dA, \quad C = -\tfrac{1}{2}\boldsymbol{\beta}_F^{\mathrm{T}}\int_{A_m} \mathbf{P}_F^{\mathrm{T}}\mathbf{P}_F\, dA\, \boldsymbol{\beta}_F. \end{aligned} \tag{8.79}$$

Equation (8.79) is in the same form as (8.42), with the present matrices defined in (8.79). As before, the vector $\boldsymbol{\beta}$ can be determined (and then eliminated) by maximizing π_{hm} with respect to $\boldsymbol{\beta}$. The element-stiffness matrix $\mathbf{k}$ and the element-load vector $\mathbf{Q}$ can then be obtained in the same form as that in (8.37) and (8.46).

8.7 Generalization

The construction of the hybrid finite-element model for the biharmonic equation can be easily generalized. Consider the equation

$$\frac{\partial^2}{\partial x_\alpha \partial x_\beta}\left(E_{\alpha\beta\lambda\theta}\frac{\partial^2 w}{\partial x_\lambda \partial x_\theta}\right) = p(x_1, x_2) \quad \text{in} \quad A \tag{8.80}$$

with boundary conditions

$$w = \bar{w} \quad \text{and} \quad \frac{\partial w}{\partial \nu} = \bar{w}_\nu \quad \text{on} \quad \partial A, \tag{8.81}$$

in which

$$E_{\alpha\beta\lambda\theta} = E_{\lambda\theta\alpha\beta} = E_{\beta\alpha\lambda\theta} = E_{\alpha\beta\theta\lambda}, \tag{8.82}$$

and $E_{\alpha\beta\lambda\theta}$ is positive definite.

Introducing the variables

$$M_{\alpha\beta} = E_{\alpha\beta\lambda\theta} w_{,\lambda\theta}, \tag{8.83}$$

eq. (8.80) becomes

$$M_{\alpha\beta,\alpha\beta} = -p(x_1, x_2), \tag{8.84}$$

which is the same as (8.55).

Then the proper functional for the hybrid finite-element model is

$$\Pi_h = \sum_{\substack{\text{all} \\ \text{elements}}} \pi_{hm},$$

where

$$\pi_{hm} = -\int_{\partial A_m}(M_{\alpha\beta}\nu_\beta \tilde{w}_{,\alpha} - Q_\alpha \nu_\alpha \tilde{w})ds + \int_{A_m}(-M_{\alpha\beta,\alpha\beta}w - pw \\ - \tfrac{1}{2}D_{\alpha\beta\lambda\theta}M_{\alpha\beta}M_{\lambda\theta})dA, \tag{8.85}$$

where Q_α is defined in the same way as (8.57), and $D_{\alpha\beta\lambda\theta}$ is related to (8.83) and is defined by the relation

$$w_{,\alpha\beta} = -D_{\alpha\beta\lambda\theta}M_{\lambda\theta}. \tag{8.86}$$

If the $M_{\alpha\beta}$ are so chosen that (8.84) is satisfied, then (8.85) is reduced to

$$\pi_{hm} = \int_{\partial A_m} (M_{\alpha\beta}\nu_\alpha \tilde{w}_{,\beta} - Q_\alpha \nu_\alpha \tilde{w})ds - \tfrac{1}{2}\int_{A_m} D_{\alpha\beta\lambda\theta}M_{\alpha\beta}M_{\lambda\theta}dA \tag{8.87}$$

for the hybrid stress model. The procedure for constructing the finite-element equations is identically the same as in the case of the biharmonic equation. In the present case the matrices $\mathbf{H}$ and $\mathbf{H}_F$ in (8.79) are defined by

$$\mathbf{H} = \int_{A_m} \mathbf{P}^T \mathbf{D} \mathbf{P} dA, \quad \mathbf{H}_F = \int_{A_m} \mathbf{P}^T \mathbf{D} \mathbf{P}_F dA, \tag{8.88}$$

where

$$\mathbf{D} = \begin{bmatrix} D_{1111} & D_{1122} & 2D_{1112} \\ D_{1122} & D_{2222} & 2D_{2212} \\ 2D_{1112} & 2D_{2212} & 4D_{1212} \end{bmatrix}. \tag{8.89}$$

8.8 Some Remarks

In constructing the hybrid elements in both secs. 8.3 and 8.6, approximations are introduced in the assumed forms for $\tilde{\phi}$ (or $\tilde{w}$) along the element boundaries and $\mathbf{V}$ (or the Ms) within the elements. The best policy in the selection of these functions is that the approximation used along the element boundaries should be of the same order as that within the elements. (Why?) For example, if a linear function is used for $\tilde{\phi}$ in (8.19), then a linear function such as that given in (8.23) is adequate for $\mathbf{V}$. There is a requirement relating the number of qs and the number of βs used in secs. 8.3 and 8.6 to insure that the final finite-element equations for the qs are solvable. It can be shown (Tong and Pian 1969) that the sufficiency condition to fulfill such a requirement is that, for each element, number of qs $- 1 \le$ number of βs for the harmonic equation; and number of qs $- 3 \le$ number of βs for the biharmonic equation.

By using the functional (8.12) and imposing different restrictions on ϕ and $\mathbf{V}$ in (8.13) one can obtain various finite-element models. For example, rather than choosing $\mathbf{V}$ to satisfy (8.18), we can choose $\mathbf{V}$ in terms of ϕ as given in (8.5), and require that $\phi = \tilde{\phi}$ over the element boundary. Then, equation (8.12) reduces to (8.4) which is the same as the displacement model discussed in ch. 2. As another example, if $\mathbf{V}$ is chosen to satisfy (8.18) as in the last section, and in addition $\mathbf{V} \cdot \boldsymbol{\nu}$ is required to be continuous across all the interelement boundaries and be equal to $\bar{\phi}_\nu$ over ∂A_ν, the so-called equilibrium model will result. Similarly, various other finite-element models for the biharmonic equation can be derived by requiring different restrictions on the Ms and w in (8.56). The behavior of some of these models is discussed by Pian and Mau (1972) and Tong and Pian (1970).

The greatest virtue of the hybrid method lies in its flexibility. There are no special constraints required between functions assumed for one element and those for another. It resolves the difficulties in seeking compatible functions for fourth-order differential equations (e.g., biharmonic equation for plate bending). And it allows one to utilize all local information in a particular problem to construct the element matrices. For instance, in a crack problem the elasticity solution near the crack tip indicates singular behavior for the stresses, with a given distribution of high-stress gradients. This known singular solution can be used to construct special element matrices around the crack tip (Tong, Pian, and Lasry 1973), while polynomials are used to construct element matrices for the other elements in regions where stress gradients are not high. The hybrid approach provides, in general, a convenient and rational way to match approximate solutions for two different regions (Tong, Mau, and Pian 1974), and therefore it is an important tool for analyzing nonhomogeneous media and discontinuous structures.

9 Selected Topics and Recent Developments

9.1 Dynamic Problems of Elastic Solids

The finite-element process described in ch. 5 for the static analysis of elastic solids can easily be extended to problems involving the dynamic behavior of structures. In such problems the finite-element idealization again leads to a set of simultaneous equations. In terms of assembled matrices these equations can be written in the form

$$\mathbf{M}\ddot{\mathbf{q}} + \mathbf{C}\dot{\mathbf{q}} + \mathbf{K}\mathbf{q} = \mathbf{Q}, \tag{9.1}$$

where $\mathbf{q}$ represents the vector of generalized coordinates for the entire structure and dots denote differentiation with respect to time. The matrices $\mathbf{M}$ and $\mathbf{C}$ are the assembled mass and damping matrices, respectively. For time-independent problems these matrices do not appear, and (9.1) reduces to the equation governing the static elastic response of the structure, where $\mathbf{K}$ is the familiar assembled stiffness matrix introduced earlier. As will be seen, the matrix $\mathbf{M}$ is symmetric, and when linear viscous damping is assumed to exist, the matrix $\mathbf{C}$ is also symmetric.

In order to derive appropriate expressions for such mass and damping matrices, a variational formulation can be used in conjunction with the finite element discretization. This formulation is in fact related to the application of Hamilton's principle to a dynamic structural system possessing a finite number of degrees of freedom. We consider the integral

$$H = \int_{t_1}^{t_2} (U - W - T)dt, \tag{9.2}$$

where T is the kinetic energy; U, the strain energy; and W, the work of the applied loads. Hamilton's principle may be written as

$$\delta H = 0, \tag{9.3}$$

and stated as follows: Among all possible time histories of displacement configurations that satisfy compatibility and geometric boundary conditions, and that also satisfy conditions at time t_1 and t_2, the actual solution makes H have a stationary value.

A simple example can be used to fix these ideas still further. Consider the spatially one-dimensional dynamic problem for the axial displacement $u(x, t)$ of a rod; this is governed by the equation

$$\frac{\partial}{\partial x}\left(AE\frac{\partial u}{\partial x}\right) + p(x) = \rho\ddot{u}, \tag{9.4}$$

where $\rho(x)$ is the mass density (mass per unit length), $p(x)$ is an applied distributed axial loading, and AE is the rod axial stiffness. It is also assumed that $u(x, t_1)$ and $u(x, t_2)$ are prescribed values at t_1 and t_2. Let the rod be divided into $n - 1$ elements. For an element between points x_i and x_{i+1}, nodal values at these points are denoted by $u(x_i, t) = q_i$ and $u(x_{i+1}, t) = q_{i+1}$. The equivalent variational statement associated with (9.4), which can also be identified with Hamilton's principle, is expressed by $H = 0$, where

$$H = \int_{t_1}^{t_2} \sum_{i=1}^{n-1} \int_{x_i}^{x_{i+1}} \left[\frac{1}{2}AE\left(\frac{\partial u}{\partial x}\right)^2 - pu - \frac{1}{2}\rho\dot{u}^2\right]dxdt. \tag{9.5}$$

If the interpolation function $\mathbf{D}(x)$ is introduced, with $u = \mathbf{D}^T\mathbf{q}_n$, where $\mathbf{q}_n$ is the generalized-coordinate vector for the nth element, then H can be written as

$$H = \int_{t_1}^{t_2} \left[\sum_{i=1}^{n-1} \tfrac{1}{2}\mathbf{q}_i^T\mathbf{k}_i\mathbf{q}_i - \mathbf{q}_i^T\mathbf{Q}_i - \tfrac{1}{2}\dot{\mathbf{q}}_i^T\mathbf{m}_i\dot{\mathbf{q}}_i\right]dt. \tag{9.6}$$

After all elements are summed, we have

$$H = \int_{t_1}^{t_2} \left[\tfrac{1}{2}\mathbf{q}^T\mathbf{K}\mathbf{q} - \mathbf{q}^T\mathbf{Q} - \tfrac{1}{2}\dot{\mathbf{q}}^T\mathbf{M}\dot{\mathbf{q}}\right]dt. \tag{9.7}$$

Setting $\delta H = 0$, integrating by parts, and noting that $\delta\mathbf{q}(t_1) = \delta\mathbf{q}(t_2) = 0$, the resulting Euler equation is $\mathbf{M}\ddot{\mathbf{q}} + \mathbf{K}\mathbf{q} = \mathbf{Q}$. It is clear that the matrices $\mathbf{K}$ and $\mathbf{Q}$ are a consequence of the strain energy and work terms respectively in (9.5), so that the matrix $\mathbf{M}$ results from consideration of the kinetic-energy term. For this example the element-mass matrix $\mathbf{m}_i$ is implied by the equation

$$\int_{x_i}^{x_{i+1}} \tfrac{1}{2}\rho\dot{u}^2 dx = \tfrac{1}{2}\dot{\mathbf{q}}_i^T\mathbf{m}_i\dot{\mathbf{q}}_i. \tag{9.8}$$

When the interpolation function used in (9.8) is the same as that used in obtaining $\mathbf{k}_i$, the mass matrix so obtained is referred to as a consistent mass matrix, and differs from the so-called lumped mass matrix. In the lumped-mass formulation a certain amount of structural mass that surrounds a given node is assumed to be concentrated or lumped at that node. The consistent element-mass matrix is usually fully populated while the lumped element-mass matrix is diagonal. As will be seen later in this section, each representation has certain advantages and disadvantages with regard to computer implementation.

To continue with the example of (9.4), we use the following linear interpolation at the element level:

$$u = u_1(1 - \xi) + u_2\xi, \tag{9.9}$$

where $\xi = (x - x_1)/\varepsilon$, $\varepsilon = x_2 - x_1$, and where u_1 and u_2 are the local nodal values of u at x_1 and x_2. To derive a consistent mass matrix for constant density over an element, we integrate:

$$\frac{1}{2} \rho \int_{x_1}^{x_2} \dot{u}^2 dx = \frac{\rho\varepsilon}{2} \int_0^1 [\dot{u}_1(1 - \xi) + \dot{u}_2\xi]^2 d\xi = \frac{\rho\varepsilon}{2}\left[\frac{\dot{u}_1^2}{3} + \frac{\dot{u}_2^2}{3} + 2\frac{\dot{u}_1\dot{u}_2}{6}\right]$$

$$= \frac{1}{2}[\dot{u}_1 \quad \dot{u}_2]\begin{bmatrix} \dfrac{\rho\varepsilon}{3} & \dfrac{\rho\varepsilon}{6} \\ \dfrac{\rho\varepsilon}{6} & \dfrac{\rho\varepsilon}{3} \end{bmatrix}\begin{Bmatrix} \dot{u}_1 \\ \dot{u}_2 \end{Bmatrix} \tag{9.10}$$

$$= \tfrac{1}{2}\dot{\mathbf{q}}_e^T \mathbf{m}_e \dot{\mathbf{q}}_e,$$

where the subscript e denotes element values. The matrix $\mathbf{m}_e$ is then given by

$$\mathbf{m}_e = \frac{\rho\varepsilon}{3}\begin{bmatrix} 1 & 1/2 \\ 1/2 & 1 \end{bmatrix}, \tag{9.11}$$

where $\rho = \bar{\rho}A$, $\bar{\rho}$ is mass per unit volume, and A is the rod cross-sectional area.

Consistent mass matrices for beam elements and triangular plane-stress elements can be derived in a similar fashion. For a beam element of length ε we can use the cubic interpolation (7.11) which was used to derive the element-stiffness matrix $\mathbf{k}$. We therefore express the deflection w in the form

$$w = \sum_{i=1}^{4} q_i f_i(x) = \mathbf{f}^T \mathbf{q}_e. \tag{9.12}$$

Recall that the element vector of generalized nodal coordinates is $\mathbf{q}_e^T = [w_1\ w_1'\ w_2\ w_2']$, and the $\mathbf{f}_i$ ($i = 1, 4$) are Hermite polynomials. The expression for the kinetic energy can be written as

$$\tfrac{1}{2}\int_0^\varepsilon \rho(x)\dot{w}^2 dx = \tfrac{1}{2}\dot{\mathbf{q}}_e^T\left(\int_0^\varepsilon \rho \mathbf{f}^T\mathbf{f}dx\right)\dot{\mathbf{q}}_e = \tfrac{1}{2}\dot{\mathbf{q}}_e^T\mathbf{m}_e\dot{\mathbf{q}}_e. \qquad (9.13)$$

The consistent mass matrix for the element is then given by

$$\mathbf{m}_e = \int_0^\varepsilon \rho\mathbf{f}^T\mathbf{f}dx = \frac{\rho\varepsilon}{420}\begin{bmatrix} 156 & 22\varepsilon & 54 & -13\varepsilon \\ & 4\varepsilon^2 & 13\varepsilon & -3\varepsilon^2 \\ & & 156 & -22\varepsilon \\ \text{sym} & & & 4\varepsilon^2 \end{bmatrix}, \qquad (9.14)$$

where again ρ is mass per unit length. The corresponding lumped mass matrix for a beam element having a uniform mass distribution can be obtained by lumping half the beam element mass onto each node, or

$$\mathbf{m}_e = \frac{\rho\varepsilon}{2}\begin{bmatrix} 1 & 0 & 0 & 0 \\ 0 & \varepsilon^2/12 & 0 & 0 \\ 0 & 0 & 1 & 0 \\ 0 & 0 & 0 & \varepsilon^2/12 \end{bmatrix}. \qquad (9.15)$$

One can even neglect the rotary inertia terms, $\varepsilon^2/12$, in (9.15) associated with the angular deflections w_1' and w_2'. In this case, however, the mass matrix becomes singular. This will make the numerical integration of (9.1) by the explicit integration scheme more cumbersome.

For the triangular plane-stress element we use the same interpolation as that used in sec. 5.3.1 for evaluating the element-stiffness matrix, i.e.,

$$\mathbf{u} = \begin{Bmatrix} u \\ v \end{Bmatrix} = \mathbf{D}\mathbf{q}_e, \qquad (9.16)$$

where

$$\mathbf{D} = \begin{bmatrix} f_1 & 0 & f_2 & 0 & f_3 & 0 \\ 0 & f_1 & 0 & f_2 & 0 & f_3 \end{bmatrix},$$

$$\mathbf{q}_e^T = [u_1\ v_1\ u_2\ v_2\ u_3\ v_3],$$

and the $f_i(x, y)$ ($i = 1, 2, 3$) are as defined in sec. 5.3.1. The expression for

9.1 Dynamic Problems of Elastic Solids

the kinetic energy of an element of thickness t, mass per unit volume $\bar{\rho}$, and area A is

$$\tfrac{1}{2} t\bar{\rho} \iint_{A_p} \dot{\mathbf{u}}^T \dot{\mathbf{u}}\, dxdy = \tfrac{1}{2}\dot{\mathbf{q}}_e^T \left(t\bar{\rho} \iint_{A_p} \mathbf{D}^T\mathbf{D}\, dxdy \right) \dot{\mathbf{q}}_e = \tfrac{1}{2}\dot{\mathbf{q}}_e^T \mathbf{m}_e \dot{\mathbf{q}}_e. \tag{9.17}$$

Thus the consistent mass matrix is given by

$$\mathbf{m}_e = t\bar{\rho} \iint_A \mathbf{D}^T\mathbf{D}\, dxdy = \frac{\bar{\rho} t A_p}{12} \begin{bmatrix} 2 & 1 & 1 & 0 & 0 & 0 \\ & 2 & 1 & 0 & 0 & 0 \\ & & 2 & 0 & 0 & 0 \\ & & & 2 & 1 & 1 \\ & & & & 2 & 1 \\ \text{sym} & & & & & 2 \end{bmatrix}. \tag{9.18}$$

The corresponding lumped mass matrix for a triangular element of uniform mass is obtained by lumping one-third of the element mass onto each node to get

$$\mathbf{m}_e = \frac{\bar{\rho} t A_p}{3} \begin{bmatrix} 1 & 0 & 0 & 0 & 0 & 0 \\ & 1 & 0 & 0 & 0 & 0 \\ & & 1 & 0 & 0 & 0 \\ & & & 1 & 0 & 0 \\ & & & & 1 & 0 \\ \text{sym} & & & & & 1 \end{bmatrix}. \tag{9.19}$$

Note that the total mass of the element is equally divided among the nodes for each translation quantity u and v.

In order to determine the damping matrix $\mathbf{C}$ we need to consider the virtual work done by damping forces. As an example consider the beam element and the case where damping forces are proportional to velocity (i.e., force $= \gamma \dot{w}$). Using the interpolation given by (9.12), we find that the virtual work done is

$$\int_0^\varepsilon \gamma \dot{w} \delta w\, dx = \delta\mathbf{q}_e^T \left(\int_0^\varepsilon \gamma \mathbf{f}^T \mathbf{f}\, dx \right) \dot{\mathbf{q}}_e = \delta\mathbf{q}_e^T \mathbf{c}_e \dot{\mathbf{q}}_e; \tag{9.20}$$

so that the element damping matrix is calculated from

$$\mathbf{c}_e = \int_0^\varepsilon \gamma \mathbf{f}^\mathrm{T} \mathbf{f} dx. \tag{9.21}$$

It should be noted that both the matrices $\mathbf{M}$ and $\mathbf{C}$ for the entire structure are assembled by using the element matrices $\mathbf{m}_e$ and $\mathbf{c}_e$ in a fashion analogous to that in obtaining $\mathbf{K}$. Once the system of equations implied by (9.1) is written out, its solution can be obtained by standard methods (Clough 1971; Pian 1975). The eigenvalue problems (free vibration) associated with the corresponding homogeneous system of (9.1) can also be treated by well-established techniques (Clough and Bathe 1972).

From the point of view of computer implementation and questions of accuracy, it is important to consider the effect of employing either the consistent or lumped mass matrix, since both formulations are used quite often in practice. Since the lumped mass matrix is often a diagonal matrix, it requires less storage in a computer and less time to generate than the corresponding sparsely populated consistent mass matrix (a consistent mass matrix requires the same amount of storage space as a stiffness matrix). The diagonal form also facilitates calculation when using an explicit numerical-integration scheme, where new vectors to be computed in a given time step are functions only of known vectors computed in previous steps (Clough and Bathe 1972; Tong and Rossettos 1975). As an example of the integration of (9.1) for the case $\mathbf{C} = 0$ by the central difference scheme, we have

$$\mathbf{M}\mathbf{q}_{n+1} = \mathbf{M}(2\mathbf{q}_n - \mathbf{q}_{n-1}) + \mathbf{K}\mathbf{q}_n(\Delta t)^2 \tag{9.22}$$

where $\mathbf{q}_n$ denotes the solution for $\mathbf{q}$ at time $t = n\Delta t$. If $\mathbf{M}$ is a diagonal matrix, the solution $\mathbf{q}_{n+1}$ can be evaluated rapidly from $\mathbf{q}_n$ and $\mathbf{q}_{n-1}$. Otherwise, the equation to be solved for $\mathbf{q}_{n+1}$ at each time step requires considerably more computation time.

A particularly useful feature and the principle advantage of the use of a consistent mass matrix is that the natural frequencies obtained in a vibration problem are proven upper bounds if the interpolation is conforming. The use of lumped mass matrices tends to lower natural frequencies, while the stiffness matrix in a compatible model will tend to increase frequencies. Since the two effects oppose each other, there may be instances where accurate frequencies are obtained because of this. Unfortunately, this rarely happens in reality except for the case where the stiffness matrix is formulated to be excessively stiff (Felippa 1966).

9.1 Dynamic Problems of Elastic Solids

In regard to the rate of convergence of mode shapes and frequencies by the finite-element method using consistent and lumped mass formulations, the following results have been established (Tong, Pian and Bucciarelli 1971). In cases where certain of the lowest-order elements are used, such as the constant-strain triangle or four-node quadrilateral elements, for second-order differential equations a lumped mass matrix provides the same order of approximation as a consistent mass matrix. However, when using higher-order elements, or for problems involving higher-order differential equations, there appears to be no systematic procedure for finding a palatable way of lumping the mass matrix to improve convergence. In these cases, the consistent mass matrix will often given the higher-order approximation. In general, all that may be recommended is that the lumped mass matrix be used for the low-order elements described above, where they provide the same order of approximation.

In order to treat problems in real structures, however, such as auto or rail vehicles, buildings, or bridges, where one must often use a gross approximation for the stiffness description, and the mass distribution itself is often distinctly concentrated in certain areas, then there is no doubt that one should use a lumped-mass approach. An example where this concept has been used to advantage concerns the structural deformation of vehicles in a crash (Rossettos and Weinstock 1974; Tong and Rossettos 1975). In these cases the vehicle can be conveniently divided into individual modules, each with its own stiffness and mass characteristics. Mass lumping becomes clear from the physical problem (e.g., engine mass, etc.), the specific engineering information one is seeking, and the desired accuracy. The modules are combined by standard finite-element methodology.

In a study of numerical instability in dynamic problems (Tong 1971) involving the explicit integration procedure, it has been shown that the time increment Δt required for stability is inversely related to the maximum frequency of the system. In the study it was shown that the lumped mass matrix leads to a lower value for the maximum frequency than the consistent-mass formulation, so that a larger time step could be used for integration. This particular advantage is no longer so clear when an implicit integration scheme is used. Implicit schemes, where expressions for the new vectors in a given time step contain the new vectors, are often unconditionally stable (Clough and Bathe 1972).

Considering the discussion in the last several paragraphs, we can say that it is not always a straightforward matter to decide which matrix formulation is to be preferred in any given situation.

9.2 Hybrid Singular and Infinite-Domain Superelements

The versatility of the finite-element method has been extensively exposed in earlier chapters. The essential features involve the utilization of local approximations and the incorporation of known characteristics of a given problem to establish the final solution for an entire system. We shall now give additional evidence of such versatility by introducing the concept of the superelement and giving three examples of such elements.

When treating a problem whose solution is smooth, the order of approximation within an element is proportional to the order of the polynomials used for the interpolation functions (Fix and Strang 1969). When the solution involves a singularity, however, the order of the approximation is dominated by the order of the singularity; the high-order elements discussed in ch. 6 cannot improve the rate of convergence (Tong and Pian 1972; Key 1966). This difficulty can be resolved by using the so-called hybrid superelement (Tong, Pian, and Lasry 1973). The next two examples will illustrate the procedure.

9.2.1 Wedge Element for Laplace's Equation

Near a wedge of angle $2(\pi - \alpha)$, the solution ϕ of Laplace's equation in polar coordinates can be taken in the form

$$\phi = \sum_{i=1}^{4} \phi_i(r, \theta), \tag{9.23}$$

where

$$\phi_1 = \sum_{j=0}^{N} \beta_{1j} R^{j+1/2} \cos\left(j + \frac{1}{2}\right) \frac{\pi}{\alpha} \theta,$$

$$\phi_2 = \sum_{j=1}^{N} \beta_{2j} R^j \sin j \frac{\pi}{\alpha} \theta,$$

$$\phi_3 = \sum_{j=1}^{N} \beta_{3j} R^j \cos j \frac{\pi}{\alpha} \theta,$$

$$\phi_4 = \sum_{j=0}^{N} \beta_{4j} R^{j+1/2} \sin\left(j + \frac{1}{2}\right) \frac{\pi}{\alpha} \theta,$$

$$R = r^{\pi/\alpha}. \tag{9.24}$$

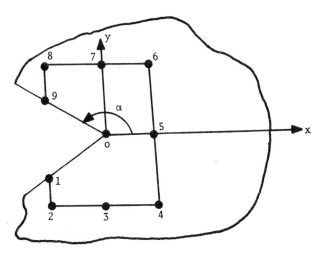

Fig. 9.1 A wedge superelement.

N is a positive integer and the β's are constants. In the above series, ϕ_1 and ϕ_2 satisfy

$$\phi = 0 \quad \text{at} \quad \theta = \pm \alpha, \tag{9.25}$$

and ϕ_3 and ϕ_4 satisfy

$$\frac{\partial \phi}{\partial \theta} = 0 \quad \text{at} \quad \theta = \pm \alpha. \tag{9.26}$$

The solution ϕ is not analytic at $r = 0$, and, in fact $\partial \phi_1 / \partial r$ and $\partial \phi_4 / \partial r$ become infinite there. In this case an improved finite-element approximation can be developed by using the series in (9.24) as the interpolation function to construct a superelement around the tip of the wedge. For the rest of the domain, where the solution has no singularity, we can use regular elements with polynomials as interpolation functions as done in previous chapters. Such a superelement is shown in fig. 9.1 with nodes 0 through 9. Using this approach, the solution for ϕ within the superelement will, in general, not be continuous with the solution for ϕ in the surrounding elements. The hybrid technique can be used to match the solutions for the two regions. Consider an example where the boundary condition on the wedge surface is

$$\phi(r, \pm \alpha) = \bar{\phi}(r). \tag{9.27}$$

As indicated in ch. 8, a hybrid functional for the present superelement can be written as

$$\pi(\phi, \bar{\phi}) = \int_{\partial A} \bar{\phi}\phi_n ds - \tfrac{1}{2} \int_A (\nabla\phi)^2 dxdy \tag{9.28}$$

where ∂A is the element boundary, $\phi_n (= \partial\phi/\partial n)$ is the normal derivative along ∂A, and $\bar{\phi}$ is that ϕ defined only on ∂A which is equal to $\bar{\phi}$ on the wedge surface. In constructing the element matrices, we choose

$$\phi = \phi_1 + \phi_2 + \bar{\phi}_3, \tag{9.29}$$

where ϕ_1 and ϕ_2 are defined in (9.24), and $\bar{\phi}_3$ has the same form as ϕ_3 with the $\bar{\beta}_{3i}$'s chosen to satisfy (9.27). Since ϕ_1 and ϕ_2 vanish at the wedge surface, ϕ as defined in (9.29) satisfies the boundary condition (9.27) and Laplace's equation. On substituting (9.29) into (9.28), using Green's theorem for the area integral, and noting that ϕ satisfies $\nabla^2\phi = 0$ within the element, we obtain

$$\pi(\phi, \bar{\phi}) = \int_1^9 \bar{\phi}\phi_n ds - \tfrac{1}{2}\int_1^9 \phi\phi_n ds + \tfrac{1}{2}\left(\int_9^0 + \int_0^1\right) \bar{\phi}\phi_n ds. \tag{9.30}$$

We may write ϕ and $\bar{\phi}$ in matrix form as follows:

$$\phi_1 + \phi_2 = \mathbf{A}(r, \theta) \begin{Bmatrix} \beta_{11} \\ \vdots \\ \beta_{1N} \\ \beta_{21} \\ \vdots \\ \beta_{2N} \end{Bmatrix} = \mathbf{A}\boldsymbol{\beta},$$

$$\frac{\partial}{\partial n}(\phi_1 + \phi_2) = (\phi_1 + \phi_2)_n = \mathbf{A}_n \boldsymbol{\beta},$$

$$\bar{\phi}_3 = \mathbf{B}(r, \theta) \begin{Bmatrix} \bar{\beta}_{31} \\ \vdots \\ \bar{\beta}_{3N} \end{Bmatrix} = \mathbf{B}\bar{\boldsymbol{\beta}}_3, \tag{9.31}$$

$$\frac{\partial \phi_3}{\partial n} = (\phi_3)_n = \mathbf{B}_n \bar{\boldsymbol{\beta}}_3$$

$$\bar{\phi} = \mathbf{L}(s) \begin{Bmatrix} q_1 \\ \vdots \\ q_9 \end{Bmatrix} = \mathbf{L}(s)\mathbf{q},$$

where $\mathbf{A}$, $\mathbf{A}_n$, $\mathbf{B}$, $\mathbf{B}_n$, and $\mathbf{L}$ are row matrices. The qs are the nodal values of $\tilde{\phi}$ along the element boundaries, and $\mathbf{L}$ is the interpolation function matrix defined only on ∂A, which is chosen so that, when the nodal values of $\tilde{\phi}$ are matched with those of the adjacent element, $\tilde{\phi}$ is the same for the two elements over their common boundaries. For example, if the elements around the superelement have a linear-varying ϕ along their element boundaries, we should have

$$\tilde{\phi} = \mathbf{L}\mathbf{q} = \begin{bmatrix} 1 - \dfrac{s}{l} & \dfrac{s}{l} \end{bmatrix} \begin{Bmatrix} q_p \\ q_{p+1} \end{Bmatrix}$$

between node p and node $p+1$, $p = 1, 2, \cdots 8$, where l is the distance between the two nodes and s the distance measured from node p. A substitution of (9.31) into (9.30) yields

$$\pi(\phi, \tilde{\phi}) = \boldsymbol{\beta}^T \mathbf{G} \mathbf{q} + \mathbf{q}^T \mathbf{Q}_1 - \tfrac{1}{2} \boldsymbol{\beta}^T \mathbf{H} \boldsymbol{\beta} + \boldsymbol{\beta}^T \mathbf{Q}_2 + c_1, \tag{9.32}$$

where

$$\begin{aligned}
\mathbf{G} &= \int_1^9 \mathbf{A}_n^T \mathbf{L}\, ds, \qquad \mathbf{Q}_1 = \int_1^9 \mathbf{L}^T (\bar{\phi}_3)_n\, ds, \\
\mathbf{H} &= \tfrac{1}{2} \int_1^9 (\mathbf{A}^T \mathbf{A}_n + \mathbf{A}_n^T \mathbf{A})\, ds, \\
\mathbf{Q}_2 &= -\int_1^9 \mathbf{A}_n^T \bar{\phi}_3\, ds + \tfrac{1}{2} \left(\int_9^0 + \int_0^1 \right) \mathbf{A}_n^T \bar{\phi}\, ds, \\
c_1 &= -\tfrac{1}{2} \int_1^9 \bar{\phi}_3 (\bar{\phi}_3)_n\, ds + \tfrac{1}{2} \left(\int_9^0 + \int_0^1 \right) (\bar{\phi}_3)_n \bar{\phi}\, ds.
\end{aligned} \tag{9.33}$$

We can eliminate $\boldsymbol{\beta}$ in (9.32) in the same manner as we did in ch. 8, with (9.32) now playing the same role as (8.27), so that we obtain

$$\pi = \tfrac{1}{2} \mathbf{q}^T \mathbf{k} \mathbf{q} - \mathbf{q}^T \mathbf{Q} + c_2, \tag{9.34}$$

where

$$\begin{aligned}
c_2 &= c_1 + \tfrac{1}{2} \mathbf{Q}_2^T \mathbf{H}^{-1} \mathbf{Q}_2, \\
\mathbf{k} &= \mathbf{G}^T \mathbf{H}^{-1} \mathbf{G} \quad \text{and} \quad \mathbf{Q} = -\mathbf{Q}_1 - \mathbf{G}^T \mathbf{H}^{-1} \mathbf{Q}_2.
\end{aligned} \tag{9.35}$$

The matrix $\mathbf{k}$ is the stiffness matrix of the superelement, and is in the same

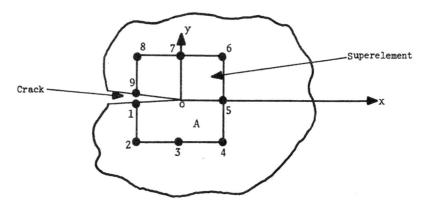

Fig. 9.2 A crack-tip superelement.

form as (8.37). This superelement is then to be used with elements whose variation of ϕ along their boundaries is the same as that of $\tilde{\phi}$.

9.2.2 Crack Element in Plane Elasticity

Consider a plane-elasticity problem with a crack as shown in fig. 9.2. For simplicity, we shall assume a traction-free crack surface. The nature of the singularity at the crack tip is known to be $\sigma_{kj} \sim 1/\sqrt{r}$ for the stresses, and $(u, v) \sim \sqrt{r}$ for the displacements, where r is the distance from the crack tip. To construct a superelement around the crack tip which accounts for singularities of all orders, the complex variable formulations of Muskhelishvili can be used (Tong, Pian, and Lasry 1973) to express the stresses and displacements in the form:

$$\begin{aligned}
\sigma_y + \sigma_x &= 4\mathrm{Re}[\phi'(\zeta)/w'(\zeta)], \\
\sigma_y - \sigma_x + 2i\sigma_{xy} &= 2\{\overline{w(\zeta)}[\phi'(\zeta)/w'(\zeta)]' + \psi'(\zeta)\}/w'(\zeta), \\
2\mu(u + iv) &= \eta\phi(\zeta) - w(\zeta)\overline{\phi'(\zeta)/w'(\zeta)} - \overline{\psi(\zeta)},
\end{aligned} \quad (9.36)$$

where

$$w(\zeta) = x + iy = \zeta^2,$$

and $i = \sqrt{-1}$. Also, $\mu = E/2(1 + \nu)$; $\eta = 3 - 4\nu$ for plane strain or $\eta = (3 - \nu)/(1 + \nu)$ for plane stress; E and ν are, respectively, Young's modulus and Poisson's ratio; ( )' denotes differentiation, and $\overline{(\ )}$ is the complex conjugate. Both ϕ and ψ are analytic functions. The argument of ζ is limited to the range $-\pi/2 < \arg \zeta \leq \pi/2$.

By the extension concept of Muskhelishvili (Bowie and Neal 1970;

Muskhelishvilli, 1953), the stress-free condition (the traction components $T_x = T_y = 0$) corresponding to

$$\phi(\zeta) + w(\zeta)\overline{\phi'(\zeta)/w'(\zeta)} + \overline{\psi(\zeta)} = 0 \tag{9.37}$$

over the crack surface can easily be satisfied by choosing*

$$\psi(\zeta) = - \overline{\phi(-\bar{\zeta})} - \overline{w(-\bar{\zeta})}\phi'(\zeta)/w'(\zeta). \tag{9.38}$$

Using (9.37) and (9.38), a hybrid functional for the element around the crack tip shown in fig. 9.2 can be written as

$$\pi = \int_{\partial A} \tilde{u}_k \sigma_{kj} \nu_j ds - \tfrac{1}{2} \int_{\partial A} u_k \sigma_{kj} \nu_j ds;$$

or, in matrix form as

$$\pi = \int_{\partial A} \mathbf{T}^T \tilde{\mathbf{u}} ds - \tfrac{1}{2} \int_{\partial A} \mathbf{T}^T \mathbf{u} ds, \tag{9.39}$$

in which

$$\mathbf{T} = \begin{Bmatrix} T_1 \\ T_2 \end{Bmatrix} = \begin{Bmatrix} \sigma_x \nu_x + \sigma_{xy} \nu_y \\ \sigma_{xy} \nu_x + \sigma_y \nu_y \end{Bmatrix},$$
$$\mathbf{u} = \begin{Bmatrix} u \\ v \end{Bmatrix} \quad \tilde{\mathbf{u}} = \begin{Bmatrix} \tilde{u} \\ \tilde{v} \end{Bmatrix}, \tag{9.40}$$

where $\tilde{\mathbf{u}}$ represents the element boundary displacements which are to be the same as those of the neighboring elements. In constructing the element-stiffness matrix for the superelement, one may assume

$$\phi(\zeta) = \sum_{j=1}^{N} b_j \zeta^j,$$
$$\psi(\zeta) = - \sum_{j=1}^{N} \left[\bar{b}_j (-1)^j + \frac{j}{2} b_j \right] \zeta^j, \tag{9.41}$$

where N is a finite integer and $b_j = \beta_j + i\beta_{N+j}$, with the βs being real con-

---

*On the imaginary axis (crack surface), $\zeta = - \bar{\zeta}$. Therefore, if $\psi(\zeta)$ is chosen as in (9.38), eq. (9.37) is identically satisfied on the crack surface.

stants and $\beta_{N+2} = 0$. (This is because $\phi = i\zeta^2$, and $\psi = 0$ make no contribution to the stresses.) From (9.36), (9.40), and (9.41) one can express

$$\mathbf{T} = \mathbf{R}\beta, \qquad \mathbf{u} = \mathbf{U}\beta, \tag{9.42}$$

where β is the column vector with components $\beta_1, \beta_2, \cdots, \beta_{2N}$, excluding β_{N+2}. The displacements $\tilde{\mathbf{u}}$ shall be taken in terms of the generalized displacements $\mathbf{q}$, namely,

$$\tilde{\mathbf{u}} = \mathbf{L}\mathbf{q}, \tag{9.43}$$

where $\mathbf{L}$ is the interpolation-function matrix defined only on the element boundary ∂A; $\mathbf{L}$ is chosen such that, when the corresponding generalized coordinates of the adjacent elements are matched, $\tilde{\mathbf{u}}$ is the same for the two elements over their common boundaries. When the surrounding elements are the regular elements of the displacement model, $\tilde{\mathbf{u}}$ is simply the boundary displacement of the surrounding elements at their common boundaries. In the case of a linear variation of $\tilde{\mathbf{u}}$ along the element boundaries, we have

$$\tilde{\mathbf{u}} = \begin{bmatrix} 1 - \dfrac{s}{l} & 0 & \dfrac{s}{l} & 0 \\ 0 & 1 - \dfrac{s}{l} & 0 & \dfrac{s}{l} \end{bmatrix} \begin{Bmatrix} u_p \\ v_p \\ u_{p+1} \\ v_{p+1} \end{Bmatrix} = \mathbf{L}\mathbf{q} \tag{9.44}$$

between node p and node $p + 1$, where l is the distance between the two nodes, (u_p, v_p) is the value of $\tilde{\mathbf{u}}$ at node p, and s is distance measured from p. A substitution of (9.42) and (9.43) into (9.39) yields

$$\pi = \beta^{\mathrm{T}}\mathbf{G}\mathbf{q} - \tfrac{1}{2}\beta^{\mathrm{T}}\mathbf{H}\beta, \tag{9.45}$$

in which*

$$\mathbf{G} = \int_{\partial A} \mathbf{R}^{\mathrm{T}}\mathbf{L}\,ds, \qquad \mathbf{H} = \tfrac{1}{2}\int_{\partial A} (\mathbf{R}^{\mathrm{T}}\mathbf{U} + \mathbf{U}^{\mathrm{T}}\mathbf{R})\,ds. \tag{9.46}$$

Since the βs can be assumed independently of those of the surrounding elements, one may eliminate the βs from (9.45) if so desired. Thus $\delta\pi = 0$ gives

$$\mathbf{H}\beta = \mathbf{G}\mathbf{q} \quad \text{or} \quad \beta = \mathbf{H}^{-1}\mathbf{G}\mathbf{q}. \tag{9.47}$$

*No integrations from node 9 to node 0 and from node 0 to node 1 are necessary, because $\mathbf{R} = 0$.

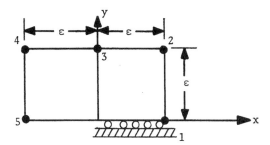

Fig. 9.3 Superelement for symmetric deformation.

Substituting for β as given in (9.47), eq. (9.45) becomes

$$\pi = \tfrac{1}{2}\mathbf{q}^T\mathbf{k}\mathbf{q}, \tag{9.48}$$

where

$$\mathbf{k} = \mathbf{G}^T\mathbf{H}^{-1}\mathbf{G} \tag{9.49}$$

is the element-stiffness matrix of the superelement.

From the series expressions for the stresses, it is seen that the stress intensity factors of mode I and mode II are related to β_1 and β_{N+1}, i.e.,

$$k_\mathrm{I} = \sqrt{2}\beta_1, \qquad k_\mathrm{II} = \sqrt{2}\beta_{N+1}. \tag{9.50}$$

Since $\boldsymbol{\beta}$ is related to the nodal displacements by (9.47), one can express k_I and k_II in the form

$$k_\mathrm{I} = \mathbf{B}_\mathrm{I}\mathbf{q}, \qquad k_\mathrm{II} = \mathbf{B}_\mathrm{II}\mathbf{q}, \tag{9.51}$$

where $\mathbf{q}^T = [u_1 \; v_1 \; u_2 \; v_2 \; \cdots]$ is the nodal displacement vector of the superelement.

In the case of symmetric deformation, it is only necessary to use half of the superelement, as shown in fig. 9.3. Its element-stiffness matrix for plane stress, with $N = 9$ in (9.41), $\nu = 0.3$, is given in table 9.1, and the corresponding vector $\mathbf{B}_\mathrm{I}$ is given in table 9.2. It has been shown that the use of such superelements to solve stress intensity factor problems in fracture mechanics is extremely accurate and efficient (Tong, Pian, and Lasry 1973; Tong and Pian 1973; Orringer, Lin, Stalk, Tong, and Mar 1976).

Table 9.1 Superelement stiffness matrix for symmetric deformation, $\nu = 0.3$.

$$k = \frac{Et}{10} \begin{bmatrix} 3.3424 & & & & & & & & & \\ -1.0221 & 1.1903 & & & & & & & & \\ 0.2349 & 1.0909 & 3.1226 & & & & & & & \\ -0.5313 & 0.7923 & 0.9932 & 4.0338 & & & & & & \\ -1.6870 & -0.1558 & -1.3593 & 0.5978 & 5.7584 & & & & & \\ 1.4594 & -0.9537 & -0.0866 & 1.1371 & -0.3201 & 6.9045 & & & & \\ -0.9502 & 0.0665 & -1.0610 & -0.4408 & -1.0757 & 0.4306 & 2.9899 & & & \\ 0.6302 & 0.1008 & 0.2787 & -0.0989 & -1.3924 & 0.0431 & -0.3169 & 1.4739 & & \\ -0.9400 & 0.0205 & -0.9371 & -0.6190 & -1.6364 & -1.4833 & 0.0970 & 0.8004 & 3.4165 & \\ -0.0530 & 0.1472 & -0.0307 & 0.0898 & -0.3234 & -1.4181 & -0.7122 & -0.3075 & 1.1193 & 1.5582 \end{bmatrix}$$

Table 9.2 Row vector for evaluation of k_1 of symmetric deformation, $\nu = 0.3$.

$\mathbf{B}_I = (1/50\varepsilon)^{1/2} \{0.1609 \quad 0.4150 \quad 0.3353 \quad 0.3735 \quad 0.2239$
$\qquad\qquad\qquad\qquad\qquad 2.6821 \quad -0.6514 \quad -0.9985 \quad 0.0686 \quad 0.1644\}$

9.2.3 Element for an Infinite Domain

In treating a problem involving an infinite domain, where the solution decays rapidly at infinity, we may approximate the problem by one with finite boundaries. That is, by choosing the boundaries far from the region of interest, we can solve the problem using regular elements in a finite domain. However, for certain problems (such as those involving wave propagation), the disturbance is often transmitted to infinity. Any truncation to a finite domain will produce a false solution.

To treat the finite domain properly, we can use the hybrid approach. We first divide the domain, by means of a curve C, into the two regions R_1 and R_2. This is shown in fig. 9.4, where R_1 is sufficiently large to enclose the area of interest, say around B, and R_2 is the rest of the domain which extends to infinity. The region R_1 is to be divided further into a finite number of elements where the regular finite-element approximations discussed in earlier chapters can be used. The region R_2 is here considered to be a single superelement. Within R_2 the asymptotic solution to ϕ will be used as the interpolation function. The hybrid technique is then used to match the solutions in R_1 and R_2. Two examples will be considered to illustrate this idea further.

(a) *A superelement for Laplace's equation.* Consider the boundary value problem

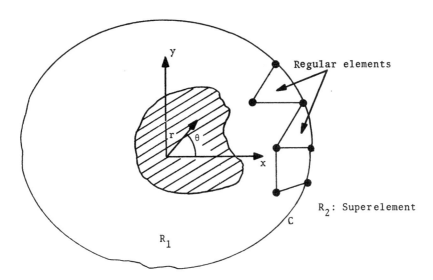

Fig. 9.4 An infinite-domain element.

$$\nabla^2 \phi = 0 \quad \text{in} \quad R_1 + R_2;$$
$$\partial \phi / \partial n = \bar{\phi}_n \quad \text{on} \quad \partial B;$$
$$\nabla \phi \to 0 \quad \text{as} \quad (x^2 + y^2)^{1/2} \to \infty, \tag{9.52}$$

where $\partial/\partial n$ is the normal derivative at the boundary. A variational functional can be written in the form

$$\pi(\phi, \tilde{\phi}) = \pi_1(\tilde{\phi}) + \pi_2(\phi, \tilde{\phi}), \tag{9.53}$$

where

$$\pi_1(\tilde{\phi}) = \tfrac{1}{2} \int_{R_1} (\nabla \tilde{\phi}) dx dy,$$
$$\pi_2(\phi, \tilde{\phi}) = \int_C \tilde{\phi} \phi_n ds - \tfrac{1}{2} \int_{R_2} (\nabla \phi)^2 dx dy, \tag{9.54}$$

in which $\phi_n = \partial \phi / \partial n$, π_1 is the regular functional for R_1, and π_2 is the hybrid functional for R_2. The finite-element matrices for R_2 can be constructed by choosing ϕ in the form*

---

*If there is circulation for a potential flow, an additional term proportional to θ is needed.

$$\phi = \sum_{j=1}^{N} (\beta_j \cos j\theta + \beta_{j+N} \sin j\theta)/r^j. \tag{9.55}$$

This form satisfies Laplace's equation, and the βs are unknown constants. Substituting into (9.54), π_2 becomes

$$\pi_2(\phi, \tilde{\phi}) = \int_C \tilde{\phi} \phi_n ds - \tfrac{1}{2} \int_C \phi \phi_n ds, \tag{9.56}$$

where the integration over the infinite domain is eliminated. We can write ϕ and ϕ_n in matrix form by using (9.55) to yield

$$\phi = \mathbf{A}(r, \theta)\boldsymbol{\beta}, \qquad \phi_n = \mathbf{A}_n \boldsymbol{\beta}.$$

Assuming $\tilde{\phi}$ in the same form as in (9.31), i.e., $\tilde{\phi} = \mathbf{L}(s)\mathbf{q}$, π_2 becomes

$$\pi_2 = \boldsymbol{\beta}^\mathrm{T} \mathbf{G} \mathbf{q} - \tfrac{1}{2} \boldsymbol{\beta}^\mathrm{T} \mathbf{H} \boldsymbol{\beta}, \tag{9.57}$$

where

$$\mathbf{G} = \int_C \mathbf{A}_n^\mathrm{T} \mathbf{L} ds,$$

$$\mathbf{H} = \tfrac{1}{2} \int_C [\mathbf{A}^\mathrm{T} \mathbf{A}_n + \mathbf{A}_n^\mathrm{T} \mathbf{A}] ds.$$

The elimination of $\boldsymbol{\beta}$ can be accomplished as was done previously in regard to (9.32), to yield

$$\pi_2 = \tfrac{1}{2} \mathbf{q}^\mathrm{T} \mathbf{k} \mathbf{q}, \tag{9.58}$$

where the hybrid stiffness matrix for the element is given again by the standard form $\mathbf{k} = \mathbf{G}^\mathrm{T} \mathbf{H}^{-1} \mathbf{G}$.

(b) Long wave in potential flow. Consider a three-dimensional region whose (x, y)-plane is that shown in fig. 9.4. B is a solid body. Under longwave theory for potential flow (Stoker 1957; Chen and Mei 1974), with waves propagating along the (x, y)-plane, the velocity potential $\phi e^{-i\omega t}$ is governed by

$$\nabla \cdot (h \nabla \phi) + \frac{\omega^2}{g} \phi = 0 \quad \text{in} \quad R_1 + R_2, \tag{9.59}$$

where $h(x, y)$ is the depth of the fluid, and g is the gravitational acceleration. The boundary condition on ∂B is $\partial \phi / \partial n = 0$. For an incident wave

$$\phi^I = -\frac{iga_0}{\omega} e^{ikr\cos(\theta-\alpha)}, \tag{9.60}$$

where $k = \omega(gh)^{-1/2}$, a_0 is a constant and α is the angle of the incident wave direction. The scattered wave potential, $\phi^S = \phi - \phi^I$ must satisfy

$$\lim_{kr\to\infty} \sqrt{r}\left(\frac{\partial}{\partial r} - ik\right)\phi^S = 0, \tag{9.61}$$

which represents the behavior of an outgoing wave. To construct the finite-element equations, we shall again regard R_2 as a superelement and divide R_1 further into regular elements as shown in fig. 9.4. As before, the hybrid functional can be written as

$$\pi(\phi, \tilde{\phi}) = \pi_1(\tilde{\phi}) + \pi_2(\phi, \tilde{\phi}), \tag{9.62}$$

where now

$$\pi_1(\tilde{\phi}) = \frac{1}{2}\int_{R_1}\left[h(\nabla\tilde{\phi})^2 - \frac{\omega^2}{g}\tilde{\phi}^2\right]dxdy,$$

$$\pi_2(\phi, \tilde{\phi}) = \int_C h(\tilde{\phi} - \phi^I)\frac{\partial\phi}{\partial n}ds + \frac{1}{2}\int_{r\to\infty} ikh(\phi - \phi^I)^2 ds$$
$$- \frac{1}{2}\int_{R_2}\left\{h[\nabla(\phi - \phi^I)]^2 - \frac{\omega^2}{g}(\phi - \phi^I)^2\right\}dxdy, \tag{9.63}$$

where $\partial/\partial n$ is the outward normal derivative along $\partial R_2 (= C)$.

The hybrid element for the infinite domain R_2 can easily be evaluated from π_2. Consider the case where h is constant in R_2. We can easily select the scattered-wave potential which satisfies the Helmholtz equation (9.59) and the radiation condition (9.61), i.e.,

$$\phi - \phi^I = \beta_0 H_0(kr) + \sum_{j=1}^{N}(\beta_j \cos j\theta + \beta_{j+N}\sin j\theta)H_j(kr), \tag{9.64}$$

in which the Hs are Hankel functions of the first kind. Using (9.64) the integration over the infinite domain R_2 can be eliminated, and π_2 becomes

$$\pi_2(\phi, \tilde{\phi}) = \int_C h(\tilde{\phi} - \phi^I)\frac{\partial(\phi - \phi^I)}{\partial n}ds + \int_C h\tilde{\phi}\frac{\partial\phi^I}{\partial n}ds$$
$$- \frac{1}{2}\int_C h(\phi - \phi^I)\frac{\partial(\phi - \phi^I)}{\partial n}ds - \int_C h\phi^I\frac{\partial\phi^I}{\partial n}ds. \tag{9.65}$$

Substituting (9.64) into (9.65) and assuming $\bar{\phi}$ in the same form as in (9.31) along C, π_2 is reduced to

$$\pi_2 = \boldsymbol{\beta}^T\mathbf{G}\mathbf{q} - \mathbf{q}^T\mathbf{Q} - \tfrac{1}{2}\boldsymbol{\beta}^T\mathbf{H}\boldsymbol{\beta} + c_3. \tag{9.66}$$

The four terms in (9.66) are derived from the four integrals in (9.65). By eliminating $\boldsymbol{\beta}$, we have

$$\pi_2 = \tfrac{1}{2}\mathbf{q}^T\mathbf{k}\mathbf{q} - \mathbf{q}^T\mathbf{Q} + c_4, \tag{9.67}$$

where

$\mathbf{k} = \mathbf{G}^T\mathbf{H}^{-1}\mathbf{G}$; c_3 and c_4 are constants.

It should be noted that if C is a circle of radius a, the third line integral can be evaluated analytically to obtain $\mathbf{H}$, i.e.,

$$\begin{aligned}\boldsymbol{\beta}^T\mathbf{H}\boldsymbol{\beta} &= \int_{r=a} ha(\phi - \phi^I)\frac{\partial(\phi - \phi^I)}{\partial r}\,d\theta \\ &= \pi a k h\left[2\beta_0^2 H_0 H_0' + \sum_{j=1}^N H_j H_j'(\beta_j^2 + \beta_{j+N}^2)\right],\end{aligned} \tag{9.68}$$

where

$$H_j = H_j(ka), \qquad H_j' = \left.\frac{dH_j}{dx}(x)\right|_{x=ka}.$$

A more general discussion of the water-wave problem and corresponding finite-element solutions can be found in Chen and Mei (1974) and in Mei and Chen (1975).

9.3 Heat Transfer and Fluid Flow

Much work has been done on the application of the finite-element method to heat conduction problems (Visser 1966; Wilson and Nickell 1966; and many others). The governing equation for the temperature distribution in a domain V is

$$\rho\left(\frac{\partial T}{\partial t} + u_i T_{,i}\right) = (c_{ij}T_{,j})_{,j} + H; \tag{9.69}$$

and the initial condition and boundary condition are, respectively,

$$T = T_0(\mathbf{x}) \quad \text{for} \quad t = 0, \tag{9.70}$$
$$a_1 T + a_2 c_{ij} T_{,j} \nu_i = \theta \quad \text{on} \quad \partial V, \tag{9.71}$$

where ρ is the density; T, the temperature; $\mathbf{u}$, the velocity vector; c_{ij} ($= c_{ji}$), the thermal-conductivity coefficient, which can be a function of spatial coordinates $\mathbf{x}$, as well as of the temperature itself; H, the heat source; T_0, a_1, a_2, and θ are known functions; and ν_i is the direction cosine of the unit normal over the surface ∂V. The present problem cannot be stated in the form of the stationary property of a functional; thus, in applying the finite-element method, we will transform the problem into a variational statement. We shall restrict ourselves to the problems of a finite domain with a prescribed velocity field $\mathbf{u}$.

We are seeking a solution such that

$$\delta \pi = \int_V \left[\left(\rho \frac{\delta T}{\delta t} + \rho u_i T_{,i} - H \right) \delta T + c_{ij} T_{,i} \delta T_{,j} \right] dV$$
$$+ \int_{\partial V} \frac{1}{a_2} (a_1 T - \theta) \delta T dS = 0, \tag{9.72}$$

where the variation is for arbitrary δT. Furthermore, over the portion of the boundary where a_2 is zero, we have the requirement that

$$a_1 T_1 = \theta \tag{9.73}$$

and

$$\delta T = 0.$$

In formulating a finite-element method, we, as usual, subdivide the region into a finite number of discrete elements V_n, each with selected node points. For each individual element the nodal values of T or also the derivatives of T are chosen as the generalized coordinates which are represented by a vector $\mathbf{q}$. Appropriate interpolations can then be constructed to approximate the temperature T over the individual elements; e.g., for the nth element,

$$T = \mathbf{f}(\mathbf{x}) \mathbf{q}_n, \tag{9.74}$$

from which

$$\delta T = \mathbf{f}(\mathbf{x}) \delta \mathbf{q}_n. \tag{9.75}$$

where $\mathbf{f(x)}$ is a row matrix of interpolation functions.

It is because the first integral in (9.72) contains a term $c_{ij}\,T_{,i}\,\delta T_{,j}$ that we must choose the interpolation functions $\mathbf{f}$ such that both T and δT are continuous over the entire domain V. This is to guarantee that the first integral is defined for the chosen T and δT in (9.74) and (9.75). The present formulation is, of course, applicable to one-dimensional, two-dimensional, and three-dimensional problems by using the appropriate interpolation functions described in ch. 6.

Realizing that the domain V is represented by individual elements V_n and then substituting (9.74) and (9.75) into (9.72), we obtain

$$\delta\pi = \sum_n \delta\mathbf{q}_n^T\left(\mathbf{m}_n\frac{d\mathbf{q}_n}{dt} + \mathbf{k}_n\mathbf{q}_n - \mathbf{Q}_n\right), \tag{9.76}$$

where $\mathbf{m}_n$, $\mathbf{k}_n$, and $\mathbf{Q}_n$ are constructed by using

$$\begin{aligned}
\delta\mathbf{q}_n^T\mathbf{m}\frac{d\mathbf{q}_n}{dt} &= \int_{V_n} \rho\,\frac{\partial T}{\partial t}\,\delta T dV, \\
\delta\mathbf{q}_n^T\mathbf{k}_n\mathbf{q}_n &= \int_{V_n} (\rho u_i T_{,i}\delta T + c_{,j}T_{,i}\partial T_{,j})dV + \int_{\bar{S}_n} \frac{a_1}{a_2} T\delta T dS, \\
\delta\mathbf{q}_n^T\mathbf{Q}_n &= \int_{\bar{S}_n} \theta\delta T dS + \int_{V_n} H\delta T dV,
\end{aligned} \tag{9.77}$$

where $\bar{S}_n$ is the portion of the boundary of V_n which is on the boundary ∂V where $a_2 \neq 0$.

With proper assembly the resulting equation is

$$\delta\pi = \delta\mathbf{q}^T\left(\mathbf{M}\frac{d\mathbf{q}}{dt} + \mathbf{Kq} - \mathbf{Q}\right) = 0 \tag{9.78}$$

for arbitrary δq. This variation, of course, is subjected to the conditions of constraint given by (9.73), and a constraining procedure as outlined in sec. 3.4 should be used. The ordinary differential equations for the determination of $\mathbf{q}$ are then

$$\mathbf{M}\frac{d\mathbf{q}}{dt} + \mathbf{Kq} = \mathbf{Q}. \tag{9.79}$$

We shall illustrate this formulation with two examples.

Example 1 Consider a simple one-dimensional problem (fig. 9.5) given by the equation

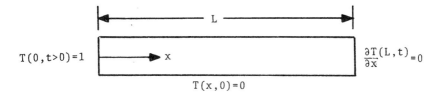

Fig. 9.5 The one-dimensional heat-transfer problem.

$$\rho\left(\frac{\partial T}{\partial t} + u\frac{\partial T}{\partial x}\right) = c\frac{\partial^2 T}{\partial x^2}. \tag{9.80}$$

For $0 \le x \le L$, the initial condition is

$$T = T_0(x) \tag{9.81}$$

at $t = 0$, and the boundary conditions for $t > 0$ are

$$T = 1 \quad \text{at} \quad x = 0,$$
$$\partial T/\partial x = 0 \quad \text{at} \quad x = L. \tag{9.82}$$

Referring to (9.69), we see that the present problem corresponds to the case where the heat source H is zero and u_1 and c_{ij} are, respectively, scalar constants u and c. The boundary conditions, by (9.71), are simply $a_1 = 1$, $a_2 = 0$, $\theta = 1$ at $x = 0$ and $a_2 = 1$, $a_1 = \theta = 0$ at $x = L$.

We divide the region into N uniform elements each of length ε ($= L/N$). The generalized coordinate q_n ($n = 1, \cdots, N + 1$) is the value of T at x_n ($=(n - 1)\varepsilon$). The interpolation function is simply a linear function. From (9.77), the element matrices are found to be

$$\mathbf{m} = \rho\varepsilon \begin{bmatrix} 1/3 & 1/6 \\ 1/6 & 1/3 \end{bmatrix},$$

$$\mathbf{k} = \frac{1}{\varepsilon} \begin{bmatrix} c - \tfrac{1}{2}\rho u\varepsilon & -c + \tfrac{1}{2}\rho u\varepsilon \\ -c - \tfrac{1}{2}\rho\varepsilon & c + \tfrac{1}{2}\rho u\varepsilon \end{bmatrix}, \tag{9.83}$$

and $\mathbf{Q}_n$ is zero for every element. By (9.73), it is required that $q_1 = 1$ and $\delta q_1 = 0$; and (9.79) becomes

$$\frac{\rho\varepsilon}{6}(\dot{q}_{n-1} + 4\dot{q}_n + \dot{q}_{n+1}) = \frac{c}{\varepsilon}\left[\left(1 + \frac{\rho u\varepsilon}{2c}\right)q_{n-1} - 2q_n + \left(1 - \frac{\rho u\varepsilon}{2c}\right)q_{n+1}\right]$$
$$\text{for} \quad n = 2, 3, \cdots, N, \tag{9.84}$$

and

$$\frac{\rho\varepsilon}{6}(\dot{q}_N + 2\dot{q}_{N+1}) = \frac{c}{\varepsilon}\left[\left(1 + \frac{\rho u \varepsilon}{2c}\right)q_N - \left(1 + \frac{\rho u \varepsilon}{2c}\right)q_{N+1}\right].$$

In practice, the values of q_n are obtained by integrating (9.84) numerically using the initial condition

$$q_n = T_0(x_n) \tag{9.85}$$

at $t = 0$. In the numerical integration procedure further approximations are usually made for the left-hand sides of (9.84) and the equations are written as

$$\rho\varepsilon\dot{q}_n = \frac{c}{\varepsilon}\left[\left(1 + \frac{\rho u \varepsilon}{2c}\right)q_{n-1} - 2q_n + \left(1 - \frac{\rho u \varepsilon}{2c}\right)q_{n+1}\right] \tag{9.86}$$
$$\text{for} \quad n = 2, 3, \cdots N,$$

and

$$\frac{\rho\varepsilon\dot{q}_{N+1}}{2} = \frac{c}{\varepsilon}\left[\left(1 + \frac{\rho u \varepsilon}{2c}\right)q_N - \left(1 + \frac{\rho u \varepsilon}{2c}\right)q_{N+1}\right].$$

Some graphical results of (9.86) are presented in fig. 9.6.

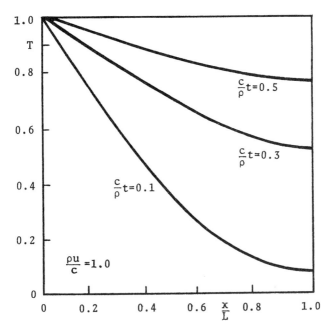

Fig. 9.6 Transient temperature distribution of the one-dimensional heat-transfer problem.

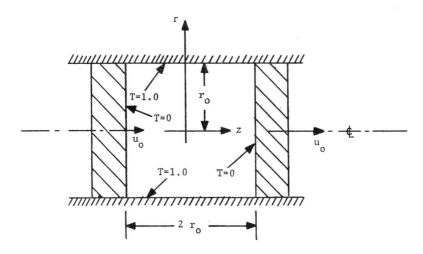

Fig. 9.7 Steady-state heat conduction in axially symmetric Stokes flow.

Example 2. We consider next the steady-state heat conduction in axially symmetric Stokes flow (fig. 9.7) for which the heat source is zero and the thermal conductivity coefficient is a scalar constant c. The governing equation in cylindrical coordinates is

$$u_r \frac{\partial T}{\partial r} + u_z \frac{\partial T}{\partial z} = c \left[\frac{1}{r} \frac{\partial}{\partial r} \left(r \frac{\partial T}{\partial r} \right) + \frac{\partial^2 T}{\partial z^2} \right], \tag{9.87}$$

and the boundary conditions are

$$\begin{aligned} T &= 0 \quad \text{at} \quad z = \pm r_0, \\ T &= 1 \quad \text{at} \quad r = r_0, \\ \partial T/\partial r &= 0 \quad \text{at} \quad r = 0. \end{aligned} \tag{9.88}$$

Referring to the boundary condition (9.71), we see that these boundary conditions correspond to $a_2 = 0$, $a_1 = \theta = 1$ at $r = r_0$; $a_2 = 1$, $a_1 = \theta = 0$ at $r = 0$; and $a_2 = \theta = 0$, $a_1 = 1$ at $z = \pm r_0$.

For the finite-element formulation we divide the region into rectangular ring elements with the four circular edges as the nodes. Since the problem is axially symmetric only the cross section in the (r, z)-plane is to be considered. In this case an appropriate interpolation function is the bilinear interpolation which can easily be constructed for a rectangular element, and is

$$T(r, z) = \sum_{i=1}^{2} \sum_{j=1}^{2} H_{0i}^{(0)}(r) H_{0j}^{(0)}(z) T(r_i, z_i), \tag{9.89}$$

where $H_{0i}^{(0)}(x)$, $i = 1, 2$, is defined in (6.12). The variational equation can then be written in the form

$$\delta \Pi = \sum_n \delta \mathbf{q}_n^T \mathbf{k}_n \mathbf{q}_n, \tag{9.90}$$

where the element matrix $\mathbf{k}_n$ is defined by

$$\delta \mathbf{q}_n^T \mathbf{k}_n \mathbf{q}_n = 2\pi \int_{z_1}^{z_2} \int_{r_1}^{r_2} \left[\left(u_r \frac{\partial T}{\partial r} + u_z \frac{\partial T}{\partial z} \right) \delta T \right. \\ \left. + c \left(\frac{\partial T}{\partial r} \frac{\partial \delta T}{\partial r} + \frac{\partial T}{\partial z} \frac{\partial \delta T}{\partial z} \right) \right] r \, dr \, dz. \tag{9.91}$$

In (9.91) u_r and u_z can be obtained, for example, from the finite-element solution of slow viscous flow developed by Tong and Fung (1971). The variational statement (9.90) can again be expressed in terms of independent coordinates $\mathbf{q}$ subject to the conditions of constraint at $z = \pm r_0$ and $r = r_0$, and then, by realizing that $\delta \mathbf{q}$ is arbitrary, we can obtain a system of algebraic equations for the determination of $\mathbf{q}$.

9.3.1. Two-Dimensional or Axially Symmetric Stokes Flow

In the case of the slow motion of a fluid, the inertial forces can be neglected when compared to the viscous forces. The governing equations for such problems are given simply, in the domain V, as

$$\nabla \cdot \mathbf{v} \tag{9.92}$$

and

$$\mu \nabla \cdot \nabla \mathbf{v} - \nabla p = 0 \tag{9.93}$$

where $\mathbf{v}$ is the velocity vector, p the pressure, and μ the coefficient of viscosity.

We shall now show that in the case of two-dimensional flow or axially symmetric flow, a different finite-element approach can be used. We introduce the stream function ψ, defined by

$$v_1 = \frac{1}{x_2^m} \frac{\partial \psi}{\partial x_2},$$
$$v_2 = -\frac{1}{x_2^m} \frac{\partial \psi}{\partial x_1}, \tag{9.94}$$

where $m = 0$ and 1, respectively, for two-dimensional flow and axially symmetric flow. It is seen that (9.92) is satisfied identically. By eliminating p, (9.93) is reduced to

$$L(x_2^m L\psi) = 0, \tag{9.95}$$

where

$$L = \frac{\partial}{\partial x_2}\left(\frac{1}{x_2^m}\frac{\partial}{\partial x_2}\right) + \frac{1}{x_2^m}\frac{\partial^2}{\partial x_1^2}, \tag{9.96}$$

and x_1 is the axis of symmetry in the case of axisymmetric flow. The boundary value of ψ and $\partial \psi/\partial \nu$ is related to that of $\mathbf{v}$ by

$$\begin{aligned}\psi &= \int_{\partial A} x_2^m v_\nu \, ds, \\ \partial \psi/\partial \nu &= -x_2^m v_s\end{aligned} \tag{9.97}$$

on ∂A, where v_ν and v_s are the normal and tangential components of $\mathbf{v}$, A is the cross-sectional area of the domain, and ∂A is the boundary curve of A. (For a multiply connected region, see Tong and Fung 1971).

Equation (9.95) is similar to the plate-bending problem in solid mechanics. It is therefore convenient to introduce

$$\begin{aligned}M_{11} &= \frac{1}{x_2^m}\frac{\partial^2 \psi}{\partial x_1^2}, \\ M_{22} &= \frac{\partial}{\partial x_2}\left(\frac{1}{x_2^m}\frac{\partial \psi}{\partial x_2}\right), \\ M_{12} &= \frac{1}{x_2^m}\frac{\partial^2 \psi}{\partial x_1 \partial x_2}, \\ Q_1 &= \frac{\partial M_{11}}{\partial x_1} + \frac{\partial M_{12}}{\partial x_2}, \\ Q_2 &= \frac{\partial M_{12}}{\partial x_1} + \frac{1}{x_2^m}\frac{\partial(x_2^m M_{22})}{\partial x_2}.\end{aligned} \tag{9.98}$$

Then (9.95) is equivalent to

$$\frac{\partial^2 M_{11}}{\partial x_1^2} + 2\frac{\partial^2 M_{12}}{\partial x_1 \partial x_2} + \frac{\partial}{\partial x_2}\left[\frac{1}{x_2^m}\frac{\partial(x_2^m M_{22})}{\partial x_2}\right] = 0 \tag{9.99}$$

and

$$\frac{1}{x_2^m}\frac{\partial(x_2^m M_{11})}{\partial x_2} = \frac{\partial M_{12}}{\partial x_1},$$
$$\frac{\partial M_{22}}{\partial x_1} = \frac{\partial M_{12}}{\partial x_2}. \tag{9.100}$$

It can be shown that the present Stokes flow problem is equivalent to a variational principle with the following expression for the functional:

$$\Pi = \sum_n \left[\int_{\partial A_n} (M_{\alpha\beta}\nu_\alpha \psi_{,\beta} - Q_\alpha \nu_\alpha \psi)\, ds \right.$$
$$\left. - \tfrac{1}{2}\int_{A_n}(M_{11}^2 + 2M_{12}^2 + M_{22}^2)x_2^m\, dx_1\, dx_2 \right], \tag{9.101}$$

where $\sum_n$ denotes the sum over all elements. Incidentally, in the case of two-dimensional flow, the governing equations are similar to those for plate bending. In the paper of Tong and Fung (1971) the Ms are assumed in the form

$$M_{11} = \beta_1 + \beta_2 x_1 + \beta_3 x_2 + \beta_4 x_1^2 + \beta_5 x_1 x_2 + \beta_6 x_2^2,$$
$$M_{12} = \beta_7 + \beta_8 x_1 + \beta_9 x_2 + \beta_{10} x_1^2 + \beta_{11} x_1 x_2 + \beta_{12} x_2^2,$$
$$M_{22} = \begin{cases} \beta_{13} + \beta_{14} x_1 + \beta_{15} x_2 + \beta_{16} x_1^2 + \beta_{17} x_1 x_2 - (\beta_4 + \beta_{11}) x_2^2 \\ \qquad \text{(for two-dimensional flow),} \\ x_2(\beta_{13} + \beta_{14} x_1 + \beta_{15} x_1^2) - \tfrac{2}{3}(\beta_4 + \beta_{11}) x_2 \\ \qquad \text{(for axial symmetric flow),} \end{cases} \tag{9.102}$$

and a new variable ϕ is used for ψ/x_2^m. The nodal values of ϕ, $\partial \phi/\partial x_1$, and $\partial \phi/\partial x_2$ are then used as generalized coordinates.

9.4 Variational Formulation of Nonlinear Problems by the Incremental Method

The finite-element method can be applied to nonlinear problems in a straightforward manner by following the variational approach. If the same type of interpolation functions described in previous chapters for the linear problem are used, one can derive a set of nonlinear algebraic finite-element equations. In practice, however, these equations are difficult to solve. The common approach is to use an incremental procedure where the problems are solved in step-by-step fashion (Turner, Dill, Martin, and Melosh 1960; Washizu, 1968).

A particularly convenient and accurate incremental procedure is formulated here by considering the second variation of the function Π. Let

$$\Pi(\mathbf{u}, p) = \int F(\mathbf{u}, \mathbf{u}_{,i}, \cdots, p)\, d\mathbf{x} \tag{9.103}$$

where F is a smooth function of the dependent vector variable $\mathbf{u}$, the partial derivatives of $\mathbf{u}$ and a loading magnitude, p. Let $\mathbf{u}_0$ be the fundamental state at a loading magnitude p_0. The change of the fundamental state, $\Delta\mathbf{u}$, due to a change of load, Δp, can be obtained from the stationary value of

$$\begin{aligned}\Delta\Pi &= \Pi(\mathbf{u}_0 + \Delta\mathbf{u}, p_0 + \Delta p) - \Pi(\mathbf{u}_0, p_0) \\ &= \Delta p P_1(\mathbf{u}_0, p_0) + P_2(\mathbf{u}_0, \Delta\mathbf{u}; p_0, \Delta p) + P_3(\mathbf{u}_0, \Delta\mathbf{u}; p_0, \Delta p)\end{aligned} \tag{9.104}$$

with respect to all kinematically admissible $\Delta\mathbf{u}$. In (9.104) $P_2(\mathbf{u}_0, \Delta\mathbf{u}; p_0, \Delta p)$ is the so-called second variation of Π, which contains all the second order terms in $\Delta\mathbf{u}$ and Δp. The functional $P_3(\mathbf{u}_0, \Delta\mathbf{u}; p_0, \Delta p)$ contains all the terms of higher order in $\Delta\mathbf{u}$ and Δp, whereas $P_1(\mathbf{u}_0, p_0)$ is independent of $\Delta\mathbf{u}$ and Δp. For small Δp the first-order approximation of $\Delta\mathbf{u}$ can be obtained from the stationary value of P_2 alone, i.e., from the variation

$$\delta P_2(\mathbf{u}_0, \Delta\mathbf{u}; p_0, \Delta p) = 0 \tag{9.105}$$

with respect to $\Delta\mathbf{u}$.

Then the first-order approximation of the fundamental state $\mathbf{u}$ at load $p_0 + \Delta p$ is simply

$$\mathbf{u} = \mathbf{u}_0 + \Delta\mathbf{u}. \tag{9.106}$$

For a hypoelastic material, the total potential energy of an elastic body is given by

$$\Pi = \sum_m \left[\int_{V_m} W(\varepsilon_{ij}) dV - \int_{V_m} \bar{F}_i u_i\, dV - \int_{S\sigma_m} \bar{T}_i u_i\, dS \right], \tag{9.107}$$

where $W(\varepsilon_{ij})$ is the strain energy function per unit volume; and for a finite strain measured with respect to undeformed state coordinates a_i, the strain displacement relation is

$$\varepsilon_{ij} = \frac{1}{2}\left(\frac{\partial u_i}{\partial a_j} + \frac{\partial u_j}{\partial a_i} + \frac{\partial u_k}{\partial a_i}\frac{\partial u_k}{\partial a_j}\right). \tag{9.108}$$

The vector $\mathbf{u}$ [$\mathbf{u} = (u_1\ \ u_2\ \ u_3)$] is the displacement satisfying the prescribed displacement boundary conditions. In (9.107) the sign $\sum_m$ indicates summation over all the elements, $\bar{F}_i$ is the prescribed body force and $\bar{T}_i$ is the prescribed boundary traction, V is the volume, and S is the surface area. Both V and S are referred to the undeformed state. The second variation of Π at a fundamental state $\mathbf{u}_0$ is simply

$$P_2 = \sum_m \left\{ \int_{V_m} \left[\frac{1}{2} (E_{ijkl})_0 \Delta \varepsilon_{ij} \Delta \varepsilon_{kl} + \frac{1}{2} S_{ij} \frac{\partial \Delta u_k}{\partial a_i} \frac{\partial \Delta u_k}{\partial a_j} \right. \right. \\ \left. \left. - \Delta \bar{F}_i \Delta u_i \right] dV - \int_{S\sigma_m} \Delta \bar{T}_i \Delta u_i dS \right\}, \tag{9.109}$$

where

$$(E_{ijkl})_0 = \left. \frac{\partial W}{\partial \varepsilon_{ij} \partial \varepsilon_{kl}} \right|_{\mathbf{u}=\mathbf{u}_0} \tag{9.110}$$

is, in general, a function of the displacement $\mathbf{u}_0$; $\Delta \varepsilon_{ij}$ is the strain increment given by

$$\Delta \varepsilon_{ij} = \frac{1}{2} \left(\frac{\partial \Delta u_i}{\partial a_j} + \frac{\partial \Delta u_j}{\partial a_i} + \frac{\partial (u_k)_0}{\partial a_i} \frac{\partial \Delta u_k}{\partial a_j} + \frac{\partial (u_k)_0}{\partial a_j} \frac{\partial \Delta u_k}{\partial a_i} \right); \tag{9.111}$$

and S_{ij} is the Kirchhoff stress defined by

$$S_{ij} = \left. \frac{\partial W}{\partial \varepsilon_{ij}} \right|_{\mathbf{u}=\mathbf{u}_0}. \tag{9.112}$$

In this problem we assume that the prescribed body force $\bar{F}_i$, boundary traction $\bar{T}_i$, and boundary displacement $\bar{u}_i$ are increasing proportionally so that they can be given in terms of a simple parameter p.

In the prescribed forces given in the foregoing paragraphs, we have considered conservative loadings of the dead weight type only. If the loading is conservative but the direction of the load distribution changes with the deformation, the terms $\int_V \bar{F}_i u_i dV$ or $\int_{(S_\sigma)_m} \bar{T}_i u_i dS$ must be changed accordingly.

9.4.1 Finite-Element Formulation and Incremental Solution Procedure

In the finite-element formulation, let $\mathbf{q}_0$ be the generalized coordinates associated with dependent variables $\mathbf{u}_0$ at the load p_0. The quantity P_2 can be written as

$$P_2 = \tfrac{1}{2} \Delta \mathbf{q}^\mathrm{T} \mathbf{K}(\mathbf{q}_0, p_0) \Delta \mathbf{q} - \Delta \mathbf{q}^\mathrm{T} \mathbf{Q}(\mathbf{q}_0, p_0) \Delta p, \tag{9.113}$$

where the $\Delta\mathbf{q}$ represent the generalized coordinates associated with the incremental variables $\Delta\mathbf{u}$. Then (9.105) becomes

$$\delta P_2 = \delta\Delta\mathbf{q}^\mathrm{T}\{\mathbf{K}(q_0, p_0)\Delta\mathbf{q} - \Delta p\mathbf{Q}(q_0, p_0)\} = 0 \tag{9.114}$$

or

$$\mathbf{K}\Delta\mathbf{q} = \Delta p\mathbf{Q}, \tag{9.115}$$

where $\mathbf{K}$ depends on $\mathbf{q}_0$ and p_0. When the finite-element formulation is based on the principle of stationary potential energy, $\mathbf{K}$ is the stiffness matrix.

Equation (9.115) is only linear in $\Delta\mathbf{q}$, hence it can be solved by many standard methods. The generalized coordinates $\mathbf{q}$ at load $p_0 + \Delta p$ are given by

$$\mathbf{q} = \mathbf{q}_0 + \Delta\mathbf{q}, \tag{9.116}$$

where $\Delta\mathbf{q}$ is obtained by solving (9.115). In (9.115), by replacing $\mathbf{q}_0$ and p_0 by $\mathbf{q}$ and $\mathbf{p}$, respectively, and repeating the procedure above, we can obtain the solution for further increments of loading.

This simple incremental procedure has one major drawback. The accuracy of (9.116) is only up to order $(\Delta p)^2$, even when $\mathbf{q}$ is a smooth function of p. Thus the accumulated error due to the successive steps may become very large. We may, however, look at (9.115) from a different point of view by writing it in the form

$$\mathbf{K}(\mathbf{q}_0, p_0)\frac{\Delta\mathbf{q}}{\Delta p} = \mathbf{Q}(\mathbf{q}_0, p_0). \tag{9.117}$$

Then $\Delta\mathbf{q}/\Delta p$ can be identified as the rate of changing $\mathbf{q}$ at load p_0, and (9.117) may be treated as a system of ordinary differential equations of the first order with p as the independent variable. The values of $\mathbf{q}$ at any load p can be obtained by numerical integration from a known initial state. The expression given by (9.116) is usually called the Euler method. There are in existence many other integration schemes which provide much better accuracy. The most commonly known schemes are versions of the Runge-Kutta method and the predictor-corrector method.

Consider the specific example of the large deflection of a shallow curved beam of uniform cross section with hinged but fixed ends subjected to a laterally distributed load $p(x) = p_0 f(x)$. It is assumed that the strains remain small and the stress-strain relations are linear. The appropriate functional is given by the total potential energy:

$$\Pi = \int_0^l \left[\frac{1}{2} EI \left(\frac{d^2 w}{dx^2} \right)^2 + \frac{1}{2} EA\varepsilon^2 + p_0 f(x) w \right] dx, \qquad (9.118)$$

where w is the deflection, E is Young's modulus, I is the cross-sectional moment of inertia, A is the cross-sectional area, and ε is the axial strain along the neutral axis. Under the assumption that the initial rise $y(x)$ is small compared with the beam length l, ε can be approximated by

$$\varepsilon = \frac{du}{dx} + \frac{1}{2}\left[\left(\frac{d(y+w)}{dx} \right)^2 - \left(\frac{dy}{dx} \right)^2 \right]. \qquad (9.119)$$

The second variation of the potential energy can be written as

$$P_2 = \int_0^l \left\{ \frac{1}{2} EI \left(\frac{d^2 \Delta w}{dx} \right)^2 + \frac{1}{2} EA \left(\frac{d\Delta u}{dx} + \frac{d(y+w)}{dx} \frac{d\Delta w}{dx} \right)^2 \right.$$
$$\left. - \frac{1}{2} N \left(\frac{d\Delta w}{dx} \right)^2 + \Delta p f(x) \Delta w \right\} dx, \qquad (9.120)$$

where the axial compressive stress resultant N is given by

$$N = -EA \left[\frac{du_0}{dx} + \frac{dw_0}{dx} \frac{dy}{dx} + \frac{1}{2} \left(\frac{dw_0}{dx} \right)^2 \right]. \qquad (9.121)$$

The formulation of the finite-element solution based on the functional given by (9.120) is straightforward. In principle, the interpolation functions should be so assumed that Δw, $d\Delta w/dx$, and Δu are continuous, i.e., at least a cubic interpolation is needed for Δw and a linear interpolation for Δu.

Problems

1 Consider the axial vibration of a uniform rod of length L governed by eq. (9.4) with $p(x) = 0$, and the boundary condition $u(0) = \partial u(L)/\partial x = 0$. Perform a finite-element vibration analysis using both the consistent mass matrix (9.11), and the corresponding lumped mass matrix. Divide the length L into two and three elements, respectively, and determine the lowest two natural frequencies ω_1 and ω_2. The exact values are:

$$\omega_1 = \frac{3\pi}{2L}\sqrt{\frac{EA}{\rho}}, \qquad \omega_2 = \frac{5\pi}{2L}\sqrt{\frac{EA}{\rho}}.$$

Free undamped sinusoidal vibrations for this problem are governed by eq. (9.1) with $\mathbf{C} = \mathbf{Q} = 0$. Proceed by setting $\mathbf{q} = \mathbf{q}^0 \sin \omega t$ so that $\ddot{\mathbf{q}} = -\omega^2 \mathbf{q}$; then eq. (9.1) gives the system of equations

$$(\mathbf{K} - \omega^2 \mathbf{M})\mathbf{q}^0 = 0.$$

These equations have nontrivial solutions when the determinant $|\mathbf{K} - \omega^2 \mathbf{M}|$ vanishes, and this yields an algebraic equation whose roots give values for the natural frequencies ω. Assume $A = E = L = \rho = 1$ for convenience.

2 Solve prob. 9.1 using quadratic interpolation to obtain the consistent mass matrix.

3 Solve prob. 9.1 when the density of the rod has the variation $\rho(x) = \rho_0(1 + \frac{x}{L})$, by finding appropriate consistent and lumped mass matrices.

4 Obtain eq. (9.30) by performing the indicated operations.

5 Determine explicitly the matrices $\mathbf{A}$, $\mathbf{A}_n$, $\mathbf{B}$, $\mathbf{B}_n$, and $\mathbf{L}$ that appear in (9.31).

6 Construct the stiffness matrix of the hybrid element shown in fig. 9.1 for Laplace's equation with the following boundary conditions on the wedge surface:

$$\frac{\partial \phi}{\partial \theta} = \begin{cases} \bar{\Phi}_1(r) & \text{at} \quad \theta = -\alpha, \\ \bar{\Phi}_2(r) & \text{at} \quad \theta = +\alpha. \end{cases}$$

7 Derive the matrices $\mathbf{m}$ and $\mathbf{k}$ in eq. (9.83) for the one-dimensional heat transfer associated with fig. 9.5.

8 Set up equations equivalent to (9.117) for the nonlinear shallow curved beam which is discussed in the text with respect to eqs. (9.118)—(9.121).

Appendix A Notation and Matrix Algebra

The use of the finite-element method involves a lot of algebraic manipulation. Matrix algebra is used to simplify the notation and writing.

Throughout this book, matrices (and vectors) are printed in boldface type. A matrix is a two-dimensional ordered array,

$$\mathbf{A} = \begin{bmatrix} a_{11} & a_{12} & \cdots & a_{1n} \\ a_{21} & a_{22} & \cdots & a_{2n} \\ \vdots & & \ddots & \vdots \\ a_{m1} & & \cdots & a_{mn} \end{bmatrix} = (a_{ij}), \tag{A.1}$$

where a_{ij}, $i = 1, 2, \cdots, m$, $j = 1, 2, \cdots, n$ are called the elements, entries, or components of $\mathbf{A}$. The matrix $\mathbf{A}$ is said to have m rows and n columns or to be of order $m \times n$. Sometimes we use the notation $\underset{m \times n}{\mathbf{A}}$ to denote the order of the matrix.

A vector is a matrix that consists of only one row or one column. In (A.1), when $n = 1$, the resulting vector is called a column vector; when $m = 1$, it is called a row vector. We usually use only one subscript for the elements of a vector, and use braces instead of square brackets to enclose the elements when we have to write out a vector explicitly:

$$\mathbf{A} = \begin{Bmatrix} a_1 \\ a_2 \\ \vdots \\ a_n \end{Bmatrix} \tag{A.2}$$

for a column vector, and

$$\mathbf{A} = \{a_1 \quad a_2 \quad \cdots \quad a_n\} \tag{A.3}$$

for a row vector. Sometimes we write $\underset{n \times 1}{\mathbf{A}}$ or $\underset{1 \times n}{\mathbf{A}}$ to indicate the order of a column or row vector.

A scalar can be considered to be a matrix of order 1×1. We do not use Αυεsubscript or bold-face type to distinguish scalars. Throughout the book, a quantity with a bar on the top will be a prescribed quantity. We now list a number of definitions and properties of matrices.

A.1 Transpose

The transpose of the matrix $\mathbf{A}$ is denoted by $\mathbf{A}^T$ and is defined as

$$\mathbf{B} = (b_{ij}) = \mathbf{A}^T, \tag{A.4}$$

where

$$b_{ij} = a_{ji}.$$

Therefore, if $\mathbf{A}$ is of order $m \times n$, then $\mathbf{A}^T$ will be of order $n \times m$. The transpose of a column vector is a row vector and vice versa.

A.2 Addition and Subtraction

Addition and subtraction can only be performed for matrices of the same order. For addition,

$$\mathbf{C} = (c_{ij}) = \mathbf{A} + \mathbf{B} \tag{A.5}$$

implies that

$$c_{ij} = a_{ij} + b_{ij}; \tag{A.6}$$

and for subtraction

$$\mathbf{C} = (c_{ij}) = \mathbf{A} - \mathbf{B} \tag{A.7}$$

implies that

$$c_{ij} = a_{ij} - b_{ij}. \tag{A.8}$$

Both $\mathbf{A}$ and $\mathbf{B}$ are of the same order, say $m \times n$. The resulting matrix $\mathbf{C}$ is also of order $m \times n$. It can be easily shown that matrix addition and subtraction are associative:

$$\mathbf{A} + \mathbf{B} + \mathbf{C} = (\mathbf{A} + \mathbf{B}) + \mathbf{C} = \mathbf{A} + (\mathbf{B} + \mathbf{C}), \tag{A.9}$$
$$\mathbf{A} + \mathbf{B} - \mathbf{C} = (\mathbf{A} + \mathbf{B}) - \mathbf{C} = \mathbf{A} + (\mathbf{B} - \mathbf{C}). \tag{A.10}$$

They are also commutative:

$$\mathbf{A} + \mathbf{B} = \mathbf{B} + \mathbf{A}, \tag{A.11}$$
$$\mathbf{A} - \mathbf{B} = -\mathbf{B} + \mathbf{A}. \tag{A.12}$$

A.3 Multiplication by a Scalar

Scalar multiplication,

$$\mathbf{B} = (b_{ij}) = \alpha \mathbf{A} = \mathbf{A}\alpha, \tag{A.13}$$

implies

$$b_{ij} = \alpha\, a_{ij} = a_{ij}\alpha. \tag{A.14}$$

Scalar multiplication is commutative.

A.4 Multiplication of Matrices

The product of two matrices is

$$\mathbf{C} = (c_{ij}) = \mathbf{AB}, \tag{A.15}$$

where $\mathbf{A}$ and $\mathbf{B}$ are of order $l \times m$ and $m \times n$, respectively. Then

$$c_{ij} = \sum_{k=1}^{m} a_{ik} b_{kj}, \tag{A.16}$$

and $\mathbf{C}$ is a matrix of order $l \times n$. For the multiplication of three matrices, we have

$$\mathbf{ABC} = (\mathbf{AB})\mathbf{C} = \mathbf{A}(\mathbf{BC}), \tag{A.17}$$

i.e., multiplication is associative. However, matrix multiplication is, in general, not commutative:

$$\mathbf{AB} \neq \mathbf{BA}. \tag{A.18}$$

A.5 Transpose of a Product

From the definition of the multiplication of two matrices, it can easily be proven that

$$(\mathbf{AB})^\mathrm{T} = \mathbf{B}^\mathrm{T}\mathbf{A}^\mathrm{T}. \tag{A.19}$$

A.6 Partitioning and Submatrices

A matrix can be partitioned into submatrices. For example,

$$\mathbf{A} = \begin{bmatrix} a_{11} & a_{12} & a_{13} & \vdots & a_{14} & a_{15} \\ a_{21} & a_{22} & a_{23} & \vdots & a_{24} & a_{25} \\ a_{31} & a_{32} & a_{33} & \vdots & a_{34} & a_{35} \\ \hdashline a_{41} & a_{42} & a_{43} & \vdots & a_{44} & a_{45} \end{bmatrix}, \tag{A.20}$$

is shown partitioned into four submatrices by the dotted lines. We may write

$$\mathbf{A} = \begin{bmatrix} \mathbf{A}_{11} & \mathbf{A}_{12} \\ \mathbf{A}_{21} & \mathbf{A}_{22} \end{bmatrix}, \tag{A.21}$$

where $\mathbf{A}_{11}_{3\times 3}$, $\mathbf{A}_{12}_{3\times 2}$, $\mathbf{A}_{21}_{1\times 3}$ and $\mathbf{A}_{22}_{1\times 2}$ themselves are matrices. In performing any matrix operation, all the rules can be first applied as if each of the submatrices were a scalar element and then carrying out any further operation in the usual way. For example, if we have $\mathbf{A}$ as given in (A.21) and

$$\mathbf{B} = \begin{bmatrix} \mathbf{B}_1 \\ \mathbf{B}_2 \end{bmatrix} \tag{A.22}$$

partitioned into two submatrices, then

$$\mathbf{AB} = \begin{bmatrix} \mathbf{A}_{11}\mathbf{B}_1 + \mathbf{A}_{12}\mathbf{B}_2 \\ \mathbf{A}_{21}\mathbf{B}_1 + \mathbf{A}_{22}\mathbf{B}_2 \end{bmatrix}. \tag{A.23}$$

It should be pointed out that the order of the submatrices has to be such as to make any subsequent multiplications, etc., meaningful.

A.7 Differentiation and Integration

If the elements of the matrix are a function of x, then

$$\frac{d\mathbf{A}}{dx} = \left(\frac{da_{ij}}{dx}\right); \tag{A.24}$$

that is, differentiation of a matrix is differentiation of every element of the matrix separately. Likewise,

$$\int \mathbf{A}\, dx = \left(\int a_{ij}\, dx\right), \tag{A.25}$$

or the integration of a matrix is the integration of each individual entry of the matrix separately.

A.8 Some Definitions for a Square Matrix

A square matrix is a matrix with the number of rows and columns the same. An element, say a_{ij}, with both of its subscripts being the same ($i = j$) is on the diagonal of the matrix. The rest of the elements are off the diagonal of the matrix.

A.8.1 Identity Matrix

An identity matrix, usually denoted by $\mathbf{I}$, is a square matrix defined as

$$\mathbf{I} = (\delta_{ij}) \tag{A.26}$$

where δ_{ij} is the Kronecker delta,

$$\begin{aligned}\delta_{ij} &= 1 \quad \text{for} \quad i = j, \\ &= 0 \quad \text{for} \quad i \neq j.\end{aligned} \tag{A.27}$$

For any matrix $\mathbf{A}$, we have

$$\mathbf{IA} = \mathbf{A} = \mathbf{AI}.$$

A.8.2 Symmetric Matrix

The matrix $\mathbf{A}$ is symmetric if

$$\mathbf{A} = \mathbf{A}^T, \tag{A.28}$$

that is, if $a_{ij} = a_{ji}$. Often in this book we write symmetric $\mathbf{A}$ out as

$$\begin{bmatrix} a_{11} & & \text{sym} \\ a_{21} & a_{22} & \\ \vdots & & \ddots \\ a_{n1} & \cdots & & a_{nn} \end{bmatrix} \quad \text{or} \quad \begin{bmatrix} a_{11} & a_{12} & \cdots & a_{1n} \\ & a_{22} & & \vdots \\ & & \ddots & \\ \text{sym} & & & a_{nn} \end{bmatrix} \tag{A.29}$$

A.8.3 Inverse

The inverse of a matrix $\mathbf{A}$ is usually denoted by $\mathbf{A}^{-1}$. The inverse satisfies

$$\mathbf{AA}^{-1} = \mathbf{A}^{-1}\mathbf{A} = \mathbf{I}. \tag{A.30}$$

A.8. Positiv4e-Semidefinite Matrix

A matrix $\mathbf{A}$ is positive semidefinite if

$$x^T A x \geq 0 \tag{A.31}$$

for all x, where A is matrix of order $n \times n$, and x is a vector of order n.

A.8.5 Positive-Definite Matrix

A matrix A is positive definite if

$$x^T A x > 0 \tag{A.32}$$

for all $x \neq 0$, and if

$$x^T A x = 0 \tag{A.33}$$

only if

$x \equiv 0$.

A.8.6 The Determinant of a Matrix; Singular Matrix

The determinant of a matrix is the determinant formed from the elements of the matrix, taken in the same order as they appear in the matrix. Thus, if $A = (a_{ij})$, $\det A = |a_{ij}|$. A matrix A is singular if the determinant of the matrix vanishes,

$$\det A = 0. \tag{A.34}$$

For a singular matrix, there exists at least one vector with at least one nonzero component (i.e., a vector that is not identically zero, $x \neq 0$), such that

$$Ax = 0; \tag{A.35}$$

x is called the zero vector of A.

A.8.7 Banded Matrix

A square matrix A is said to be banded if all its nonzero entries lie within a band near its principal diagonal a_{ii}. For example,

$$A = \begin{bmatrix} a_{11} & a_{12} & a_{13} & 0 \\ a_{21} & a_{22} & a_{23} & a_{24} \\ 0 & a_{32} & a_{33} & a_{34} \\ 0 & 0 & a_{43} & a_{44} \end{bmatrix} \tag{A.36}$$

is said to have a band width of 4. Its left semiband width is 2 (count a_{12} and a_{22}) and its right semiband width is 3 (count a_{22}, a_{23}, and a_{24}).

A.8.8 Diagonal Matrix

A diagonal matrix is a banded matrix of unit band width, so that it has nonzero entries only on the principal diagonal.

A.8.9 Triangular matrix

An upper triangular matrix has all its nonzero entries on and above the diagonal,

$$\mathbf{A} = \begin{bmatrix} a_{11} & a_{12} & \cdots & a_{1n} \\ & a_{22} & \cdots & a_{2n} \\ & & \ddots & \vdots \\ 0 & & & a_{nn} \end{bmatrix}. \tag{A.37}$$

A lower triangular matrix has all its nonzero entries on and below the diagonal,

$$\mathbf{A} = \begin{bmatrix} a_{11} & & & 0 \\ a_{21} & a_{22} & & \\ \vdots & & \ddots & \\ a_{n1} & \cdots & & a_{nn} \end{bmatrix}. \tag{A.38}$$

A.8.10 Zero or Null matrix

A zero matrix has all zero entries, and is usually denoted by

$$\mathbf{A} = 0$$

A.9 Cyclic Permutation

By a cyclic permutation of a group of p distinct integers, say, $i_1, i_2, \cdots, i_p$, we mean a transformation that assigns i_2 to i_1, i_3 to i_2, $\cdots$, i_p to i_{p-1}, and i_1 to i_p. For example, the cyclic permutation of 1, 2, 3 is 2, 3, 1, that of 2, 3, 1 is 3, 1, 2; and that of 3, 1, 2 is 1, 2, 3.

Problems

1 Show that if $\mathbf{K}$ is symmetric and positive semidefinite,

$\det \mathbf{K} = 0$.

2 Show that if **K** is symmetric and positive definite,

det **K** $\neq$ 0,

and that all the eigenvalues of **K** are positive.

Appendix B Rectangular Elements

For the rectangular elements having edges parallel to the x- and y-axes, such as shown in fig. B.1, the transformation from (x, y)- to (ξ, η)-coordinates is simply

$$x = x_1 + \frac{a}{2} + \frac{a}{2}\xi,$$
$$y = y_1 + \frac{b}{2} + \frac{b}{2}\eta. \tag{B.1}$$

The limits for ξ and η are ± 1. The Jacobian (det **J**) is equal to $ab/4$, and $ds/d\zeta = 1/2$.

The generalized-coordinate vector is

$$\mathbf{q}^* = \begin{Bmatrix} \mathbf{q}_u \\ \mathbf{q}_v \end{Bmatrix}, \tag{B.2}$$

in which

$$\mathbf{q}_u = \begin{Bmatrix} u_1 \\ u_2 \\ \vdots \end{Bmatrix}, \qquad \mathbf{q}_v = \begin{Bmatrix} v_1 \\ v_2 \\ \vdots \end{Bmatrix}.$$

The displacement, strain, and stress vectors are

$$\mathbf{u} = \begin{Bmatrix} u \\ v \end{Bmatrix} = \begin{bmatrix} \mathbf{f}^T & 0 \\ 0 & \mathbf{f}^T \end{bmatrix} \begin{Bmatrix} \mathbf{q}_u \\ \mathbf{q}_v \end{Bmatrix} = \mathbf{D}^* \mathbf{q}^*, \tag{B.3}$$

$$\mathbf{e} = \begin{Bmatrix} e_x \\ e_y \\ e_{xy} \end{Bmatrix} = \begin{bmatrix} \dfrac{\partial \mathbf{f}^T}{\partial x} & 0 \\ 0 & \dfrac{\partial \mathbf{f}^T}{\partial y} \\ \dfrac{\partial \mathbf{f}^T}{\partial y} & \dfrac{\partial \mathbf{f}^T}{\partial x} \end{bmatrix} \begin{Bmatrix} \mathbf{q}_u \\ \mathbf{q}_v \end{Bmatrix} = \mathbf{E}^* \mathbf{q}^*, \tag{B.4}$$

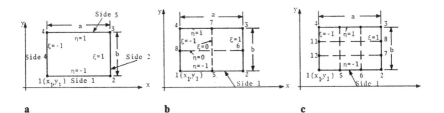

Fig. B.1 Rectangular elements.

$$\boldsymbol{\sigma} = \begin{Bmatrix} \sigma_x \\ \sigma_y \\ \sigma_{xy} \end{Bmatrix} = \mathbf{CE}^*\mathbf{q}^*, \tag{B.5}$$

where **C** is the elastic-coefficient matrix given in sec. 5.1.3, and where

$$\begin{aligned} \frac{\partial \mathbf{f}^T}{\partial x} &= \frac{2}{a} \frac{\partial \mathbf{f}^T}{\partial \xi}, \\ \frac{\partial \mathbf{f}^T}{\partial y} &= \frac{2}{b} \frac{\partial \mathbf{f}^T}{\partial \eta}. \end{aligned} \tag{B.6}$$

The interpolation functions **f** are given in (6.23), (6.25), and (6.28) for four-, eight-, and twelve-node elements, respectively. It is a common practice to approximate also the distributed loading in terms of interpolation functions

$$\begin{aligned} \bar{\mathbf{F}} &= \begin{Bmatrix} \bar{F}_x \\ \bar{F}_y \end{Bmatrix} = \begin{bmatrix} \mathbf{f}^T & 0 \\ 0 & \mathbf{f}^T \end{bmatrix} \begin{Bmatrix} \bar{\mathbf{F}}_x \\ \bar{\mathbf{F}}_y \end{Bmatrix} = \mathbf{D}^* \begin{Bmatrix} \bar{\mathbf{F}}_x \\ \bar{\mathbf{F}}_y \end{Bmatrix}, \\ \bar{\mathbf{T}} &= \begin{Bmatrix} \bar{T}_x \\ \bar{T}_y \end{Bmatrix} = \begin{bmatrix} \mathbf{g}^T & 0 \\ 0 & \mathbf{g}^T \end{bmatrix} \begin{Bmatrix} \bar{\mathbf{T}}_x \\ \bar{\mathbf{T}}_y \end{Bmatrix}, \end{aligned} \tag{B.7}$$

where $\bar{\mathbf{F}}_x$ and $\bar{\mathbf{F}}_y$ are nodal values of $\bar{F}_x$ and $\bar{F}_y$, respectively, $\bar{\mathbf{T}}_x$ and $\bar{\mathbf{T}}_y$ are nodal values of $\bar{T}_x$ and $\bar{T}_y$, respectively, and **g** is an appropriate one-dimensional interpolation function given in sec. 6.2.

For the case where material is homogeneous within the element, we have from (6.95)

$$\mathbf{k}_n^* = t \begin{bmatrix} \frac{b}{a} C_{11} \mathbf{A}_{\xi\xi} + C_{13}(\mathbf{A}_{\xi\eta} + \mathbf{A}_{\xi\eta}^T) + \frac{a}{b} C_{33} \mathbf{A}_{\eta\eta} & \frac{b}{a} C_{13} \mathbf{A}_{\xi\xi} + C_{12} \mathbf{A}_{\xi\eta} + C_{33} \mathbf{A}_{\xi\eta}^T + \frac{a}{b} C_{23} \mathbf{A}_{\eta\eta} \\ \text{sym} & \frac{b}{a} C_{33} \mathbf{A}_{\xi\xi} + C_{23}(\mathbf{A}_{\xi\eta} + \mathbf{A}_{\xi\eta}^T) + \frac{a}{b} C_{22} \mathbf{A}_{\eta\eta} \end{bmatrix}, \tag{B.8}$$

which is in the same form as (5.74), and

$$\mathbf{Q}_n^* = \begin{Bmatrix} \mathbf{Q}_x^* \\ \mathbf{Q}_y^* \end{Bmatrix}, \tag{B.9}$$

where

$$A_{\xi\xi} = \int_{-1}^{1}\int_{-1}^{1} \frac{\partial \mathbf{f}}{\partial \xi} \frac{\partial \mathbf{f}^T}{\partial \xi} d\xi d\eta,$$

$$A_{\eta\eta} = \int_{-1}^{1}\int_{-1}^{1} \frac{\partial \mathbf{f}}{\partial \eta} \frac{\partial \mathbf{f}^T}{\partial \eta} d\xi d\eta, \qquad (B.10)$$

$$A_{\xi\eta} = \int_{-1}^{1}\int_{-1}^{1} \frac{\partial \mathbf{f}}{\partial \xi} \frac{\partial \mathbf{f}^T}{\partial \eta} d\xi d\eta,$$

$$\mathbf{Q}_x^* = tab\, \mathbf{M}\bar{\mathbf{F}}_x + tl_j\, \mathbf{R}\bar{\mathbf{T}}_x,$$
$$\mathbf{Q}_y^* = tab\, \mathbf{M}\bar{\mathbf{F}}_y + tl_j\, \mathbf{R}\bar{\mathbf{T}}_y, \qquad (B.11)$$

$$\mathbf{M} = \tfrac{1}{4}\int_{-1}^{1}\int_{-1}^{1} \mathbf{f}\mathbf{f}^T d\xi d\eta, \qquad (B.12)$$

and

$$\mathbf{R} = \tfrac{1}{2}\int_{-1}^{1} \mathbf{f}\bigg|_{\text{side}\,j} \mathbf{g}^T d\zeta, \qquad (B.13)$$

in which the tractions are prescribed over side j. The quantity l_j is the length of the side of the element (i.e., $l_j = a$ or b), and $d\zeta$ is integrated counter-clockwise along side j in (ξ, η)-coordinates.

The $\mathbf{k}_n$ and $\mathbf{Q}_n$ associated with the generalized-coordinate vector

$$\mathbf{q}^T = \{u_1 \quad v_1 \quad u_2 \quad v_2 \quad \cdots\} \qquad (B.14)$$

are given by

$$\mathbf{k}_n = [k_{ij}],$$
$$\mathbf{Q}_n = \{Q_i\}, \qquad (B.15)$$

in which

$$k_{ij} = k^*_{I(i)I(j)},$$
$$Q_i = Q^*_{I(i)}, \qquad (B.16)$$

where

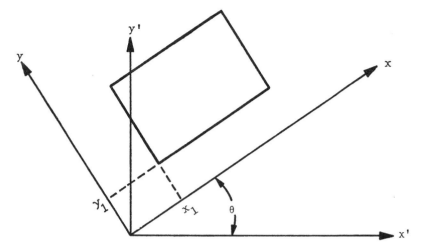

Fig. B.2 A rotation.

$$I(2i - 1) = i$$
$$I(2i) = i + \text{NNPE}, \qquad i = 1, 2, \cdots, \text{NNPE}. \tag{B.17}$$

Throughout this book, NNPE means "number of nodes per element."

For an element as shown in fig. B.2, the element matrices $\mathbf{k}'_n$ and $\mathbf{Q}'_n$ are related to $\mathbf{k}_n$ and $\mathbf{Q}_n$ by

$$\mathbf{k}'_n = \mathbf{T}^{\mathrm{T}} \mathbf{k}_n \mathbf{T},$$
$$\mathbf{Q}'_n = \mathbf{T}^{\mathrm{T}} \mathbf{Q}_n, \tag{B.18}$$

in which

$$\mathbf{T} = \begin{bmatrix} \mathbf{s} & & & 0 \\ & \mathbf{s} & & \\ & & \ddots & \\ 0 & & & \mathbf{s} \end{bmatrix}, \tag{B.19}$$

where

$$\mathbf{s} = \begin{bmatrix} \cos\theta & \sin\theta \\ -\sin\theta & \cos\theta \end{bmatrix}; \tag{B.20}$$

the number of submatrices $\mathbf{s}$ in $\mathbf{T}$ is equal to NNPE.

B.1 Four-Node Element (Fig. B.1a)

We have NNPE = 4 and

$$\mathbf{q}_u^T = \{u_1 \quad u_2 \quad u_3 \quad u_4\},$$
$$\mathbf{q}_v^T = \{v_1 \quad v_2 \quad v_3 \quad v_4\}.$$
(B.21)

The interpolation function **f** is given in (6.23) and for a linear distribution of boundary traction, we have that

$$\mathbf{g} = \frac{1}{2} \begin{Bmatrix} 1 - \zeta \\ 1 + \zeta \end{Bmatrix}.$$

Here $\mathbf{k}_n^*$ is an 8×8 matrix and $\mathbf{Q}_n^*$ is an 8-component vector. The values of the matrices defined in (B.10) through (B.12) are

$$\mathbf{A}_{\xi\xi} = \begin{bmatrix} \frac{1}{3} & -\frac{1}{3} & -\frac{1}{6} & \frac{1}{6} \\ & \frac{1}{3} & \frac{1}{6} & -\frac{1}{6} \\ & & \frac{1}{3} & -\frac{1}{3} \\ \text{sym} & & & \frac{1}{3} \end{bmatrix},$$
(B.22)

$$\mathbf{A}_{\eta\eta} = \begin{bmatrix} \frac{1}{3} & \frac{1}{6} & -\frac{1}{6} & -\frac{1}{3} \\ & \frac{1}{3} & -\frac{1}{3} & -\frac{1}{6} \\ & & \frac{1}{3} & \frac{1}{6} \\ \text{sym} & & & \frac{1}{3} \end{bmatrix},$$
(B.23)

$$\mathbf{A}_{\xi\eta} = \frac{1}{4} \begin{bmatrix} 1 & 1 & -1 & -1 \\ -1 & -1 & 1 & 1 \\ -1 & -1 & 1 & 1 \\ 1 & 1 & -1 & -1 \end{bmatrix},$$
(B.24)

and

$$\mathbf{M} = \begin{bmatrix} \frac{1}{9} & \frac{1}{18} & \frac{1}{36} & \frac{1}{18} \\ & \frac{1}{9} & \frac{1}{18} & \frac{1}{36} \\ & & \frac{1}{9} & \frac{1}{18} \\ \text{sym} & & & \frac{1}{9} \end{bmatrix};$$
(B.25)

and

$$\mathbf{\bar{F}}_x^T = \{F_{x_1} \quad F_{x_2} \quad F_{x_3} \quad F_{x_4}\},$$
$$\mathbf{\bar{F}}_y^T = \{F_{y_1} \quad F_{y_2} \quad F_{y_3} \quad F_{y_4}\},$$
(B.26)

where the F_xs and F_ys have the values $\bar{F}_x$ and $\bar{F}_y$ at the nodes. If side j on which the tractions are prescribed has nodes l, m, then

$$\bar{\mathbf{T}}_x = \begin{Bmatrix} \bar{T}_{x_l} \\ \bar{T}_{x_m} \end{Bmatrix}, \qquad \bar{\mathbf{T}}_y = \begin{Bmatrix} \bar{T}_{y_l} \\ \bar{T}_{y_m} \end{Bmatrix}, \tag{B.27}$$

where the $\bar{T}_x$s and $\bar{T}_y$s are the known values of T_x and T_y at nodes l and m (counting counterclockwise), and where

$$\begin{aligned} \text{row } l \text{ of } \mathbf{R} &= \{\tfrac{1}{3} \quad \tfrac{1}{6}\}, \\ \text{row } m \text{ of } \mathbf{R} &= \{\tfrac{1}{6} \quad \tfrac{1}{3}\}. \end{aligned} \tag{B.28}$$

For example, say side 2 has prescribed traction $\bar{T}_x = (\bar{T}_x)_2$ at node 2 and $\bar{T}_x = (\bar{T}_x)_3$ at node 3; then

$$\mathbf{R}\bar{\mathbf{T}}_x = \begin{bmatrix} 0 & 0 \\ \tfrac{1}{3} & \tfrac{1}{6} \\ \tfrac{1}{6} & \tfrac{1}{3} \\ 0 & 0 \end{bmatrix} \begin{Bmatrix} (\bar{T}_x)_2 \\ (\bar{T}_x)_3 \end{Bmatrix}. \tag{B.29}$$

Or if side 4 has prescribed traction $\bar{T}_x = (\bar{T}_x)_4$ at node 4 and $\bar{T}_x = (\bar{T}_x)_1$ at node 1, then

$$\mathbf{R}\bar{\mathbf{T}}_x = \begin{bmatrix} \tfrac{1}{6} & \tfrac{1}{3} \\ 0 & 0 \\ 0 & 0 \\ \tfrac{1}{3} & \tfrac{1}{6} \end{bmatrix} \begin{Bmatrix} (\bar{T}_x)_4 \\ (\bar{T}_x)_1 \end{Bmatrix}. \tag{B.30}$$

If T_x is constant, that is, if, $(\bar{T}_x)_4 = (\bar{T}_x)_1$, (B.29) and (B.30) are reduced to the case of constant traction discussed in sec. 5.32.

Appendix C Triangular Elements With Straight Edges

For triangular elements, such as shown in Fig. C.1, the transformation between (x, y)- and $(\zeta_1, \zeta_2, \zeta_3)$-coordinates is given in (6.45) and (6.46):

$$\begin{Bmatrix} 1 \\ x \\ y \end{Bmatrix} = \begin{bmatrix} 1 & 1 & 1 \\ x_1 & x_2 & x_3 \\ y_1 & y_2 & y_3 \end{bmatrix} \begin{Bmatrix} \zeta_1 \\ \zeta_2 \\ \zeta_3 \end{Bmatrix} \tag{6.45}$$

and

$$\begin{Bmatrix} \zeta_1 \\ \zeta_2 \\ \zeta_3 \end{Bmatrix} = \frac{1}{2\Delta} \begin{bmatrix} a_1 & b_1 & c_1 \\ a_2 & b_2 & c_2 \\ a_3 & b_3 & c_3 \end{bmatrix} \begin{Bmatrix} 1 \\ x \\ y \end{Bmatrix}. \tag{6.46}$$

Recall that

$\Delta = \frac{1}{2}(x_2 y_3 - x_3 y_2 + x_3 y_1 - x_1 y_3 + x_1 y_2 - x_2 y_1),$
$a_1 = x_2 y_3 - x_3 y_2, \quad a_2 = x_3 y_1 - x_1 y_3, \quad a_3 = x_1 y_2 - x_2 y_1,$
$b_1 = y_2 - y_3, \quad b_2 = y_3 - y_1, \quad b_3 = y_1 - y_2,$
$c_1 = x_3 - x_2, \quad c_2 = x_1 - x_3, \quad c_3 = x_2 - x_1.$

The displacement, the strain, the stress, and the distributed loading vector can be written in the same form as (B.3), (B.4), (B.5), and (B.7), with the interpolation functions defined in (6.31), (6.33) and (6.35). In this case, both

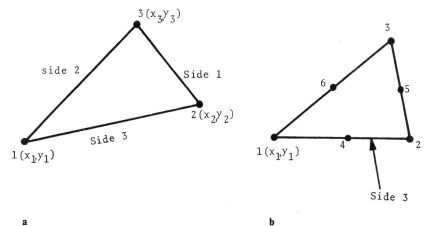

Fig. C.1 Triangular elements: (a) constant strain; (b) linear strain.

$\partial \mathbf{f}^T/\partial x$ and $\partial \mathbf{f}^T/\partial y$ can be expressed in the form

$$\frac{\partial \mathbf{f}^T}{\partial x} = \frac{\partial \zeta_1}{\partial x}\frac{\partial \mathbf{f}^T}{\partial \zeta_1} + \frac{\partial \zeta_2}{\partial x}\frac{\partial \mathbf{f}^T}{\partial \xi_2} + \frac{\partial \zeta_3}{\partial x}\frac{\partial \mathbf{f}^T}{\partial \zeta_3} = \zeta \mathbf{X},$$
$$\frac{\partial \mathbf{f}^T}{\partial y} = \frac{\partial \zeta_1}{\partial y}\frac{\partial \mathbf{f}^T}{\partial \zeta_1} + \frac{\partial \zeta_2}{\partial y}\frac{\partial \mathbf{f}^T}{\partial \zeta_2} + \frac{\partial \zeta_3}{\partial y}\frac{\partial \mathbf{f}^T}{\partial \zeta_3} = \zeta \mathbf{Y},$$
(C.1)

where $\mathbf{X}$, $\mathbf{Y}$ are constants that only depend on the spatial coordinates of the nodes, and where the ζ on the right side of the equations are functions of ζ_1, ζ_2, and ζ_3. The matrices $\mathbf{X}$, $\mathbf{Y}$ will be written down explicitly later for different cases (see eqs. (C.8) and (C.13) for constant-strain and linear-strain triangle, respectively). From (6.97) and (6.98) we have

$$\mathbf{k}_n^* = t\triangle \begin{bmatrix} \mathbf{X} & 0 \\ 0 & \mathbf{Y} \\ \mathbf{Y} & \mathbf{X} \end{bmatrix}^T \begin{bmatrix} C_{11}\mathbf{H} & & \text{sym} \\ C_{21}\mathbf{H} & C_{22}\mathbf{H} & \\ C_{31}\mathbf{H} & C_{32}\mathbf{H} & C_{33}\mathbf{H} \end{bmatrix} \begin{bmatrix} \mathbf{X} & 0 \\ 0 & \mathbf{Y} \\ \mathbf{Y} & \mathbf{X} \end{bmatrix},$$
(C.2)

and

$$\mathbf{Q}_n^* = \begin{Bmatrix} \mathbf{Q}_x \\ \mathbf{Q}_y \end{Bmatrix},$$
(C.3)

where

$$\mathbf{H} = 2\int_0^1\int_0^{1-\eta} \zeta\zeta^T d\xi\partial\eta,$$
(C.4)

$$\mathbf{Q}_x = t\triangle \mathbf{M}\bar{\mathbf{F}}_x + tl_j\mathbf{R}\bar{\mathbf{T}}_x,$$
$$\mathbf{Q}_y = t\triangle \mathbf{M}\bar{\mathbf{F}}_y + tl_j\mathbf{R}\bar{\mathbf{T}}_y,$$
(C.5)

$$\mathbf{M} = 2\int_0^1\int_0^{1-\eta} \mathbf{f}\mathbf{f}^T d\xi d\eta,$$
(C.6)

and

$$\mathbf{R} = \int_0^1 \mathbf{f}\Big|_{\text{side}j} \mathbf{g}^T d\zeta,$$
(C.7)

where $\zeta = \zeta_3$ if $j = 1$, $\zeta = \zeta_1$ if $j = 2$, or $\zeta = \zeta_2$ if $j = 3$, with side j the side with prescribed tractions.

The matrix $\mathbf{k}_n$ and the vector $\mathbf{Q}_n$ associated with $\mathbf{q} = \{u_i \quad v_1 \quad u_2 \quad v_2 \quad \cdots\}$ can be obtained from (B.15) and (B.16).

C.1 Constant-Strain Triangle (Fig. C.1a)

We have

$$\zeta = 1,$$
$$\mathbf{X} = \{b_1 \quad b_2 \quad b_3\}, \tag{C.8}$$
$$\mathbf{Y} = \{c_1 \quad c_2 \quad c_3\}.$$

The matrix $\mathbf{k}_n^*$ is 6×6 and $\mathbf{Q}_n^*$ is 6×1. The matrices defined in (C.3) through (C.7) are

$$\mathbf{H} = 1, \tag{C.9}$$

$$\mathbf{M} = \frac{1}{12} \begin{bmatrix} 2 & 1 & 1 \\ 1 & 2 & 1 \\ 1 & 1 & 2 \end{bmatrix}, \tag{C.10}$$

$$\bar{\mathbf{F}}_x^T = \{F_{x_1} \quad F_{x_2} \quad F_{x_3}\},$$
$$\bar{\mathbf{F}}_y^T = \{F_{y_1} \quad F_{y_2} \quad F_{y_3}\}, \tag{C.11}$$

and

$$\bar{\mathbf{T}}_x = \begin{Bmatrix} T_{x_l} \\ T_{x_m} \end{Bmatrix}$$
$$\bar{\mathbf{T}}_y = \begin{Bmatrix} T_{y_l} \\ T_{y_m} \end{Bmatrix}, \tag{C.12}$$

and the rows of $\mathbf{R}$ are as given in (B.28); (T_{x_l}, T_{y_l}) and (T_{x_m}, T_{y_m}) are the given nodal values of l and m which are nodes (counting counterclockwise) on the side j where tractions are prescribed.

C.2 Linear-Strain Triangle (Fig. C.1b)

We have

$$\zeta^T = \{\zeta_1 \quad \zeta_2 \quad \zeta_3\},$$

$$\mathbf{X} = \frac{1}{2\Delta} \begin{bmatrix} 3b_1 & -b_2 & -b_3 & 4b_2 & 0 & 4b_3 \\ -b_1 & 3b_2 & -b_3 & 4b_1 & 4b_3 & 0 \\ -b_1 & -b_2 & 3b_3 & 0 & 4b_2 & 4b_1 \end{bmatrix}, \tag{C.13}$$

$$\mathbf{Y} = \frac{1}{2\Delta} \begin{bmatrix} 3c_1 & -c_2 & -c_3 & 4c_2 & 0 & 4c_3 \\ -c_1 & 3c_2 & -c_3 & 4c_1 & 4c_3 & 0 \\ -c_1 & -c_2 & 3c_3 & 0 & 4c_2 & 4c_1 \end{bmatrix}.$$

The matrix $\mathbf{k}_n^*$ is 12×12 and $\mathbf{Q}_n^*$ is 12×1. The matrices defined in (C.3) through (C.7) are

$$\mathbf{H} = \frac{1}{12}\begin{bmatrix} 2 & 1 & 1 \\ 1 & 2 & 1 \\ 1 & 1 & 2 \end{bmatrix}, \tag{C.14}$$

$$\mathbf{M} = \frac{1}{180}\begin{bmatrix} 6 & & & & & \\ -1 & 6 & & & \text{sym} & \\ -1 & -1 & 6 & & & \\ 0 & 0 & -4 & 32 & & \\ -4 & 0 & 0 & 16 & 32 & \\ 0 & -4 & 0 & 16 & 16 & 32 \end{bmatrix}, \tag{C.15}$$

$$\begin{aligned}\bar{\mathbf{F}}_x^T &= \{F_{x_1} \quad F_{x_2} \quad \cdots \quad F_{x_6}\}, \\ \bar{\mathbf{F}}_y^T &= \{F_{y_1} \quad F_{y_2} \quad \cdots \quad F_{y_6}\},\end{aligned} \tag{C.16}$$

and

$$\begin{aligned}\bar{\mathbf{T}}_x^T &= \{T_{x_l} \quad T_{x_m} \quad T_{x_n}\}, \\ \bar{\mathbf{T}}_y^T &= \{T_{y_l} \quad T_{y_m} \quad T_{y_n}\},\end{aligned} \tag{C.17}$$

and the rows of $\mathbf{R}$ are

$$\begin{aligned}\text{row } l \text{ of } \mathbf{R} &= \tfrac{1}{30}[4 \quad 2 \quad -1], \\ \text{row } m \text{ of } \mathbf{R} &= \tfrac{1}{30}[2 \quad 16 \quad 2], \\ \text{row } n \text{ of } \mathbf{R} &= \tfrac{1}{30}[-1 \quad 2 \quad 4],\end{aligned} \tag{C.18}$$

where l, m, n are the nodal numbers of the side where traction is prescribed, with m the node at the midpoint of the side.

Appendix D Variational Methods

D.1 The Calculus of Variations

We will work our way into the calculus of variations by means of a simple example whose solution is known. Consider the curve in fig. D.1. What function $y = y(x)$ (often called the extremal function) will minimize the length of the curve AB? The answer is, of course, the straight line $y = ax + b$. Now, the length l of curve AB is

$$l = \int_{x_A}^{x_B} \sqrt{1 + (y')^2}\, dx \quad \text{where } (\ )' = \frac{d(\ )}{dx}. \tag{D.1}$$

We therefore wish to minimize the integral l as given by (D.1). This integral is called a functional, and its value depends on the function $y(x)$. Before we proceed with this minimization, it is helpful to consider the slightly more general problem of a functional $I(y)$ given by

$$I(y) = \int_{x_A}^{x_B} f(x, y, y')\, dx. \tag{D.2}$$

We wish to determine $y = y(x)$ so that I is stationary (i.e., $\delta I = 0$, which

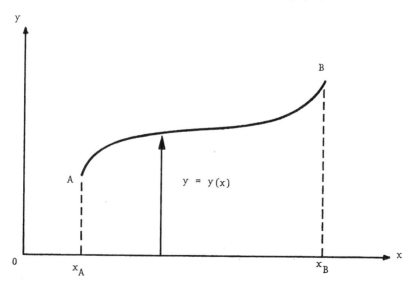

Fig. D.1 The function $y(x)$ is to minimize length of the curve AB.

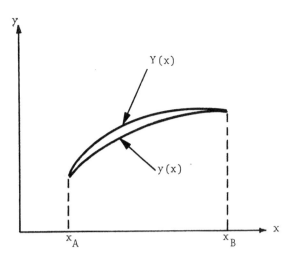

Fig. D.2 Neighboring functions.

means that a small variation about the correct $y(x)$ leads to zero change in δI). This is the problem of the calculus of variations. The method yields the differential equation from which we determine that particular y which makes $\delta I = 0$. If $y(x)$ is the solution being sought, and $Y(x)$ is some neighboring function with the same end values (fig. D.2), we define the variation of y as $\delta y = Y - y$, and note that $\delta y = 0$ at x_A and x_B. Note also that

$$\delta\left(\frac{dy}{dx}\right) = \delta y' = Y' - y' = \frac{d}{dx}(Y - y) = \frac{d}{dx}(\delta y). \tag{D.3}$$

Equation (D.3) shows that the operators δ and d/dx are commutative, so that the derivative of the variation is the same as the variation of the derivative. It can also be shown that one can use the δ operator just as one uses the d operator; so that, for example,

$$\begin{aligned}\delta x^n &= nx^{n-1}\delta x,\\ \delta(y')^2 &= 2y'\delta y'.\end{aligned} \tag{D.4}$$

Now, since $Y = y + \delta y$, we have

$$I(Y) = \int_{x_A}^{x_B} f(x, y + \delta y, y' + \delta y')dx, \tag{D.5}$$

$$\delta I = I(Y) - I(y) = \int_{x_A}^{x_B} \delta f\, dx. \tag{D.6}$$

From here on, we adopt a formal approach (see Forray 1968 or Hildebrand 1962 for the more traditional procedure): we expand in Taylor series and truncate for small δy to obtain an expression for δf. Hence,

$$\begin{aligned}
\delta f &= f(x, Y, Y') - f(x, y, y') \\
&= f(x, y + \delta y, y' + \delta y') - f(x, y, y') \\
&= \text{Taylor-series expansion} = f(x, y, y') + \frac{\partial f}{\partial y}\delta y + \frac{\partial f}{\partial y'}\delta y' + \cdots \\
&\quad - f(x, y, y') \\
&= \frac{\partial f}{\partial y}\delta y + \frac{\partial f}{\partial y'}\delta y' + \text{higher order terms.}
\end{aligned} \quad (D.7)$$

Note that in the preceding and in what follows $\delta x = 0$, since the independent variable x is not varied. The function of x, namely $y(x)$, is what is varied. With (D.7), eq. (D.6) gives

$$\delta I = I(Y) - I(y) = \int_{x_A}^{x_B} \left(\frac{\partial f}{\partial y}\delta y + \frac{\partial f}{\partial y'}\delta y' \right) dx. \quad (D.8)$$

Equation (D.8) is then integrated by parts to obtain

$$\delta I = \frac{\partial f}{\partial y'}\delta y \bigg|_{x_A}^{x_B} - \int_{x_A}^{x_B} \left[\frac{d}{dx}\left(\frac{\partial f}{\partial y'}\right) - \frac{\partial f}{\partial y} \right] \delta y\, dx. \quad (D.9)$$

The integrated term in (D.9) vanishes because $\delta y = 0$ at $x = x_A, x_B$. For arbitrary δy in the interval (x_A, x_B), $\delta I = 0$ in (D.9) if the quantity in brackets vanishes, and this yields the equation

$$\frac{d}{dx}\left(\frac{\partial f}{\partial y'}\right) - \frac{\partial f}{\partial y} = 0. \quad (D.10)$$

Equation (D.10) is called the Euler equation of the variational problem; it is the one whose solution is that $y = y(x)$ which makes $\delta I = 0$.

This procedure can be generalized to higher derivatives inside the integral in (D.2), to obtain a higher-order ordinary differential equation for y as our Euler equation. Generalization to more than one dependent variable leads to more than one Euler equation, and generalization to more than one independent variable leads to partial differential equations as our Euler equations.

We now get back to our example where the functional is given by (D.1), so that $I = l, f = (1 + y'^2)^{1/2}$. It is found that in this case, $\delta I = 0$ leads to the Euler equation

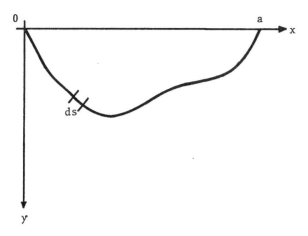

Fig. D.3 The hanging-chain problem.

$$\frac{d}{dx}\left[\frac{y'}{(1 + y'^2)^{1/2}}\right] = 0. \tag{D.11}$$

Integration gives $y'\sqrt{1 + (y')^2}$ = constant $\to y' = c_1 \to y = c_1 x + c_2$, where c_1, c_2 are constants.

D.1.1 Lagrange Multipliers

In this next example we introduce Lagrange multipliers. To find the equilibrium position of a hanging, inextensional chain, we stipulate that this be the position of minimum potential energy. Referring to fig. D.3, the potential energy Π and the fixed length of chain l are given by (ρ is the density per unit length):

$$\Pi = -\int_0^a \rho y ds,$$
$$l = \int_0^a ds = \int_0^a [1 + (y')^2]^{1/2} dx.$$

We have boundary conditions at $x = 0$ and a; and, in addition, a side condition or constraint—namely, a fixed length of chain l. To handle this side condition, instead of minimizing Π, we minimize an expression Φ, where $\Phi = \Pi + \lambda l$ and λ is the Lagrange multiplier which is to be determined (Forray 1968). Then we require that $\delta\Phi = 0$, where

$$\Phi = -\int_0^a (\rho y - \lambda)ds = -\int_0^a (\rho y - \lambda)\sqrt{1 + (y')^2}\,dx. \tag{D.12}$$

We next solve the Euler equation for this problem, wherein

$$f = -(\rho y + \lambda)\sqrt{1 + (y')^2}.$$

After some algebra, we obtain

$$y' = \left(\frac{\rho y - \lambda}{\rho c_1} - 1\right)^{1/2}, \quad \text{or} \quad y = \frac{\lambda}{\rho} + c_1 \cosh\left(\frac{x - c_2}{c_1}\right). \tag{D.13}$$

The three constants c_1, c_2, and λ are evaluated with the aid of the two boundary conditions and the constraint of fixed length.

In some problems, λ may have a physical interpretation. The Lagrange-multiplier approach is very useful in conjunction with the Rayleigh-Ritz method, which we will outline in the following section (Forray 1968; Hildebrand 1962), when certain assumed functions do not satisfy the boundary conditions exactly, and are better treated as additional constraints.

D.2 The Rayleigh-Ritz Method (An Approximate Method)

As we have seen in the previous section, the problem to be solved by the calculus of variations is to find $y(x)$ such that the functional $I(y)$ is rendered stationary. It was found that the correct $y(x)$ satisfies the Euler equation associated with the variational problem. Another approach, leading to an approximate solution for $y(x)$, is available. Rather than satisfying the Euler equation directly, a mathematical expression (such as a truncated power series) is assumed for the unknown function $y(x)$. The functional I after integration becomes a function of, say, the n parameters $a_1, a_2, \cdots, a_n$ (that are the coefficients of the power series). Then I is made stationary by a proper selection of the a_i ($i = 1, 2, \cdots, n$), thereby leading to an approximate solution. Mathematically,

$$\delta I = \sum_{i=1}^{n} \frac{\partial I}{\partial a_i} \delta a_i = 0. \tag{D.14}$$

For arbitrary variations of the parameters a_i, eq. (D.14) implies that

$$\partial I/\partial a_i = 0, \quad i = 1, 2, \cdots, n. \tag{D.15}$$

Equations (D.15) yield n equations for the n unknown parameters $a_1, a_2, \cdots, a_n$. This is the Rayleigh-Ritz method. In variational problems of solid mechanics, for instance, where I is the potential energy and y is a displacement, the Euler equation usually (not invariably!) turns out to be the equilibrium equation. With more than one dependent and independent variable, we obtain, of course, more than one equilibrium equation. Also, the stationary value of I is *usually* a minimum. The quantity $y(x)$ must also be selected so that it satisfies the displacement boundary conditions of the particular problem; that is, $y(x)$ must be kinematically admissible. (It may be remarked at this point that in the finite-element method, the parameters which play the same role as the coefficients, $a_1, a_2, \cdots, a_n$ of the Rayleigh-Ritz method are normally the unknown nodal values at the nodes of an element.)

D.3 Boundary-Value Problems and Variational Methods

The functional I, together with the variational statement $\delta I = 0$, can be given broader meaning by noting that any boundary-value problem (differential equations and boundary conditions) is equivalent to an appropriate variational statement. That is, for a properly chosen functional I, the statement $\delta I = 0$ leads to Euler equations which are in fact the same equations as those in the boundary-value problem. Consequently, the solution of such problems can also be obtained by the approximate method indicated, once the variational statements are formulated. Some examples will now be given to expand further on these ideas.

D.3.1 First Example

This example will show how the variational statement can be derived from the differential equation. Consider the problem of a uniform bar of length L, under a distributed axial load $p = p(x)$. The axial displacement u is governed by

$$AEu_{xx} + p(x) = 0, \tag{D.16}$$

where AE is the axial stiffness of the rod. Note also the notation $u_x = du/dx$, etc. We proceed by multiplying (D.16) by the variation δu and integrating:

$$\int_0^L \left(AE \frac{d^2u}{dx^2} + p\right) \delta u\, dx = \int_0^L AE \frac{du_x}{dx} \delta u\, dx + \int_0^L p\delta u\, dx = 0. \tag{D.17}$$

The first integral on the right can be integrated by parts:

$$\int_0^L AE \frac{du_x}{dx} \delta u\, dx = AE \int_0^L \delta u\, d(u_x) = AEu_x \delta u \Big|_0^L - \int_0^L AEu_x \delta u_x\, dx, \quad (D.18)$$

where we used the rule

$$\int_0^L U dV = UV \Big|_0^L - \int_0^L V dU,$$

where $U = \delta u$ and $V = u_x$, and where we have interchanged δ and d/dx, in order to write

$$dU = \frac{d\delta u}{dx} dx = \delta \frac{du}{dx} dx = \delta u_x dx.$$

With (D.18), the integral in (D.17) is

$$AEu_x \delta u \Big|_0^L - \int_0^L (AEu_x \delta u_x - p\delta u)\, dx = 0. \quad (D.19)$$

Noting that $\delta \tfrac{1}{2}(u_x)^2 = u_x \delta u_x$, we may write (D.19) as

$$AEu_x \delta u \Big|_0^L - \delta \int_0^L \left[\frac{AE}{2}(u_x)^2 - pu \right] dx = 0. \quad (D.20)$$

The integrated term in (D.20) vanishes if we impose at $x = 0, L$ the condition that either

$$u = u_0 \quad \text{or} \quad AEu_x = 0, \quad (D.21)$$

where u_0 is a prescribed constant, so that $\delta u_0 = 0$. Under conditions (D.21), eq. (D.20) yields

$$\delta \int_0^L \left[\frac{AE}{2} \left(\frac{du}{dx} \right)^2 - pu \right] dx = 0 \quad (D.22)$$

(i.e., $\delta I = 0$, where I is the functional under the integral). Equation (D.22) is the variational statement whose Euler equation is (D.16). That is, if u satisfies (D.16) and the end conditions (D.21), the integral (D.22) will be stationary. The condition $AEu_x = 0$ at $x = L$ is a so-called natural boundary condition.

The condition $u = u_0$ is referred to as a geometric or rigid boundary condition. From solid mechanics, AEu_x is the axial force; and the integrated term in (D.20) represents the work done by this force through the displacement δu. If u is prescribed (so that $\delta u = 0$) or there is no axial constraint (i.e., $AEu_x = 0$) at $x = L$, the work done vanishes. If there were a spring of modulus k at $x = L$, the end condition would be

$$AE\left(\frac{du}{dx}\right)_{x=L} = -k(u)_{x=L}. \tag{D.23}$$

Multiplying (D.23) by δu yields

$$AE\frac{du}{dx}\delta u = -ku\delta u = -\delta\left(\frac{1}{2}ku^2\right) \quad \text{at} \quad x = L. \tag{D.24}$$

Since this term would not vanish in (D.20), eq. (D.22) is

$$\delta\left\{I + \left(\frac{1}{2}ku^2\right)_{x=L}\right\} = 0 \quad \text{where} \quad I = \int_0^L \left[\frac{AE}{2}(u_x)^2 - pu\right]dx. \tag{D.25}$$

If instead of a spring there is an applied force F at $x = L$, then the end condition would be

$$AE\frac{du}{dx} = F \quad \text{at} \quad x = L, \tag{D.26}$$

and instead of (D.25), eq. (D.22) would be

$$\delta\{I - (Fu)_{x=L}\} = 0 \quad \text{or} \quad \delta\left\{\int_0^L \left[\frac{AE}{2}(u_x)^2 - pu\right]dx - (Fu)_{x=L}\right\} = 0. \tag{D.27}$$

The bracketed expression in (D.27) is the potential energy of the bar; the first term is the strain energy, and the other terms are work done by the distributed force p and the end load F. In solid mechanics, the variational statement or principle is often a minimum principle. For the present example, the variational statement is that of all displacements $u(x)$ satisfying the geometric (or rigid) boundary conditions, the correct one also minimizes the potential energy (or equivalently, also satisfies equilibrium). It is instructive to start from the potential energy Π, minimize it (i.e., render it stationary), and obtain the differential equation [i.e., the Euler equation, which, for this problem is the equilibrium equation (D.16)] and natural boundary condition. In other words, we can work the problem just treated in reverse. The potential energy is

$$\Pi = \int_0^L \left[\frac{AE}{2}(u_{,x})^2 - pu \right] dx - Fu \bigg|_{x=L}. \qquad (D.28)$$

We now seek to make $\delta \Pi = 0$, and therefore take variations on u:

$$\delta \Pi = \int_0^L [AEu_{,x}\delta u_{,x} - p\delta u] dx - F\delta u \bigg|_{x=L} = 0. \qquad (D.29)$$

Integrating by parts, the first term is

$$\int_0^L AEu_{,x}\delta u_{,x} dx = \int_0^L AEu_{,x} \frac{d\delta u}{dx} dx = \int_0^L AEu_{,x} d(\delta u)$$

$$= AEu_{,x}\delta u \bigg|_0^L - \int_0^L AEu_{,xx}\delta u \, dx \qquad (D.30)$$

$$= AEu_{,x}\delta u \bigg|_{x=L} - AEu_{,x}\delta u \bigg|_{x=0} - \int_0^L AEu_{,xx}\delta u \, dx.$$

By using (D.30), and combining terms, (D.29) gives

$$\delta \Pi = -\int_0^L [AEu_{,xx} + p]\delta u \, dx + (AEu_{,x} - F)\delta u]_{x=L} \qquad (D.31)$$
$$- AEu_{,x}\delta u]_{x=0} = 0.$$

The last term in (D.31) vanishes if u satisfies prescribed geometric boundary conditions at $x = 0$, so that $\delta u(0) = 0$. In what remains, in order for $\delta \Pi$ to vanish for arbitrary δu, the expressions in brackets and parentheses must vanish, yielding

$$AEu_{,xx} + p = 0 \quad \text{for} \quad 0 < x < L, \qquad (D.32)$$
$$AEu_{,x} = F \quad \text{at} \quad x = L. \qquad (D.33)$$

Equation (D.32) is the equilibrium equation (or Euler equation) and equation (D.33) is the natural boundary condition. We comment at this point that in approximate methods (such as the Rayleigh-Ritz and finite-element methods), if $\delta\Pi = 0$ is satisfied approximately, then (D.32) and (D.33) are satisfied approximately, since they are a consequence of $\delta\Pi = 0$.

An example of a higher-order equation is the deflection y of a beam of stiffness EI caused by a transverse load of intensity $q(x)$, for which

$$\frac{d^2}{dx^2}(EI\, d^2y/dx^2) - q(x) = 0. \tag{D.34}$$

Again, we multiply by δy and integrate from 0 to L. We will now need to integrate the following term by parts, noting that the δ and d/dx operators commute:

$$\int_0^L \frac{d^2}{dx^2}\left(EI\frac{d^2y}{dx^2}\right)\delta y\, dx = \int_0^L \frac{d}{dx}(EIy'_{xx})'_x \delta y\, dx = \int_0^L \delta y\, d(EIy'_{xx})'_x$$

$$= (EIy'_{xx})'_x \delta y \bigg|_0^L - \int_0^L (EIy'_{xx})'_x \delta(y'_x)\, dx.$$

Integrate again by parts the integral

$$\int_0^L (EIy'_{xx})'_x \delta(y'_x)\, dx = \int_0^L \delta(y'_x)\, d(EIy'_{xx}) = EIy'_{xx}\delta(y'_x)\bigg|_0^L - \int_0^L EIy'_{xx}\delta(y'_{xx})\, dx,$$

where we have used

$$\frac{d}{dx}\delta(y'_x) = \delta\frac{d}{dx}(y'_x) = \delta(y'_{xx}).$$

Using these results, integration of (D.34) leads to

$$(EIy'_{xx})'_x \delta y\bigg|_0^L - EIy'_{xx}\delta(y'_x)\bigg|_0^L + \int_0^L [EIy'_{xx}\delta(y'_{xx}) - q\delta y]\, dx = 0. \tag{D.35}$$

The integrated term vanishes if at $x = 0, L$ we prescribe

$$y = y_0 \quad \text{or} \quad (EIy'_{xx})'_x = 0 \tag{D.36a}$$

and

$$y' = y'_0 \quad \text{or} \quad EIy'_{xx} = 0. \tag{D.36b}$$

Then (D.35) yields [noting that $\delta(y'_{xx})^2 = 2y'_{xx}\delta y'_{xx}$],

$$\delta \int_0^L \left[\frac{EI}{2} (y'_{xx})^2 - qy \right] dx = 0. \tag{D.37}$$

Equation (D.37) is the variational statement $\delta \Pi = 0$, where Π is the potential energy of our beam problem, and (D.36a, b) are boundary conditions.

D.3.2 Second Example

In our next example, we consider problems described by the partial differential equation

$$\partial^2 w/\partial x^2 + \partial^2 w/\partial y^2 + w + p = 0 \quad \text{or} \quad w_{,xx} + w_{,yy} + w + p = 0. \tag{D.38}$$

Here we multiply (D.38) by the variation $\delta w(x, y)$ and integrate over $x_1 \leq x \leq x_2$, $y_1 \leq y \leq y_2$. We can then write

$$\int_{x_1}^{x_2}\int_{y_1}^{y_2} (w_{,xx} + w_{,yy})\delta w\, dx\, dy + \delta \int_{x_1}^{x_2}\int_{y_1}^{y_2} (\tfrac{1}{2}w^2 + pw)\, dx\, dy = 0. \tag{D.39}$$

In the first integral, integrate by parts first with respect to x. The first term gives

$$\int_{x_1}^{x_2}\int_{y_1}^{y_1} w_{,xx}\delta w\, dx\, dy = \int_{y_1}^{y_2} w_{,x}\delta w \bigg|_{x_1}^{x_2} dy - \int_{x_1}^{x_2}\int_{y_1}^{y_2} w_{,x}\delta w_{,x}\, dx\, dy$$

$$= \int_{y_1}^{y_2} w_{,x}\delta w \bigg|_{x_1}^{x_2} dy - \delta \int_{x_1}^{x_2}\int_{y_1}^{y_2} \tfrac{1}{2}(w_{,x})^2\, dx\, dy.$$

Integration with respect to y is done similarly, and then (D.39) can be written as

$$\delta \int_{x_1}^{x_2}\int_{y_1}^{y_2} [-\tfrac{1}{2}(w_{,x}^2 + w_{,y}^2) + \tfrac{1}{2}w^2 + pw]\, dx\, dy + \int_{y_1}^{y_2} w_{,x}\delta w \bigg|_{x_1}^{x_2} dy$$
$$+ \int_{x_1}^{x_2} w_{,y}\delta w \bigg|_{y_1}^{y_2} dx = 0. \tag{D.40}$$

The result in (D.40) can be written in a form independent of the coordinate system:

$$\delta \iint_A \left[\frac{1}{2}(\nabla w)^2 - \frac{1}{2}w^2 + pw \right] dA - \oint_C \frac{\partial w}{\partial n} \delta w\, ds = 0, \tag{D.41}$$

where A is the area of the region and C is its closed boundary of arbitrary contour. The term $\partial w/\partial n$ is the derivative normal to the boundary, and is associated with the natural boundary condition.

As indicated previously, the Rayleigh-Ritz method is a procedure for obtaining approximate solutions to problems expressed in variational form. In the case when a function $y(x)$ is to be determined, the usual approach is to represent $y(x)$ as a linear combination of suitable functions, such as

$$y = c_1\phi_1(x) + c_2\phi_2(x) + \cdots + c_n\phi_n(x), \tag{D.42}$$

and to determine the parameters c_i so that the associated functional is stationary. The functions ϕ_i are chosen to satisfy the geometric (rigid) boundary conditions, and if the ϕ_i form a complete set, (such as is formed, e.g., by the terms in a power or trigonometric series) the corresponding infinite series ($n \to \infty$) will converge to the exact solution.

D. 3.3 Third Example

The foregoing method is illustrated by the example of the deflection y of a string of length L under a constant tension T and acted on by a distributed load q. It is fixed at $x = 0$ and L so that $y(0) = y(L) = 0$. In this case, the variational statement is expressed as

$$\delta I = 0 \quad \text{where} \quad I = \int_0^L [\tfrac{1}{2}T(y')^2 + qy]\,dx. \tag{D.43}$$

Assume for y the three-term expansion

$$y \approx c_1 \sin \frac{\pi x}{L} + c_2 \sin \frac{2\pi x}{L} + c_3 \sin \frac{3\pi x}{L}. \tag{D.44}$$

Substitute (D.44) into I, and perform the integration, making use of the fact that integrals of $\cos n\pi x/L \cos m\pi x/L$, where m and n are integers, vanish when $m \neq n$, to obtain

$$I = \frac{T}{4}\frac{\pi^2}{L}(c_1^2 + 4c_2^2 + 9c_3^2) + q\frac{L}{\pi}\left(2c_1 + \frac{2}{3}c_3\right). \tag{D.45}$$

Three equations for the unknowns c_i are obtained from

$$\partial I/\partial c_1 = 0, \quad \partial I/\partial c_2 = 0, \quad \partial I/\partial c_3 = 0, \tag{D.46}$$

so that

$$c_1 = -4qL^2/\pi^3 T, \quad c_2 = 0, \quad c_3 = -4qL^2/27\pi^3 T, \tag{D.47}$$

and the approximate solution for y is

$$y \approx -\frac{4qL^2}{\pi^3 T}\left(\sin\frac{\pi x}{L} + \frac{1}{27}\sin\frac{3\pi x}{L}\right), \tag{D.48}$$

which, on evaluation at certain values of x, turns out to be a good approximation to the exact solution given by $y = -qx(L-x)/2T$. The exact solution is obtained by solving the differential equation directly. This, of course, is the Euler equation $Ty'' - q = 0$, which can be obtained by writing down the appropriate expression for δI and integrating by parts to give

$$\delta I = Ty'\delta y\Big|_0^L - \int_0^L (Ty'' - q)\delta y\, dx = 0. \tag{D.49}$$

Since the geometric boundary conditions are satisfied by (D.44), the integrated term in (D.49) would vanish and (D.49) becomes equivalent to

$$\int_0^L (Ty'' - q)\delta y\, dx = 0. \tag{D.50}$$

In fact, if the expression (D.44) is substituted directly into (D.50), integration yields

$$\left(c_1 + \frac{4qL^2}{\pi^3 T}\right)\delta c_1 + 4c_2\,\delta c_2 + \left(9c_3 + \frac{4qL^2}{3\pi^3 T}\right)\delta c_3 = 0. \tag{D.51}$$

The terms multiplying the arbitrary variations δc_1, δc_2, and δc_3 must vanish, giving the same values for c_1, c_2, and c_3 as (D.47). Now instead of (D.44) we may choose a polynomial as an approximation for y, such as

$$y \approx x(L-x)(c_1 + c_2 x + \cdots + c_n x^{n-1}), \tag{D.52}$$

and again, on taking $\partial I/\partial c_i = 0$ ($i = 1, 2, \cdots, n$), we obtain n equations for the n constants.

D.3.4 Fourth Example

We next illustrate an application to a partial differential equation by considering St-Venant torsion in a rectangular region. The appropriate functional is

$$I(u) = \int_{-b}^{b}\int_{-a}^{a}\left[\left(\frac{\partial u}{\partial x}\right)^2 + \left(\frac{\partial u}{\partial y}\right)^2 - 4u\right]dxdy, \tag{D.53}$$

where $u = 0$ on the boundary of the rectangle ($x = \pm a$, $y = \pm b$). For the single-term approximation, $u \approx c_i\phi_i(x, y)$, the geometric (rigid) boundary condition is satisfied by taking

$$\phi_1(x, y) = (a^2 - x^2)(b^2 - y^2), \tag{D.54}$$

where the origin of coordinates is $x = y = 0$. In this case $\partial I/\partial c_1 = 0$ yields $c_1 = \frac{5}{4}(a^2 + b^2)$.

D.4 The Method of Weighted Residuals (an Approximate Method)

In addition to the variational approaches described in the preceding paragraphs, one can construct a variational statement directly by taking a differential equation, squaring it, and integrating over a given region. For instance, consider the problem of finding a $u(x)$ that satisfies the differential equation $L(u) = 0$ and given boundary conditions at the end points a, b. Then u can be regarded as that function which satisfies the boundary conditions, and renders I stationary, where I is given by

$$I = \int_a^b [L(u)]^2 dx. \tag{D.55}$$

If we assume that $u = \phi_0 + \sum_{i=1}^{n} c_i\phi_i$, where ϕ_0 satisfies the boundary conditions, and that the ϕ_i ($i = 1, 2, \cdots, n$) satisfy homogeneous boundary conditions, then $\delta I = 0$ leads to $\partial I/\partial c_j = 0$, ($j = 1, 2, \cdots, n$), which is equivalent to

$$\int_a^b L\left(\phi_0 + \sum_{i=1}^{n} c_i\phi_i\right)L(\phi_j)dx = 0. \tag{D.56}$$

If $L(\phi_j)$ is replaced by ϕ_j itself, then (D.56) leads to a technique known as the Galerkin method (Kantorovich and Krylov 1958), where in our example, $L(\phi_0 + \sum_{i=1}^{n} c_i\phi_i)$ is made orthogonal to the functions $\phi_j(j = 1, 2, \cdots, n)$ to yield n equations for $c_1, c_2, \cdots, c_n$. Note also that $\delta I = 0$ with I given by (D.55) is a least-square-error approach. Both methods are members of a class of techniques known as the method of *weighted residuals* (Forray 1968; Hildebrand 1962).

References

Chapter 1

Argyris, J. H., and Kelsey, S. 1960. *Energy Theorems and Structural Analysis.* Butterworth, London.

Babuska, I. 1971. Error-Bounds for Finite Element Method. *Numerische Math.* 16: 322–333.

Berke, L., Bader, R. M., Mykytow, W. J., Przemieniecki, J. S., and Shirk, M. H., eds. 1968. *Proc. 2nd Conf. on Matrix Meth. in Struct. Mech.* AFFDL-TR-68-150. Wright-Patterson AFB.

Besseling, J. F. 1963. The Complete Analogy Between the Matrix Equations and Continuous Field Equations of Structural Analysis. *Colloque International des Techniques de Calcul Analogique et Numérique de l'Aeronautique,* Liège, pp. 223–242.

Birkhoff, G., and Garabedian, H. I. 1960. Smooth Surface Interpolation *J. Math. Phys.* 39: 258–268.

Birkhoff, G., Schultz, M. H., and Varga, R. S. 1968. Piecewise Hermite Interpolation in One and Two Variables with Applications to Partial Differential Equations. *Numerische Math.* 11: 232–256.

Clough, R. W. 1960. The Finite Element Method in Plane Stress Analysis. *Proc. Am. Soc. Civil Engrs.* 87: 345–378.

Courant, R. 1943. Variational Methods for the Solution of Problems of Equilibrium and Vibrations. *Bull. Am. Math. Soc.* 49: 1–23.

Ergatoudis, I., Irons, B. M., and Zienkiewicz, O. C. 1968. Curved, Isoparametric "Quadrilateral" Elements for Finite Element Analysis. *Int. J. Solids Structures* 4, no. 1: 31–42.

Fix. G., and Strang, G. 1969. Fourier Analyses of the Finite Element Method in Ritz-Galerkin Theory. *Studies in Appl. Math.* 48: 265–273.

Fraeijs de Veubeke, B. 1964. Upper and Lower Bounds in Matrix Structural Analysis. *Agardograph 72,* Pergamon, Oxford, pp. 165–201.

Gallagher, R. H. 1964. *A Correlation Study of Methods of Matrix Structural Analysis.* Pergamon, Oxford.

Herrmann, L. R. 1965. A Bending Analysis for Plates. *Proc. 1st Conf. on Matrix Meth. in Struct. Mech.* AFFDL-TR-66-80. Wright-Patterson AFB, pp. 577–604.

Irons, B. M. 1966. Engineering Application of Numerical Integration in Stiffness Methods. *AIAA J.* 4: 2035–2037.

Jones, R. E. 1964. A Generalization of the Direct-Stiffness Method of Structural Analysis. *AIAA J.* 2, no. 5: 821–826.

Livesley, R. K. 1964. *Matrix Methods of Structural Analysis.* Pergamon, Oxford.

Melosh, R. J. 1963. Basis for Derivation of Matrices for the Direct Stiffness Method. *AIAA J.* 1, no. 7: 1631–1637.

Oden, J. T. 1969. A General Theory of Finite Elements. I, Topological Considerations. II, Applications. *Int. J. Numerical Meth. in Eng.* 1: 205–221, 247–259.

Pian, T. H. H. 1964. Derivation of Element Stiffness Matrices by Assumed Stress Distributions. *AIAA J.* 2, no. 7: 1333–1336.

Pian, T. H. H., and Tong, P. 1969. Basis of Finite Element Methods for Solid Continua. *Int. J. Numerical Meth. in Eng.* 1, no. 1: 3–28.

Pian, T. H. H. 1971a. Formulations of Finite Element Methods for Solid Continua. *Recent Advances in Matrix Methods in Structural Analysis and Design.* R. H. Gallagher, Y. Yamada and J. T. Oden, eds. University of Alabama Press, Huntsville.

Pian, T. H. H. 1971b. Variational Formulations of Numerical Methods in Solid Continua. *Computer-Aided Engineering.* G. M. L. Gladwell, ed. University of Waterloo Press, Waterloo, Canada, pp. 421–448.

Pian T. H. H., and Tong, P. 1972. Finite Element Methods in Continuum Mechanics. Chapter in *Advances in Applied Mechanics*, vol. 12, Academic Press, New York.

Prager, W. 1967. Variational Principles of Linear Elastostatics for Discontinuous Displacements, Strains and Stresses in *Recent Progress in Applied Mechanics.* Chapter in *The Folke-Odquist Volume.* B. Broberg, J. Hult, and F. Niordson, eds. Almquist and Wiksell, Stockholm, pp. 463–474.

Prager, W. 1968. Variational Principles for Elastic Plates with Relaxed Continuity Requirements. *Int. J. Solids Structures* 4, no. 9: 837–844.

Przemieniecki, J. S., Bader, R. M., Bozich, W. F., Johnson, J. R., and Mykytow, W. J., eds. 1965. *Proc. 1st Conf. on Matrix Meth. in Struct. Mech.* AFFDL-TR-66-80. Wright-Patterson AFB.

Strang, W. G., and Fix, G. 1973. *An Analysis of the Finite Element Method.* Prentice-Hall, Englewood Cliffs, N. J.

Szabo, B. A., and Lee, G. C. 1969. Derivation of Stiffness Matrices for Problems in Elasticity by Galerkin's Method. *Int. J. Numerical Meth. in Eng.* 1: 301–310.

Tong, P., and Pian, T. H. H. 1967. The Convergence of Finite Element Method in Solving Linear Elastic Problems. *Int. J. Solids Structures* 3: 865–879.

Tong, P., and Pian, T. H. H. 1969. A Variational Principle and the Convergence of the Finite Element Method Based on Assumed Stress Distribution. *Int. J. Solids Structures* 5: 436–472.

Tong, P. 1970. New Displacement Hybrid Finite Element Model for Solid Continua. *Int. J. Numerical Meth. in Eng.* 2: 78–83.

Turner, M. J., Clough, R. J., Martin, H. C., and Topp, L. J. 1956. Stiffness and Deflection Analysis of Complex Structures. *J. Aero. Sci.* 23, no. 9: 805–823.

Zienkiewicz, O. C. 1971. *The Finite Element Method in Engineering Science.* McGraw-Hill, New York.

Chapter 2

Bliss, G. A. 1944. *Calculus of Variations.* Mathematical Association of America. Reprint of 1925 edition.

Courant, R. 1943. Variational Methods for the Solution of Problems of Equilibrium and Vibrations. *Bull. Am. Math. Soc.* 49: 1–23.

Courant, R., and Hilbert, D. 1953. *Methods of Mathematical Physics.* vol. 1. Interscience, New York.

Kellogg, O. D. 1929. *Foundations of Potential Theory.* Springer, Berlin.

Chapter 3

Davidon, W. C. 1959. *Variable Metric Method for Minimization*. Research and Development Report ANL-5990, revised. Argonne National Laboratory, US Atomic Energy Commission.

Fletcher, R., and Powell, M. D. J. 1963. "A Rapidly Convergent Descent Method for Minimization. *Computer J.* 6: 163–168.

Fletcher, R., and Reeves, C. M. 1964. Function Minimization by Conjugate Gradients. *Computer J.* pp. 149–154.

Fox, R. L., and Kapoor, M. P. 1969. A Minimization Method for the Solution of the Eigenproblem Arising in Structural Dynamics. *Proc. 2nd Conf. on Matrix Meth. in Struct. Mech.* AFFDL TR-68-150. Wright-Patterson AFB, pp. 271–306.

Hellen, T. K. 1969. *A Front Solution for Finite Element Techniques*. Central Electricity Generating Board, RD/B/N1459, England, October.

Irons, B. M. 1970. A Frontal Solution Program for Finite Element Analysis. *Int. J. Numer. Meth. in Eng.* 2, no. 1: 5–32.

Isaacson, E., and Keller, H. B. 1966. *Analysis of Numerical Methods*. Wiley, New York.

Kotanchik, J. J., and Berg, B. A. 1969. *STACUSS 1: A Discrete Element Program for the Static Analysis of Single Layer Curved Stiffened Shells Subjected to Mechanical and Thermal Loads*. SAMSO TR 70–125, ASRL TR 146–9, December.

Luk, C. H. 1969. *Finite Element Analysis for Liquid Sloshing Problems*. AFOSR 69-1504 TR, ASRL TR 144-3, May.

Ralston, A. 1965. *A First Course in Numerical Analysis*. McGraw-Hill, New York.

Chapter 4

Irons, B. M. 1970. A Frontal Solution Program for Finite Element Analysis. *Int. J. Num. Meth. in Eng.* 2, no. 1: 5–32.

Chapter 5

Fung, Y. C. 1965. *Foundations of Solid Mechanics*. Prentice-Hall, Englewood Cliffs, N. J.

Landau, L. P., and Lifshitz, E. M. 1959. *Theory of Elasticity*. Addison-Wesley, Cambridge, Mass.

Love, A. E. H. A. 1963. *A Treatise on the Mathematical Theory of Elasticity*. Dover, New York.

Sechler, E. E. 1952. *Elasticity in Engineering*. Wiley, New York.

Sokolnikoff, I. S. 1956. *Mathematical Theory of Elasticity*. Second edition. McGraw-Hill, New York.

Timoshenko, S., and Goodier, N. 1970. *Theory of Elasticity*. McGraw-Hill, New York.

Chapter 6

Ahmad, S., Irons, B. M., and Zienkiewicz, O. C. 1969. Curved Thick Shell and Membrane

Elements with Particular Reference to Axisymmetric Problems. *Proc. 2nd Conf. Matrix Meth. in Struct. Mech.* AFFDL-TR-66-150. Wright-Patterson AFB, pp. 539–572.

Bogner, F. K., Fox, R. L., and Schmit, L. A., Jr. 1965. The Generation of Inter-Element-Compatible Stiffness and Mass Matrices by the Use of Interpolation Formulas. *Proc. 1st Conf. Matrix Methods in Struct. Mech.* AFFDL-TR-66-80. Wright-Patterson AFB, pp. 397–444.

Collatz, L. 1966. *The Numerical Treatment of Differential Equations.* Third edition. Springer, Berlin.

Davis, P., and Rabinowitz, P. 1956. Abscissas and Weights for Gaussian Quadratures of High Order. *J. Res. Nat. Bur. Stand.* 56: RP 2645.

Ergatoudis, I., Irons, B. M., and Zienkiewicz, O. C. 1968. Curved, Isoparametric, Quadrilateral Elements for Finite Element Analysis. *Int. J. Solids Structures* 4, no. 1: 31–42.

Felippa, C. A. 1966. *Refined Finite Element Analysis of Linear and Nonlinear Two-Dimensional Structures.* SESM Report 63-2. Department of Civil Engineering, University of California, Berkeley.

Felippa, C. A., and Clough, R. W. 1970. The Finite Element Method in Solid Mechanics. *Numerical Solution of Field Problems in Continuum Physics.* SIAM-AMS Proc. vol. II. G. Birkhoff and R. S. Varga, eds. American Mathematical Society, pp. 210–252.

Hammer, P. C., Marlowe, O. J., and Stroud, A. H. 1956. *Numerical Integration over Simplexes and Cones.* Math. Tables and other Aids to Computation, vol. 10.

Hammer, P. C., and Stroud, A. H. 1958. *Numerical Evaluation of Multiple Integrals.* Math. Tables and Other Aids to Computation, vol. 12.

Hildebrand, F. B. 1956. *Introduction to Numerical Analysis.* McGraw-Hill, New York.

Irons, B. M. 1966. Engineering Application of Numerical Integration in Stiffness Methods. *AIAA J.* 4: 2035–2037.

Irons, B. M. 1969. Economical Computer Techniques for Numerically Integrated Finite Elements. *Int. J. Numerical Meth. in Eng.* 1: 201–203.

Irons, B. M. 1971. Quadrature Rules for Brick Based Finite Elements. *Int. J. Numerical Meth. in Eng.* 3: 293.

Kopal, Z. 1961. *Numerical Analysis.* Second edition. Chapman & Hall, London.

Zienkiewicz, O. C., Irons, B. M., Ergatoudis, J., Ahmad, S., and Scott, F. C. 1969. Isoparametric and Associated Element Families for Two-and-Three-Dimensional Analysis. Chapter in *Finite Element Methods in Stress Analysis.* I. Holand and K. Bell, eds. TPIR (Technical University of Norway), Trondheim.

Zienkiewicz, O. C. 1971. *The Finite Element Method in Engineering Science.* McGraw-Hill, New York.

Chapter 7

Abel, J. F., and Desai, C. S. 1972. Comparison of Finite Elements for Plate Bending. *J. of the Structural Division, Proc. ASCE* 98, no. ST9: 2143–2148.

Ahmad, S., Irons, B. M., and Zienkiewicz, O. C. 1970. Analysis of Thick and Thin Shell Structures by Curved Finite Elements. *Int. J. Numerical Meth. in Eng.* 3, no. 2: 275–290.

Bell, K. 1969. A Refined Triangular Plate Bending Finite Element. *Int. J. Numerical Meth. in Eng.* 1, no. 1: 101–122 (also 1, no. 4: 395, and 2, no. 1: 146–147).

Bogner, F. K., Fox, R. L., and Schmit, L. A., Jr. 1965. The Generation of Inter-Element-Compatible Stiffness and Mass Matrices by the Use of Interpolation Formulas. *Proc. 1st. Conf. Matrix Meth. in Struct. Mech.* AFFDL-TR-66-80. Wright-Patterson AFB, pp. 397–444.

Clough, R. W., and Tocher, J. L. 1965. Finite Element Stiffness Matrices for Analysis of Plate Bending. *Proc. 1st. Conf. Matrix Meth. in Struct. Mech.* AFFDL-TR-66-80. Wright-Patterson AFB. pp. 515–546.

Clough, R. W., and Felippa, C. A. 1968. A Refined Quadrilateral Element for Analysis of Plate Bending *Proc. 2nd. Conf. Matrix Meth. in Struct. Mech.* AFFDL-TR-68-150. Wright-Patterson AFB. pp. 399–440.

Cook, R. D. 1974. *Concepts and Applications of Finite Element Analysis.* Wiley, New York.

Cowper, G. R., Kosko, E., Lindberg, G. M., and Olson, M. D. 1969. Static and Dynamic Applications of a High-Precision Triangular Plate Element. *AIAA J.* 7, no. 10: 1957–1965.

Holand, I., and Bell, K., eds. 1972. *Finite Element Methods in Stress Analysis.* Tapir (Technical University of Norway), Trondheim.

Mau, S. T., Pian, T. H. H., and Tong, P. 1973. Vibration Analysis of Laminated Plates and Shell by a Hybrid Stress Element. *AIAA J.* 2: 1450–1452.

Olsen, M.D. 1970. Some Flutter Solutions Using Finite Elements. *AIAA J.* 8, no, 4: 747–752.

Pawsey, S. F., and Clough, R. W. 1971. Improved Numerical Integration of Thick Shell Finite Elements. *Int. J. Numerical Meth. in Eng.* 3, no. 4: 575–586.

Pian, T. H. H. 1964. Derivation of Element Stiffness Matrices by Assumed Stress Distributions. *AIAA J.* 2, no. 7: 1333–1336.

Pian, T. H. H., and Tong, P. 1969. Basis of Finite Element Methods for Solid Continua. *Int. J. Numerical Meth. in Eng.* 1, no. 1: 3–28.

Pian, T. H. H., and Mau, S. T. 1972. Some Recent Studies in Assumed Stress Hybrid. Models. Chapter in *Advances in Computational Methods in Structural Mechanics and Design.* J. T. Oden, R. W. Clough, and Y. Yamamoto, eds. University of Alabama Press, Huntsville, pp. 87–106.

Rossettos, J. N., Tong, P., and Perl, E. 1972. *Finite Element Analysis of the Strength and Dynamic Behavior of Filamentary Composite Structures.* Final Report, NASA grant no. NGR 22-011-073, November.

Rossettos, J. N., and Tong, P. 1974. Finite Element Analysis of Vibration and Flutter of Cantilever Anisotropic Plates. *J. App. Mech. (Trans. ASME*, ser. E) 41, no. 4: 1075–1080.

Strang, W. G., and Fix, G. 1973. *An Analysis of the Finite Element Method.* Prentice-Hall, Englewood Cliffs, N. J.

Tong, P. 1969. Exact Solution of Certain Problems by the Finite Element Method. *AIAA J.* 7, no. 1: 179–180.

Zienkiewicz, O. C., Taylor, R. L., and Too, J. M. 1971. Reduced Integration Technique in General Analysis of Plates and Shells *Int. J. Numerical Meth. in Eng.* 3, no. 2: 275–290.

Chapter 8

Chen, H. S., and Mei, C. C. 1974. Oscillations and Wave Forces in a Man-Made Harbor in the Open Sea. *Proc. Tenth Naval Hydrodynamic Symposium.* M.I.T., Cambridge, Mass., June.

Mau, S. T., Pian, T. H. H., and Tong, P. 1973. Vibration Analysis of Laminated Plates and Shell by a Hybrid Stress Element. *AIAA J.* 2: 1450–1452.

Mau, S. T., and Tong, P. 1974. Calculation of Mechanical Impedance by the Finite Element Hybrid Model. *AIAA J.* 12, no. 2: 249–250.

Pian, T. H. H. 1964. Derivation of Element Stiffness Matrices by Assumed Stress Distribution. *AIAA J.* 2, no. 7: 1333–1336.

Pian, T. H. H., and Tong, P. 1972. Finite Element Methods in Continuum Mechanics. Chapter in *Advances in Applied Mechanics*, vol. 12. C. S. Lih, ed. Academic Press, New York, pp. 1–58.

Rossettos, J. N., and Tong, P. 1974. Finite Element Analysis of Vibration and Flutter of Cantilever Anisotropic Plates. *J. App. Mech.* (*Trans. ASME*, ser. E) 41, no. 4: 1075–1080.

Tong, P., and Pian, T. H. H. 1969. A Variational Principle and the Convergence of the Finite Element Method Based on Assumed Stress Distribution. *Int. J. Solids Structures* 5: 436–472.

Tong, P. 1970. New Displacement Hybrid Finite Element Model for Solid Continua. *Int. J. Numerical Meth. in Eng.* 2: 78–83.

Tong, P. and Pian, T. H. H. 1970. Bounds of Influence Coefficients by the Assumed Stress Method. *Int. J. Solids Structures*, 6: 1429–1432.

Tong, P., and Fung, Y. C. 1971. Slow Particulate Flow and Its Application to Biomechanics. *J. App. Mech.* (*Trans. ASME*, ser. E) 38: 721–728.

Tong, P. 1972. *Hybrid Finite Element Method.* I.E.I.—N.I. B72-17. Institute of Information Lab. Report, Pisa, Italy.

Tong, P., Pian, T. H. H., and Lasry, S. 1973. A Hybrid Element Approach to Crack Problems in Plane Elasticity. *Int. J. Numerical Meth. in Eng.* 7: 297–308.

Tong, P., Mau, S. T., and Pian, T. H. H. 1974. Derivation of Geometric Stiffness and Mass Matrices for Finite Element Hybrid Models. *Int. J. Solids Structures*, 10: 919–932.

Chapter 9

Archer, J. S. 1963. Consistent Mass Matrix for Distributed Mass Systems. *Proc. ASCE, J. Structural Div.* 39, no. 4: 161–178.

Bowie, O. L., and Neal, D. M. 1970. A Modified Mapping-Collocation Technique for Accurate Calculation of Stress-intensity Factors. *Int. J. Fracture Mech.* 6: 199–206.

Chen, H. S., and Mei, C. C. 1974. Oscillations and Wave Forces in a Man-Made Harbor in the Open Sea. *Proc. Tenth Naval Hydrodynamic Symposium.* M.I.T., Cambridge, Mass., June.

Clough, R. W. 1971. Analysis of Structural Vibrations and Dynamic Response. Chapter in *Recent Advances in Matrix Methods of Structural Analysis and Design.* R. H. Gallagher, Y. Yamada, and J. T. Oden, eds. University of Alabama Press, Huntsville, pp. 441–486.

Clough, R. W., and Bathe, K. J. 1972. Finite Element Analysis of Dynamic Response. Chapter in *Advances in Computational Methods in Structural Mechanics and Design.* J. T. Oden, R. W. Clough, and Y. Yamamoto, eds. University of Alabama Press, Huntsville, pp. 153–180.

Felippa, C. A. 1966. "Refined Finite Element Analysis of Linear and Nonlinear Two-Dimensional Structures." Ph.D. Dissertation, Dept. of Civil Eng., University of California, Berkeley.

Fix, G., and Strang, G. 1969. Fourier Analysis of the Finite Element Method in Ritz-Galerkin Theory. *Studies in Appl. Math.* 48: 265–273.

Key, S. W. 1966. A Convergence Investigation of the Direct Stiffness Method. Ph.D. Thesis, University of Washington, Seattle.

Mei, C. C., and Chen, H.S. 1975. Hybrid Element for Water Waves. *Proc. ASCE Symposium on Modeling Techniques.* San Francisco, September.

Muskhelishvilli, N. I. 1953. *Some Basic Problems of the Mathematical Theory of Elasticity.* Nordhoff, Groningen, Holland.

Orringer, O., Lin, K. Y., Stalk, G., Tong, P., and Mar, J. W. 1976. Application of the Assumed-Stress Hybrid Method to Finite Element Analysis in Fracture Mechanics. *Proc. ASCE, J. Structural Div.*, to be published.

Pian, T. H. H. 1975. *Finite Element Methods in Structural Dynamics.* AIAA Selected Reprint Series., vo. 17.

Rossetos, J. N., and Weinstock, H. 1974. Finite Element 3-Dimensional Dynamic Frame Model for Vehicle Crashworthiness Prediction. *Proc. Third Int. Conference on Vehicle System Dynamics.* V.P.I., Blacksburg, Virginia, August.

Stoker, J. J. 1957. *Water Waves.* Interscience, New York.

Tong, P. 1971. On the Numerical Problems of Finite Element Methods. Chapter in *Computer-Aided Engineering.* G. M. L. Gladwell, ed. University of Waterloo Press, Waterloo, Canada, pp. 539–559.

Tong, P., and Fung, Y. C. 1971. Slow Particulate Flow and Its Application to Biomechanics. *J. Appl. Mech. (Trans. ASME,* ser. E) 38: 721–728.

Tong, P., Pian, T. H. H., and Bucciarelli. L. L. 1971. Mode Shapes and Frequencies by the Finite Element Method using Consistent and Lumped Masses. *J. Computers and Structures.* 1: 623–638.

Tong, P. and Pian, T. H. H. 1972. On the Convergence of the Finite Element Method for Problems with a Singularity. *Int. J. Solids Structures.* 9: 313–321.

Tong, P. and Pian, T. H. H. 1973. Application of the Finite Element Method to Mixed Mode Fracture. *Proc. 10th Annual Meeting of the Society of Engineering Science.* North Carolina State University, Raleigh, November.

Tong, P. Pian, T. H. H., and Lasry, S. J. 1973. A Hybrid Element Approach to Crack Problems in Plane Elasticity. *Int. J. Numerical Meth. in Eng.* 7: 297–308.

Tong, P., and Rossettos, J. N. 1975. *Modular Approach to Structural Simulation for Vehicle Crashworthiness Prediction.* Final Report DOT-TSC-NHTSA-74-7, Department of Transportation, March.

Turner, M. J., Dill, E. H., Martin, H. C., and Melosh, R. J. 1960. Large Deflection Analysis of Complex Structures Subjected to Heating and External Loads. *J. Aero. Space Sci.* 27: 97–106.

Visser, W. 1966. A Finite Element Method for the Determination of Nonstationary Temperature Distribution and Thermal Deformation. *Proc. 1st Conf. Matrix Meth. in Struct. Mech.* AFFDL-TR-66-80. Wright-Patterson AFB, pp. 925–943.

Washiyu, K. 1968. *Variational Methods in Elasticity and Plasticity*. Pergamon, London.

Wilson, E. L., and Nickell, R. E. 1966. Application of the Finite Element Method to Heat Conduction Analysis. *Nuclear Eng. Des.* 4: 276–286.

Appendix D

Forray, M. J. 1968. *Variational Calculus in Science and Engineering*. McGraw-Hill, New York.

Hildebrand, F. B. 1962. *Methods of Applied Mathematics*. Prentice-Hall, Englewood Cliffs, N.J.

Kantorovich, L. V., and Krylov, V. I. 1958. *Approximate Methods of Higher Analysis*. Interscience, New York.

Index

Abel, J. F., 222
Addition, matrix, 284
Ahmad, S., 223
Algebra, matrix, 283–290
Algebraic equations, 13, 34, 36, 49
  linear, 82
  for three-dimensional problems, 66
Antisotropic material, 132
Approximate methods, 309–310
Approximations, finite-element, 13, 14, 37
  local, 23
  low-order differentiability of, 23
  order of, 256
Argyris, J. H., 1
Assembled finite-element model, 5, 6
Assembling
  of element matrices, 104
  example of, 74–78
  of finite elements, 254
  of master matrices, 72
  with one-dimensional array, 121–126
Asymptotic solution, 264
Automobiles, dynamic problems of, 255
Axes, principal, 235
Axisymmetric problems, 136

Back substitution, 85, 89, 115, 116
Beams
  on elastic foundation, 39, 211, 212
  shallow curved, 281
  large deflection of, 279
Beam bending, 9, 208–214
Beam element, mass matrix for, 251, 252
Bell, K., 221, 222, 223
Bending element, compatible triangular, 225
Bernoulli-Euler beam theory, 208
Bessling, J. F., 2
Biharmonic equation, 228, 239
  element-stiffness matrix for, 241–245
  hybrid finite-element model for, 246
Biharmonic operator, 216
Bogner, F. K., 218
Boundary
  common, 15, 49
  curved, 172, 205–207
  interelement, 5, 11, 19, 23, 161, 177, 214, 217, 242
  natural, 42–44
  prescribed, 143
  rigid, 43–44
  tractions, 10
Boundary conditions, 266
  for biharmonic equation, 239
  displacement, 277
  geometric, 249, 308, 313, 314
  for heat conduction problems, 268–269, 273
  for Laplace equation, 228
  mixed, 62–63, 157–160
  natural, 209, 225
  rigid, 55, 78, 79, 104, 209, 225, 308, 314
Boundary element, 44
Boundary nodes, 44
Boundary surface, 136
Boundary-value problems and variational methods, 306–311
Boundary values, at element level, 9
Bowie, O. L., 260
Bridges, dynamic problems of, 255
Bucciarelli, L. L., 255
Buildings, dynamic problems of, 255

Calculus of variations, 301–305
Cartesian coordinates, 127, 128
Central difference scheme, 254
Centroid
  of element, 235
  of triangle, 171, 172
Chen, H. S., 228, 266
Cholesky method, 87, 88, 89, 202
Clough, R. W., 1, 221, 223
Coefficients
  direct influence, 53
  variable, 63–65
Collatz, L., 161
Common storage, 126. *See also* Computers; Storage
Compatibility, 45, 208, 249
  of adjacent elements, 15
  conditions of, 214, 215, 243
  of generalized coordinates, 161
  interelement, 209, 221
  of neighboring elements, 138
  requirements, 211
Compatibility conditions
  generalized coordinates, 22
  along interelement boundaries, 21

Index

Compatibility conditions (continued)
  for neighboring elements, 19
Completeness, 45, 208, 209
  conditions for, 211
  order of, 177
  of polynomials, 20, 23, 161
  requirements for, 214
Computation, large-scale, 35
Computer programs, 83
Computers
  core storage for, 101 (see also Storage)
  high-speed digital, 72, 104, 161
  implementation of, 254
  single-array, 126
  storage of symmetric banded matrix in, 111–113
  transformation of stiffness matrix in, 160
Computer searches, 109
Computer time, 104, 161
Condensation
  static, 94
  substructure, 100
Conductivity, thermal, 273
Conjugate-gradient method, 92
Constant curvature states, 209
Constraining, 78–83
  master matrices, 55, 104
  with one-dimensional array, 121
Constraint
  application of, 36
  conditions of, 122, 123, 160, 231
Continuous function, 10
  piecewise, 37
Continuum
  elements in, 5
  solid, finite-element equations for, 127
  subregion of, 17
Convergence, 23–24
  condition for, 24–36
  cubic interpolation for, 211
  of iterative methods, 91
  monotonic, 214
  rate of, 255, 256
Cook, R. D., 223
Coordinates
  area, 170
  Cartesian, 127, 128
  cylindrical, 135, 273
  for Gaussian quadrature, 192
  generalized, 5, 6, 9, 10, 13, 19, 22, 24, 31, 34, 45, 46, 72, 139, 214, 236, 249
    compatibility of, 161
    degree numbers of, 121

  unknown, 55
  global, 155–157 (see also Global system)
  local, 31
  normalized, 162
  rotation of, 155
  nodal, 57
  rotation of, 130
  transformation, 14, 157–160, 172–179, 183
  triangular, 170, 190, 235
Core requirement, 114
Core storage, for computers, 66, 93, 101, 105, 115, 116
Couples, 127
Courant, R., 1
Cowper, G. R., 222
Crack problem, 248
Crack tip, 260
  element around, 261
Cubic element, 179–181
Cubic functions, 164
Curve, minimized length of, 301

Damping, linear viscous, 249
Damping forces, 253
Davis, P., 189
Deformation
  plane-stress, 6
  shear, 223
  strain, 130, 215
  symmetric, superelement for, 263
  transverse shear, 223
Degrees of freedom, 2–3, 14, 46
  in approximation theories, 8
  and behavior of elements, 5
  constrained, 55, 80
  of finite-element approximations, 13
  finite number of, 249
  of global system, 105, 106–107, 147
  and interelement compatibility, 209
  local numbers for, 147
  for triangular element, 222
  in two-dimensional domain, 15
  of two-node element, 33
Density, mass, 250
Desai, C. S., 222
Differentiability, 8
Differential equations
  Euler, 213, 216, 224, 241, 303, 305, 307, 308, 313
  partial, 311–312, 313
Differential-equation problems, 211
Differentiation, 286
Direction cosines, 128

Direction, principal, 130
Direct methods, for equations of large order, 82, 85–90
Discontinuous structures, 248
Discretization, finite-element, 36, 56
Discontinuities, 222
Displacement
 axial, 250
 boundary, 10, 262
 nodal values of, 149
 for plane-stress or -strain problem, 193
 strain, 130, 141–142
Displacement field, 14
 global, 4
 local, 4
Displacement method, 1
Displacement model, 15–19, 247
 polynomials for, 19–23
Divergence theorem, 42, 229
Domain
 subdivision of, 44
 triangular elements of, 44
Dynamic memory allocation, 126
Dynamic problems
 elastic solids, 249–255
 numerical instability in, 255
 one-dimensional, 250

Edge
 boundary-displacement matrix for, 244
 curved, 205
Eigenvalue, 92, 290
 problems, 213, 254
Elastic-coefficient matrix, 132, 137, 292
Elastic foundation, beam on, 39, 211, 212
Elasticity
 adjoining elements, 9
 linear, 131
 Young's modulus of, 133
Elastic material, 127
 hypoelastic material, 277
Elastic solids
 dynamic problems of, 249–255
 static analysis of, 249
Elements
 boundary, 44
 conforming, 214
 of continuum, 5
 cubic, 179–181
 curved, 178, 205
 eight-node, 178–179
 four-node, 295–296
 general quadrilateral, 178, 239
 hexahedral, 151–155, 184, 199–200
 higher-order, 162, 222
  three-dimensional, 197
  two-dimensional, 193–197
 hybrid, 247
 infinite-domain, 265
 interior, 44
 irregular, 172
 neighboring, 45, 138, 217, 242
 plate-bending finite, 214
 properties, 210
 rectangular, 48–50 (*see also* Rectangular elements)
 rectangular ring, 273
 square, 166–170
 tetrahedral, 148–150, 181–182, 185, 200
 triangular, 6, 45, 46–48, 239 (*see also* Triangular elements)
 two-dimensional, 193–197. *See also* Superelements
Element-stiffness matrix. *See* Matrix, element-stiffness
Elimination, forward, 84, 116
Elimination process, 112
Elliptic equations, second-order, 63
Energy
 complementary, 12
 kinetic, 248, 250, 252
 minimum potential, 136
 potential, 12, 41, 50, 51, 304, 308, 309
 stationary potential, 279
 strain, 249, 250, 308
Equations
 algebraic, 13, 34, 36, 66, 82
 biharmonic, 228, 239, 241–245, 246
 constrained set of, 35–36
 elliptic, 63
 equilibrium, 4, 126–130
 Euler, 213, 216, 224, 241, 303, 305, 307, 308, 313
 finite-element, 50
 fourth-order, 208
 harmonic, 228, 240
 Helmholtz, 267
 higher-order, 310
 hybrid, 229
 Laplace, 10, 40, 228, 256, 264
 partial differential, 311–312, 313
 Poisson's, 24, 40, 62, 237–239
 second-order elliptic, 63
 stress, 127–130
 Sturm-Liouville, 28, 36
 system of, 34

Equations (continued)
  temperature distribution, 7
  in triangular form, 84
Equilibrium, at node, 4
Equilibrium model, 248
Ergatoudis, I., 176
Error
  of finite-element approximations, 13
  in Gaussian quadrature, 192
  for Hammer's formula, 190, 201
  and interpolation functions, 186, 187
  roundoff, 82, 90
  truncation, 90
Euler differential equations, 213, 216, 224, 241, 303, 305, 307, 308, 313
Euler method
Extension concept, 260

Factoring, triple, 119–121, 124
Factorization, 85–90, 120
Felippa, C. A., 223
Finte-element formulation, 138–140
Finite-element methods, 1, 309
  and computers, 72
  and nonlinear problems, 276–280
Finite-element models, 5, 6, 14, 247
Finite elements
  mesh arrangement of, 14 (see also Mesh arrangement)
  nodal points of, 14
  one-dimensional, 8
  shape of, 14
Fix, G., 17, 23, 256
Fletcher, R., 93
Fluid flow and heat transfer, 268–278
Flutter problems, 213
Force, 31
  damping, 253
Force fields, nonconservative, 213
Force method, 1
Forray, M. J., 303, 314
Forward elimination, 84, 116
Forward substitution, 89
Fox, R. L., 93, 218
Fracture mechanics, 263
Fraeijs de Veubeke, B., 2
Frequencies, natural, 254
Front solution, 114–117
Functions
  admissible, 41
  analytic, 260
  bilinear, 49, 166
  biquadratic, 181
  continuous, 10
  cubic, 164
  extremal, 301
  Hankel, 267
  Hermite, 243
  interpolation, 48, 49, 63, 227 (see also Interpolation)
  linear, 17
  neighboring, 302
  piecewise-continuous, 37
  piecewise-smooth, 8
  pyramid, 18
  quadratic, 45
  shape, 161
  smooth, 163, 169
  strain energy, 277
  trilinear, 151
Functional, 24, 26, 51
  extremum of, 10, 11, 37, 43, 67
  general hybrid, 240
  hybrid, 246, 258, 261
  local, 23
  and Stokes flow problem, 276
  for variational statements, 208
Fung, Y. C., 137, 228, 276

Galerkin method, 2, 8, 314
Gallagher, R. H., 2
Gaussian elimination, 83–85, 98–99, 100, 101, 123
  banded matrices and, 110–113
  varieties of, 113–117
Gaussian quadrature, 186–189, 190, 191, 192, 199–200
Gauss-Seidel iteration, 19
Generalization, 246–247
Generalized load, 47
Geometry, rectangular, 221
Global coordinates, 155–157
Global nodal number, 44
Global subscripts, 33
Global system, 73, 156
  degrees of freedom of, 105
Global values, 36
Gradients, high-stress, 248
Green's theorem, 258

Hamilton's principle, 249
  variational statement for, 250
Hammer, P. C., 190
Hammer's formula, 201
Hanging-chain problem, 304
Harmonic operator, 216

Index

Heat conduction problems, 38, 268
  steady-state, 273
Heat transfer and fluid flow, 268–278
  one-dimensional, 281
  problem, 271
Helmholtz equation, scattered-wave potential for, 267
Hermite functions, 243
Hermite polynomials, 252
Hermann, L. R., 2
Hexahedral elements, 184
  element matrices for, 151–155
  interpolation function for, 199–200
Hexahedrons, 14
  finite-element equations for, 66
  interpolation functions for, 184
Higher-order elements, 162, 222
Hildebrand, F. B., 161, 303, 314
Holand, I., 221, 222
Hooke's law, 4
  generalized, 132
Hybrid element, for infinite domain, 267
Hybrid model, of finite-element method, 228
  and biharmonic equation, 246
Hybrid stress model, 233, 241
Hybrid stress plate elements, 223
Hybrid superelement, 256
Hypoelastic material, 277. *See also* Elastic foundation; Elastic material

Implicit integration scheme, 255
Incremental method, variational formulation of nonlinear problems by, 276
Incremental solution procedure, 278–280
Inertia, rotary, 252
Infinite domain, element for, 264–268
Initial conditions in heat conduction problems, 268–269
Integrals, weighted, 26
Integration, 286
Integration for Gaussian quadrature, 192
  implicit, 255
  numerical, 162, 185–193, 279
    in three dimensions, 191
    in two dimensions, 189
  by parts, 306
Integration formula, 190
Integration points for hexahedral elements, 199–200
Integration stations for triangular elements, 197
Interior elements, 44

Interpolation, 9, 36
  completeness property of, 179
  cubic, 171–172, 211
  first-order, 164
  Hermite, 19, 162, 164
  Lagrange's formula, 163
  linear, 25, 26, 28, 170, 211, 251
  nine-node, 168
  nodal values, 12, 16
  in one dimension, 162
  order for, 195
  quadratic, 29, 171, 224
  second-order, 164
  sixteen-node, 169
  in three dimensions, 179
  three-node, 170
  for triangular plane-stress element, 252
  twelve-node, 168
  in two dimensions, 166
  zeroth-order, 163–164
  zeroth-order Hermite, 186–187
Interpolation function, 14, 48, 49, 63, 138, 145, 161–207
  and compatibility conditions, 243
  displacement-model, 19–23
  first-order Hermitian, 210
  for hexahedral elements, 199
  for higher-order three-dimensional elements, 197
  linear, 163
  for one-dimensional problem, 18
  and order of approximation, 256
  and order of completeness, 177
  for plate bending, 222
  polynomial as, 13
  for right-triangular elements, 170
  three-dimensional, 66
  for triangular elements, 196–197 (*see also* Triangular elements)
  trilinear, 151
  for two-dimensional problem, 18
Invariance, geometric, 21–22
Invariants, stress, 129
Inverse of matrix, 287
Irons, B. M., 161, 176, 223
Isaacson, E., 91
Isotropic material, 133, 203
  stress-strain relation for, 134, 136
Iteration
  Gauss-Seidel, 91
  Jacobi's, 91
  successive, 91
Iterative method, 82, 83, 90

Jacobian and Jacobian matrix, 173, 176–177, 291
Jacobi's iteration, 91
Jones, R. E., 2

Kantorovich, L. V., 314
Kapoor, M. P., 93
Keller, H. B., 91
Key, S. W., 256
Kinetic energy, 248, 250, 252
Kirchhoff hypothesis, 215
Kirchhoff stress, 278
Kopal, Z., 161
Kronecker delta, 143, 162
Krylov, V. I., 314

Lagrange multiplier, 230, 304–305
Lagrange's interpolation formula, 163
Landau, L. P., 132
Laplace equation, 10, 40, 228, 256, 264
  superelement for, 264
  two-dimensional, 228
  wedge element for, 256
Large systems, 66
  assembly for, 72–103
Lasry, S. J., 228, 248, 263
Lee, G. C., 2
Lifshitz, E. M., 132
Lin, K. Y., 263
Livesley, R. K., 1
Load
  generalized, 72
  work of applied, 249
Local system, 156
Long-wave theory, for potential flow, 266
Love, A. E. H. A., 132
Lumped-mass approach, 255

Mar, J. W., 263
Marlowe O., 190
Martin, H. L., 1
Matrix
  aerodynamic-damping, 213
  assembled, 249 (see also Assembling)
  banded, 88, 111, 288–289
  boundary-displacement, 244
  consistent, 254
  consistent mass, 251, 252, 253, 280, 281
  constrained degrees of freedom, 81
  constrained stiffness, 82
  damping, 249, 253
  definition of, 283
  determinant of, 288
  diagonal, 86, 289
  elastic-coefficient, 132, 137, 292
  element, 32, 33, 51, 59–62, 72, 204
    evaluation of, 140
  element-stiffness, 1, 31, 54, 57, 122, 139, 212, 219, 220, 221, 226, 232–237, 241–245, 251, 252, 261, 277
  equilibrium, 5
  factorized, 124–125
  force, 213
  of force-displacement relations, 3, 5
  hybrid stiffness, 266
  identity, 287
  interpolation-function, 262
  Jacobian, 173, 176–177
  lower-triangular, 86, 124
  lumped mass, 251, 252, 253, 254, 280, 281
  master, 32, 54–55, 73 100
    constraining, 104
  master stiffness, 55, 72, 158
    constrained, 80
  multiplication, 201, 285
  null, 289
  plane-stress deformation, 6
  positive-definite, 88, 287–288, 289, 290
  rectangular-element, 53, 235
  rotation, 156
  singular, 288
  singular, 288
  sparse, 85, 88, 105, 120
  square, 287
  symmetric, 67
  stiffness, 259
    superelement, 264
  submaster, 94, 95, 100, 114
    constrained, 100
    symmetric, 104, 287, 290
  system, 32
  tetrahedral-element, 149–150 (see also Tetrahedral elements)
  triangular, 83, 289
  triangular-element, 51–53, 58–59 (see also Triangular elements)
  zero, 289
Mau, S. T., 223, 228, 248
Mei, C. C., 228, 266
Melosh, R. J., 2
Membranes, deflection of, 40
Mesh arrangement, 14, 68, 69
  rectangular, 70
  triangular, 70
Minimization techniques, 51, 92

Minimum potential energy, principle of, 1–2, 136
Minimum principle, 308
 for Poisson's equation, 41
Mixed method, 1
Models, assembled finite-element, 5, 6, 14
Multiplier, Lagrange, 230, 304–305
Muskhelishvilli, N. I., 260, 261

Natural boundary, 42–44
Natural boundary conditions, 42
Neal, D. M., 260
Neighboring elements, 45, 138, 217, 242
Neighboring functions, 302
Newton-Cotes quadrature, 186
Nickell, R. E., 268
Nodal coordinates, 57
Nodal forces, equivalent, 139
Nodal load, equivalent, 54
Nodal points, 138
Nodal values
 algebraic equations for, 34
 system of equations for, 29
Nodes, 5, 6
 boundary, 9, 44
 common, 10, 49
 generalized forces at, 26
 interior, 22–23
 in two-dimensional domain, 15
Nonconvergence, 170
Nonhomogeneous beams, 211
Nonhomogeneous media, 154, 248
Nonuniform beams, 211
Normals, rotation of, 223
Numbers
 global, 46, 75
 global degree, 72, 73, 75, 106, 108, 121
 global nodal, 44, 106
 local, 46, 75
 local degree, 72, 75, 108, 121
 local nodal, 46
Numerical-integration scheme, explicit, 254

Oden, J. T., 2
Olsen, M. D., 213
One-dimensional arrays, 117–119, 121–126
One-dimensional problems, 13
 exact solutions to, 213
Operator
 beta, 302
 biharmonic, 216
Orringer, O., 263

Orthotropic material, 132, 152
Overrelaxation, method of, 91

Parallelepiped, rectangular, 183
Partitioning, 285–286
Pascal triangle, 22
Pawsey, S. F., 223
Perl, E., 223
Permutation, cyclic, 289
Pian, T. H. H., 2, 14, 15, 17, 23, 213, 228, 242, 248, 256, 263
Piecewise-continuous functions, 37
Piecewise-linear representation, 17
Piecewise-smooth functions, differentiability of, 8
Plane elasticity, crack element in, 260–264
Plane stress, deformation behavior, 6
Plate bending, 9, 214–227
 problem, 275
Poisson's equation, 24, 40, 62, 237–239
 finite-element method for, 40–71
 minimum principle for, 41
Poisson's ratio, 133
Polygons, 14, 44
Polyhedrons, 14
Polynomials
 approximation of functions by, 13
 bicubic, 169
 complete, 13
  first-order, 47
  linear, 170
  quadratic, 167, 170
 cubic, 39, 210
 Hermite, 162, 252
 for interpolation functions, 19
 linear, 47, 149
  complete, 170
 order of, 256
 order of completeness of, 161
 quadratic, 163
  complete, 167, 170
 triquadratic, 181
Potential energy
 minimum, 1–2, 136
 second variation of, 280
 stationary, 137
Potential fields, 40
Powell, M. D., 93
Prager, W., 2
Predictor-corrector method, 279
Principal stresses, 129, 130
Problems
 axisymmetric, 136

Problems (continued)
  bending of beams and plates, 223–227
  boundary-value, 306–311
  differential-equation, 211
  dynamic, 249–255
  eigenvalue, 213, 254
  Gaussian elimination, 102–103, 123
  heat conduction, 38, 268
  one-dimensional, 13, 213
  plane-strain, 138, 142
  plane-stress, 142
  plate-bending, 9, 214–227
  Poisson's equation, 69–71
  two-dimensional, 214–227
  for variational statement, 39

Quadrature formulas, 186
Quadrature, Gaussian, 186–189, 190, 191, 192, 199–200
Quadrilateral elements, 233, 243, 245
  four-node, 255
  general, 239
  interpolation functions for, 195–196
  transformation, 174–175
Quadrilaterals, 14, 44

Rabinowitz, P., 189
Radiation condition, 267
Rail vehicles, dynamic problems of, 255
Rayleigh-Ritz method, 2, 8, 16, 23, 25, 305, 306, 309, 312
  problems, 39
Rectangles, 14, 48–50
Rectangular block, 183
Rectangular elements, 51, 53, 63, 70, 291–296
  compatible, 218
  conformable, 223
  degrees of freedom, 108
  matrix for, 53, 235
  noncompatible, 218
  nonconforming, 215
  plane strain, 144–149
  plane stress, 144–149
  transformation of, 173–175
Rectangular region, St-Venant torsion in, 313
Rectangular ring elements, 273
Reissner principle, 225
Residuals, weighted, 314
Rigid boundary condition, 43–44
Rigidity, torsional, 70
Rod, axial vibration of, 280

Rossettos, J. N., 213, 223, 228, 255
Rotary inertia, 252
Rotated coordinate system, 128
Rotation, 123
  coordinate, 155–157
  matrix, 156
  of rectangular element, 294
Roundoff, 82, 83, 90
Runge-Kutta method, 279

St-Venant torsion, 1, 69, 313
Scalar, multiplication by, 285
Schmit, L. A., 218
Search process, 106
Semiband, 88, 104–110
  left, 109
  maximum, 109, 110
  storing, 111
  width, 188, 121
Shear deformation, transverse, 223
Shear strains, transverse, 215
Simpson's rule, 187
Singularity
  at crack tip, 260
  order of, 256
  in transformation, 190
Sokolnikoff, I. S., 132
Solid mechanics, 308
  applications to, 127–160
  plate-bending problem in, 275
Solutions
  compatible, 11
  computed, 82
Sparseness, matrix, 85, 88, 105, 120
Square element, 166–170
  four-node, 166
  eight-node, 167
  twelve-node, 168
Stalk, G., 263
Static-condensation method, 94
Stationary conditions, 232, 238, 301
Stationary potential energy, principle of, 137
Stationary value, 249
Stations
  for Gaussian quadrature, 188
  for Hammer's formula, 190
Stiffness, 31
  axial, 250
  matrix, 259
  plate flexural, 215
Stoker, J. J., 266
Stokes flow

axially symmetric, 273, 274–276
  problem, 276
  two-dimensional, 274
Storage
  common, 126
  of element matrices, 104
  external, 113–117
  requirements, 161
  of symmetric banded matrix, 111–113.
    See also Computers; Core storage
Strain, 130
  constant, for triangular elements, 297
  linear, for triangular elements, 297
  plane, 133, 134, 142
    triangular element for, 140
Strain-displacement relations, 141–142
Strang, G., 17, 23, 256
Stress
  axial compressive, 280
  components of, 127
  Kirchhoff, 278
  normal, 127, 215
  plane, 129, 133, 134, 137, 148
    triangular element for, 140, 252
  principal, 129, 130
  shear, 127, 130
  two-dimensional, 129
Stress gradients, 222
Stress intensity factors, 263
Stress method, hybrid, 223
Stress-strain relations, linear, 132–133
Stroud, A. H., 190
Structural analysis, matrix methods of, 1
Structural system, dynamic, 248
Structures, theory of, 2–3
Sturm-Liouville equation, 28, 36
Submatrices, 285–286
Substitution
  back, 85, 89, 115, 116
  forward, 89
Substructure method, 94–103
Substructures, 94, 101, 114
Subtraction, in matrix algebra, 284
Superelement
  element-stiffness matrix for, 261
  crack-tip, 260
  hybrid singular, 256–268
  infinite-domain, 256–268
  for Laplace's equation, 264
  stiffness matrix of, 259–260
  for stress intensity factor problems, 263
  for symmetric deformation, 263
  wedge, 257

Szabo, B. A., 2

Taylor, R. L., 223
Taylor series, 19, 37, 47, 163, 303
  expansion, 14
  truncated, 167
Temperature, 40
  distribution, 7
Tetrahedral element, 148–150
  interpolation function for, 200
  right, 181–182
Tetrahedron, 14
  finite-element equations for, 65
  interpolation functions for, 184
Thin plate, 215
Tocher, J. L., 221
Tong, P., 2, 14, 15, 17, 23, 213, 223, 228,
  242, 248, 255, 256, 263, 276
Too, J. M., 223
Topp, L. J., 1
Torsion
  of rods, 40
  St-Venant, 69, 313
Traction
  boundary, 10, 14
  prescribed boundary, 143
Transformation
  coordinates, 14, 123, 157–160, 172–179,
    183
  to global degree numbers, 73, 75, 76, 108
  isoparametric, 14
  nonlinear, 177, 205
  order of, 195
  quadrilateral-element, 174–175
  rectangular-element, 291
  singularities in, 190
  triangular-element, 175, 297
Transpose
  of matrix, 284
  of product, 285
Trapezoidal rule, 187
Triangle. 14, 44
  constant-strain (CST), 140, 142, 299
  as degenerate quadrilateral, 191
  linear-strain, 299–300
  Pascal, 22
Triangular elements, 44–45, 46–48, 51, 53,
  63, 297–300
  compatible, 239
  compatible higher-order, 222
  conformable, 223
  constant-strain, 255
  degrees of freedom of, 108, 222

finite-element equations for, 64–65
interpolation functions for, 196–197
mesh of, 70
with one curved side, 204
right, 170–172
and rigid boundary conditions, 56
transformation of, 175–176
for plane strain and stress, 140
element-stiffness matrix for, 252
Trilinear functions, 151
Triple-factoring method, 88, 89
Truncation, 90
to finite domain, 264
Taylor series, 167
Truss, pin-joint, 3
Turner, M. J., 1
Two-dimensional array, 118
Two-dimensional elements, higher-order, 193–197
Two-dimensional problems, bending of plates, 214–227

Uniform mass distribution, 252

Variable band width, 118
Variable-metric method, 93
Variational formulations, 249
of Muskhelishvilli, 260
of nonlinear problems, 276–280
Variational functionals, stationary value of, 16
Variational methods, 301–314
for boundary-value problems, 306–311
Variational principles, 2, 7
application of, 66–67
extrema of, 7
Variational statement
of bending of plates, 216
of boundary-value problems, 25
for finite-element procedure, 36
functionals for, 208
of Hamilton's principle, 250
of problems, 39
for temperature distribution, 8
Variations, calculus of, 301–305
Vector
applied-load, 72
body-force, 137
column, 283
consistent element-loading, 139, 143
constrained load, 80, 82
definition of, 283
displacement, 155, 193

element-force, 31
element generalized-coordinate, 54
element generalized-load, 54
element-loading, 146, 158, 219, 245
generalized-coordinate, 78, 97, 194, 204
unit normal, 40
Vibration
axial, 280
free, 254
Virtual work, principle of, 1, 213, 253
Visser, W., 268

Wave, outgoing, 267
Wave front, 116
Wave potential, scattered, 267
Wave propagation, 264
Weighted residuals, method of, 314
Weighting factors, 186
for Gaussian quadrature, 188, 192
for Hammer's formula, 190, 201
for hexahedral elements, 199–200
for triangular elements, 197
Weinstock, H., 255
Wilson, E. L., 268

Young's modulus of elasticity, 133

Zienkiewicz, O. C., 176, 223

A CATALOG OF SELECTED
DOVER BOOKS
IN SCIENCE AND MATHEMATICS

CATALOG OF DOVER BOOKS

Engineering

DE RE METALLICA, Georgius Agricola. The famous Hoover translation of greatest treatise on technological chemistry, engineering, geology, mining of early modern times (1556). All 289 original woodcuts. 638pp. 6¾ x 11. 0-486-60006-8

FUNDAMENTALS OF ASTRODYNAMICS, Roger Bate et al. Modern approach developed by U.S. Air Force Academy. Designed as a first course. Problems, exercises. Numerous illustrations. 455pp. 5⅜ x 8½. 0-486-60061-0

DYNAMICS OF FLUIDS IN POROUS MEDIA, Jacob Bear. For advanced students of ground water hydrology, soil mechanics and physics, drainage and irrigation engineering and more. 335 illustrations. Exercises, with answers. 784pp. 6⅛ x 9¼. 0-486-65675-6

THEORY OF VISCOELASTICITY (Second Edition), Richard M. Christensen. Complete consistent description of the linear theory of the viscoelastic behavior of materials. Problem-solving techniques discussed. 1982 edition. 29 figures. xiv+364pp. 6⅛ x 9¼. 0-486-42880-X

MECHANICS, J. P. Den Hartog. A classic introductory text or refresher. Hundreds of applications and design problems illuminate fundamentals of trusses, loaded beams and cables, etc. 334 answered problems. 462pp. 5⅜ x 8½. 0-486-60754-2

MECHANICAL VIBRATIONS, J. P. Den Hartog. Classic textbook offers lucid explanations and illustrative models, applying theories of vibrations to a variety of practical industrial engineering problems. Numerous figures. 233 problems, solutions. Appendix. Index. Preface. 436pp. 5⅜ x 8½. 0-486-64785-4

STRENGTH OF MATERIALS, J. P. Den Hartog. Full, clear treatment of basic material (tension, torsion, bending, etc.) plus advanced material on engineering methods, applications. 350 answered problems. 323pp. 5⅜ x 8½. 0-486-60755-0

A HISTORY OF MECHANICS, René Dugas. Monumental study of mechanical principles from antiquity to quantum mechanics. Contributions of ancient Greeks, Galileo, Leonardo, Kepler, Lagrange, many others. 671pp. 5⅜ x 8½. 0-486-65632-2

STABILITY THEORY AND ITS APPLICATIONS TO STRUCTURAL MECHANICS, Clive L. Dym. Self-contained text focuses on Koiter postbuckling analyses, with mathematical notions of stability of motion. Basing minimum energy principles for static stability upon dynamic concepts of stability of motion, it develops asymptotic buckling and postbuckling analyses from potential energy considerations, with applications to columns, plates, and arches. 1974 ed. 208pp. 5⅜ x 8½.
0-486-42541-X

METAL FATIGUE, N. E. Frost, K. J. Marsh, and L. P. Pook. Definitive, clearly written, and well-illustrated volume addresses all aspects of the subject, from the historical development of understanding metal fatigue to vital concepts of the cyclic stress that causes a crack to grow. Includes 7 appendixes. 544pp. 5⅜ x 8½. 0-486-40927-9

CATALOG OF DOVER BOOKS

ROCKETS, Robert Goddard. Two of the most significant publications in the history of rocketry and jet propulsion: "A Method of Reaching Extreme Altitudes" (1919) and "Liquid Propellant Rocket Development" (1936). 128pp. 5⅜ x 8½. 0-486-42537-1

STATISTICAL MECHANICS: PRINCIPLES AND APPLICATIONS, Terrell L. Hill. Standard text covers fundamentals of statistical mechanics, applications to fluctuation theory, imperfect gases, distribution functions, more. 448pp. 5⅜ x 8½.
0-486-65390-0

ENGINEERING AND TECHNOLOGY 1650–1750: ILLUSTRATIONS AND TEXTS FROM ORIGINAL SOURCES, Martin Jensen. Highly readable text with more than 200 contemporary drawings and detailed engravings of engineering projects dealing with surveying, leveling, materials, hand tools, lifting equipment, transport and erection, piling, bailing, water supply, hydraulic engineering, and more. Among the specific projects outlined-transporting a 50-ton stone to the Louvre, erecting an obelisk, building timber locks, and dredging canals. 207pp. 8⅜ x 11¼.
0-486-42232-1

THE VARIATIONAL PRINCIPLES OF MECHANICS, Cornelius Lanczos. Graduate level coverage of calculus of variations, equations of motion, relativistic mechanics, more. First inexpensive paperbound edition of classic treatise. Index. Bibliography. 418pp. 5⅜ x 8½. 0-486-65067-7

PROTECTION OF ELECTRONIC CIRCUITS FROM OVERVOLTAGES, Ronald B. Standler. Five-part treatment presents practical rules and strategies for circuits designed to protect electronic systems from damage by transient overvoltages. 1989 ed. xxiv+434pp. 6⅛ x 9¼. 0-486-42552-5

ROTARY WING AERODYNAMICS, W. Z. Stepniewski. Clear, concise text covers aerodynamic phenomena of the rotor and offers guidelines for helicopter performance evaluation. Originally prepared for NASA. 537 figures. 640pp. 6⅛ x 9¼.
0-486-64647-5

INTRODUCTION TO SPACE DYNAMICS, William Tyrrell Thomson. Comprehensive, classic introduction to space-flight engineering for advanced undergraduate and graduate students. Includes vector algebra, kinematics, transformation of coordinates. Bibliography. Index. 352pp. 5⅜ x 8½. 0-486-65113-4

HISTORY OF STRENGTH OF MATERIALS, Stephen P. Timoshenko. Excellent historical survey of the strength of materials with many references to the theories of elasticity and structure. 245 figures. 452pp. 5⅜ x 8½. 0-486-61187-6

ANALYTICAL FRACTURE MECHANICS, David J. Unger. Self-contained text supplements standard fracture mechanics texts by focusing on analytical methods for determining crack-tip stress and strain fields. 336pp. 6⅛ x 9¼. 0-486-41737-9

STATISTICAL MECHANICS OF ELASTICITY, J. H. Weiner. Advanced, self-contained treatment illustrates general principles and elastic behavior of solids. Part 1, based on classical mechanics, studies thermoelastic behavior of crystalline and polymeric solids. Part 2, based on quantum mechanics, focuses on interatomic force laws, behavior of solids, and thermally activated processes. For students of physics and chemistry and for polymer physicists. 1983 ed. 96 figures. 496pp. 5⅜ x 8½.
0-486-42260-7

CATALOG OF DOVER BOOKS

Mathematics

FUNCTIONAL ANALYSIS (Second Corrected Edition), George Bachman and Lawrence Narici. Excellent treatment of subject geared toward students with background in linear algebra, advanced calculus, physics and engineering. Text covers introduction to inner-product spaces, normed, metric spaces, and topological spaces; complete orthonormal sets, the Hahn-Banach Theorem and its consequences, and many other related subjects. 1966 ed. 544pp. 6⅛ x 9¼. 0-486-40251-7

ASYMPTOTIC EXPANSIONS OF INTEGRALS, Norman Bleistein & Richard A. Handelsman. Best introduction to important field with applications in a variety of scientific disciplines. New preface. Problems. Diagrams. Tables. Bibliography. Index. 448pp. 5⅜ x 8½. 0-486-65082-0

VECTOR AND TENSOR ANALYSIS WITH APPLICATIONS, A. I. Borisenko and I. E. Tarapov. Concise introduction. Worked-out problems, solutions, exercises. 257pp. 5⅜ x 8¼. 0-486-63833-2

AN INTRODUCTION TO ORDINARY DIFFERENTIAL EQUATIONS, Earl A. Coddington. A thorough and systematic first course in elementary differential equations for undergraduates in mathematics and science, with many exercises and problems (with answers). Index. 304pp. 5⅜ x 8½. 0-486-65942-9

FOURIER SERIES AND ORTHOGONAL FUNCTIONS, Harry F. Davis. An incisive text combining theory and practical example to introduce Fourier series, orthogonal functions and applications of the Fourier method to boundary-value problems. 570 exercises. Answers and notes. 416pp. 5⅜ x 8½. 0-486-65973-9

COMPUTABILITY AND UNSOLVABILITY, Martin Davis. Classic graduate-level introduction to theory of computability, usually referred to as theory of recurrent functions. New preface and appendix. 288pp. 5⅜ x 8½. 0-486-61471-9

ASYMPTOTIC METHODS IN ANALYSIS, N. G. de Bruijn. An inexpensive, comprehensive guide to asymptotic methods–the pioneering work that teaches by explaining worked examples in detail. Index. 224pp. 5⅜ x 8½ 0-486-64221-6

APPLIED COMPLEX VARIABLES, John W. Dettman. Step-by-step coverage of fundamentals of analytic function theory–plus lucid exposition of five important applications: Potential Theory; Ordinary Differential Equations; Fourier Transforms; Laplace Transforms; Asymptotic Expansions. 66 figures. Exercises at chapter ends. 512pp. 5⅜ x 8½. 0-486-64670-X

INTRODUCTION TO LINEAR ALGEBRA AND DIFFERENTIAL EQUATIONS, John W. Dettman. Excellent text covers complex numbers, determinants, orthonormal bases, Laplace transforms, much more. Exercises with solutions. Undergraduate level. 416pp. 5⅜ x 8½. 0-486-65191-6

RIEMANN'S ZETA FUNCTION, H. M. Edwards. Superb, high-level study of landmark 1859 publication entitled "On the Number of Primes Less Than a Given Magnitude" traces developments in mathematical theory that it inspired. xiv+315pp. 5⅜ x 8½. 0-486-41740-9

CATALOG OF DOVER BOOKS

TENSOR CALCULUS, J.L. Synge and A. Schild. Widely used introductory text covers spaces and tensors, basic operations in Riemannian space, non-Riemannian spaces, etc. 324pp. 5⅜ x 8¼. 0-486-63612-7

ORDINARY DIFFERENTIAL EQUATIONS, Morris Tenenbaum and Harry Pollard. Exhaustive survey of ordinary differential equations for undergraduates in mathematics, engineering, science. Thorough analysis of theorems. Diagrams. Bibliography. Index. 818pp. 5⅜ x 8½. 0-486-64940-7

INTEGRAL EQUATIONS, F. G. Tricomi. Authoritative, well-written treatment of extremely useful mathematical tool with wide applications. Volterra Equations, Fredholm Equations, much more. Advanced undergraduate to graduate level. Exercises. Bibliography. 238pp. 5⅜ x 8½. 0-486-64828-1

FOURIER SERIES, Georgi P. Tolstov. Translated by Richard A. Silverman. A valuable addition to the literature on the subject, moving clearly from subject to subject and theorem to theorem. 107 problems, answers. 336pp. 5⅜ x 8½. 0-486-63317-9

INTRODUCTION TO MATHEMATICAL THINKING, Friedrich Waismann. Examinations of arithmetic, geometry, and theory of integers; rational and natural numbers; complete induction; limit and point of accumulation; remarkable curves; complex and hypercomplex numbers, more. 1959 ed. 27 figures. xii+260pp. 5⅜ x 8½. 0-486-63317-9

POPULAR LECTURES ON MATHEMATICAL LOGIC, Hao Wang. Noted logician's lucid treatment of historical developments, set theory, model theory, recursion theory and constructivism, proof theory, more. 3 appendixes. Bibliography. 1981 edition. ix + 283pp. 5⅜ x 8½. 0-486-67632-3

CALCULUS OF VARIATIONS, Robert Weinstock. Basic introduction covering isoperimetric problems, theory of elasticity, quantum mechanics, electrostatics, etc. Exercises throughout. 326pp. 5⅜ x 8½. 0-486-63069-2

THE CONTINUUM: A CRITICAL EXAMINATION OF THE FOUNDATION OF ANALYSIS, Hermann Weyl. Classic of 20th-century foundational research deals with the conceptual problem posed by the continuum. 156pp. 5⅜ x 8½. 0-486-67982-9

CHALLENGING MATHEMATICAL PROBLEMS WITH ELEMENTARY SOLUTIONS, A. M. Yaglom and I. M. Yaglom. Over 170 challenging problems on probability theory, combinatorial analysis, points and lines, topology, convex polygons, many other topics. Solutions. Total of 445pp. 5⅜ x 8½. Two-vol. set.
Vol. I: 0-486-65536-9   Vol. II: 0-486-65537-7

Paperbound unless otherwise indicated. Available at your book dealer, online at **www.doverpublications.com**, or by writing to Dept. GI, Dover Publications, Inc., 31 East 2nd Street, Mineola, NY 11501. For current price information or for free catalogues (please indicate field of interest), write to Dover Publications or log on to **www.doverpublications.com** and see every Dover book in print. Dover publishes more than 500 books each year on science, elementary and advanced mathematics, biology, music, art, literary history, social sciences, and other areas.